Methods of Teaching Agriculture

Methods of Teaching Agriculture

Third Edition

L. H. Newcomb
Professor of Agricultural Education
Price Chair in Teaching, Advising, and Learning
Senior Associate Dean, College of Food, Agricultural,
and Environmental Sciences

J. David McCracken
Professor Emeritus of Agricultural Education

J. Robert Warmbrod
Distinguished University Professor Emeritus
Agricultural Education

M. Susie Whittington
Associate Professor of Agricultural Education
Teacher Education in Agricultural Education

Department of Human & Community Resource Development
College of Food, Agricultural, and Environmental Sciences
The Ohio State University

Upper Saddle River, New Jersey 07458

Library of Congress Cataloging-in-Publication Data

Methods of teaching agriculture / L.H. Newcomb . . . [et al.].—3rd ed.
    p. cm.
    Rev. ed. Of: Methods of teaching agriculture / L.H. Newcomb, 2nd ed. 1993.
    ISBN 0-13-113418-3
    1. Agriculture—Study and teaching. I. Newcomb, L. H. II. Newcomb, L. H. Methods of teaching agriculture.

S531.M48 2004
630'.7'1—dc21
                                              2003042542

Editor-in-Chief: Stephen Helba
Executive Editor: Debbie Yarnell
Editorial Assistant: Jonathan Tenthoff
Managing Editor: Mary Carnis
Production Editor: Amy Hackett/Carlisle Publishers Services
Production Liaison: Janice Stangel
Director of Manufacturing and Production: Bruce Johnson
Manufacturing Buyer: Cathleen Petersen
Creative Director: Cheryl Asherman
Cover Design Coordinator: Miguel Ortiz
Marketing Manager: Jimmy Stephens
Printer/Binder: Phoenix Book Tech

Copyright © 2004 by Pearson Education, Inc., Upper Saddle River, New Jersey, 07458.
Pearson Prentice Hall. All rights reserved. Printed in the United States of America. This publication is protected by Copyright and permission should be obtained from the publisher prior to any prohibited reproduction, storage in a retrieval system, or transmission in any form or by any means, electronic, mechanical, photocopying, recording, or likewise. For information regarding permission(s), write to: Rights and Permissions Department.

Pearson Prentice Hall™ is a trademark of Pearson Education, Inc.
Pearson® is a registered trademark of Pearson plc
Prentice Hall® is a registered trademark of Pearson Education, Inc.

Pearson Education LTD.
Pearson Education Singapore, Pte. Ltd
Pearson Education, Canada, Ltd
Pearson Education–Japan

Pearson Education Australia PTY, Limited
Pearson Education North Asia Ltd
Pearson Educación de Mexico, S.A. de C.V.
Pearson Education Malaysia, Pte. Ltd

10 9 8 7 6 5 4 3 2 1
0-13-113418-3

Contents

PREFACE VII

PART I
FOUNDATIONS FOR METHODS IN AGRICULTURAL EDUCATION

1 FACTORS INFLUENCING DECISIONS ABOUT TEACHING *3*

2 PRINCIPLES OF TEACHING AND LEARNING *25*

3 PLANNING THE COURSE OF STUDY *49*

4 LEARNING AS PROBLEM SOLVING *72*

PART II
METHODS FOR TEACHING AND LEARNING

5 PLANNING FOR INSTRUCTION *85*

6 GROUP TEACHING TECHNIQUES *112*

7 INDIVIDUALIZED TEACHING TECHNIQUES   *150*

8 MANAGING STUDENT BEHAVIOR   *186*

PART III
APPLICATION OF LEARNING

9 APPLICATION OF LEARNING: THE LABORATORY   *213*

10 APPLICATION OF LEARNING: SUPERVISED AGRICULTURAL EXPERIENCE   *243*

11 APPLICATION OF LEARNING: FFA   *267*

PART IV
TEACHING SPECIAL POPULATIONS

12 TEACHING LEARNERS WITH SPECIAL NEEDS   *289*

13 TEACHING ADULTS   *309*

PART V
EVALUATION OF LEARNING

14 EVALUATION OF LEARNING   *327*

APPENDIX A
   SAMPLE UNIT OF INSTRUCTION   *367*

APPENDIX B
   METHODS IN TEACHING AGRICULTURE   *384*

INDEX   *387*

Preface

What is public education in agriculture like? What factors influence the effectiveness of teaching and learning? Why should problem solving be a major emphasis in teaching agriculture? How does one decide what content to teach, when to teach it, and for how long? How does one plan for effective teaching? What are some good techniques that every teacher of agriculture should have the ability to use? How does a teacher develop an appropriate classroom climate by managing student behavior? Agriculture teachers need to have students apply what is learned; how can learning be applied in the laboratory, in supervised agricultural experience programs, and in the FFA organization? What special provisions are needed in teaching students who are disadvantaged and disabled, or in teaching adult students? What techniques are helpful in assessing the extent to which students have learned?

Competence in teaching methods, along with competence in the technical subject matter, is essential to be effective as a teacher of agriculture. This book has been designed to be of use in preservice and in-service education courses. The authors assembled from a broad range of teacher education institutions solicited course syllabi, which served as the basis for the content of this book. Part I deals with the foundations for methods of teaching agriculture; Part II, the methods for teaching and learning; Part III, the application of learning; Part IV, the special needs of unique populations; and Part V, the procedures for evaluating teaching and learning. The book should prove useful as a text or reference in courses related to teaching methods, introduction or orientation to agricultural education, course or program planning, supervised agricultural experience, youth organizations or FFA, laboratory management, teaching students who are disadvantaged or disabled, teaching adults, and evaluation of learning.

The authors acknowledge the contributions of Carrie Schlechter for reading each chapter and keying in edits, Nancy Ray for creating and redesigning graphics, and Will Waidelich for supplying supporting documentation.

PART I Foundations for Methods in Agricultural Education

Teachers of agriculture need an understanding of the foundations on which effective teaching is built. Educational theory is only important as it is applied by teachers in improving instruction for students.

Chapter 1 provides an orientation to the reader concerning the purposes of public school education in agriculture, the clientele served, subject matter organization and content, and principles of learning. This background is important in developing and improving knowledge and skill in planning, delivering, and evaluating instruction. The reader who studies this introductory chapter will develop an understanding of the nature of public school education in agriculture.

One who desires to improve as a teacher will want to learn how to apply principles of teaching and learning in schools. How can a teacher use these principles in instruction? The reader of Chapter 2 will be advised on how to apply principles of learning related to organization and structure of subject matter, readiness, motivation, student involvement, student success, reinforcement and reward, directed learning, problem solving, supervised practice, and transfer of learning. Research indicates that the implementation of these principles in teaching is correlated with the extent to which desired learning outcomes are achieved.

Organization and structure of the subject matter results in increased learning. The document that describes the organization and structure of subject matter for the entire program is the course of study. The course of study is a blueprint or design for instruction, providing a rationale for course content, a vehicle for communicating what students will be taught, a basis for lesson planning, improved student learning, and a means of securing instructional resources. The teacher's responsibility for course development is outlined. A step-by-step procedure for course of study development is provided. Teachers who study and apply the instruction in Chapter 3 will enhance their ability to organize and structure the content for their courses. The need for continual updating of the course of study is stressed.

Problem solving as an approach to teaching and learning has been emphasized throughout the history of public school education in agriculture. People tend to learn through logical thought processes that approximate the problem-solving approach to teaching. In Chapter 4, the relationship between the principles of teaching and learning and the problem-solving approach to teaching is described. This chapter sets the stage for Part II, "Methods for Teaching and Learning."

1

Factors Influencing Decisions about Teaching

> POSITION AVAILABLE—TEACHER OF AGRICULTURE
>
> Public school system needs teacher of agriculture with expertise in group and individualized instruction. Must be competent in teaching youth, adults, and students who are disadvantaged and handicapped. Knowledge of and the ability to apply principles of teaching and learning expected. Application of learning by students must be managed in the laboratory, in FFA organization activities, and in programs of supervised practice. Problem-solving teaching is expected. Teacher responsible for course of study development, lesson planning, and evaluation of student learning. Thorough understanding of agricultural education in the public schools a prerequisite. Preference to applicants committed to serve in a community as an agriculturalist and an educator. Submit résumé to the Superintendent of Schools.

Being a teacher of agriculture in the public schools is challenging. One is responsible for much more than classroom and laboratory instruction. However, the primary task of any teacher is helping students learn. In the process of helping students learn, teachers plan, deliver, and evaluate instruction. The extent to which those who are taught acquire new knowledge, skills, and attitudes is determined primarily by two factors. The first is the expertise of teachers in the subject area. The second is their knowledge, understanding, and ability to put into practice what is known about teaching and learning. Prospective and practicing teachers of agriculture need to develop further their competence to plan, deliver, and evaluate instruction. Reading, study, and instruction about methods of teaching agriculture can provide that competence.

OBJECTIVES

Effective teachers consider several factors when making decisions about teaching techniques and strategies. Teaching does not take place in a vacuum. After reading this chapter, you will be able to make decisions about instruction within a context that includes

1. The purposes and objectives to be achieved through the instructional program; in this case, the purposes of public school education in agriculture.
2. The clientele taught—their interests, aspirations, experiences, and characteristics.
3. The organization and content of the subject matter.
4. The psychology of learning—what is known about some basic principles of teaching and learning.
5. The knowledge and skill of the teacher not only in the subject matter but also in planning, delivering, and evaluating instruction.

INTERRELATIONSHIP OF THE FIVE FACTORS

Thoughtful consideration of the factors influencing decisions about instruction indicated in Figure 1-1 reveals two important ideas. First, it is clear that the five factors, while influencing instructional strategies and techniques directly, are interrelated and mutually interdependent. Purposes and objectives of instructional programs are not derived in isolation from the clientele who are to be taught. Likewise, purposes and objectives influence directly the subject matter or content that will be taught. Knowledge of the principles of teaching and learning indicates how subject matter is best organized to optimize learning. In a similar manner, characteristics of learners influence decisions about teaching techniques.

A second important idea that becomes evident when the five factors influencing decisions about teaching are considered is that in any particular teaching situation, four of the five factors are relatively fixed. In real situations teachers are confronted with clearly defined instructional programs designed to achieve stated objectives for a specifically identified clientele at a particular time. In addition, the storehouse of what is known about teaching and learning at a given time is relatively stable. Consequently, the factor influencing decision making about instruction that is most flexible and potentially responsive to change is the knowledge and skill of the teacher.

A basic idea undergirds this book: A teacher's knowledge and skills, both in the content to be taught and in the psychology of teaching and learning, have a major influence on instructional decision making and, in turn, on learn-

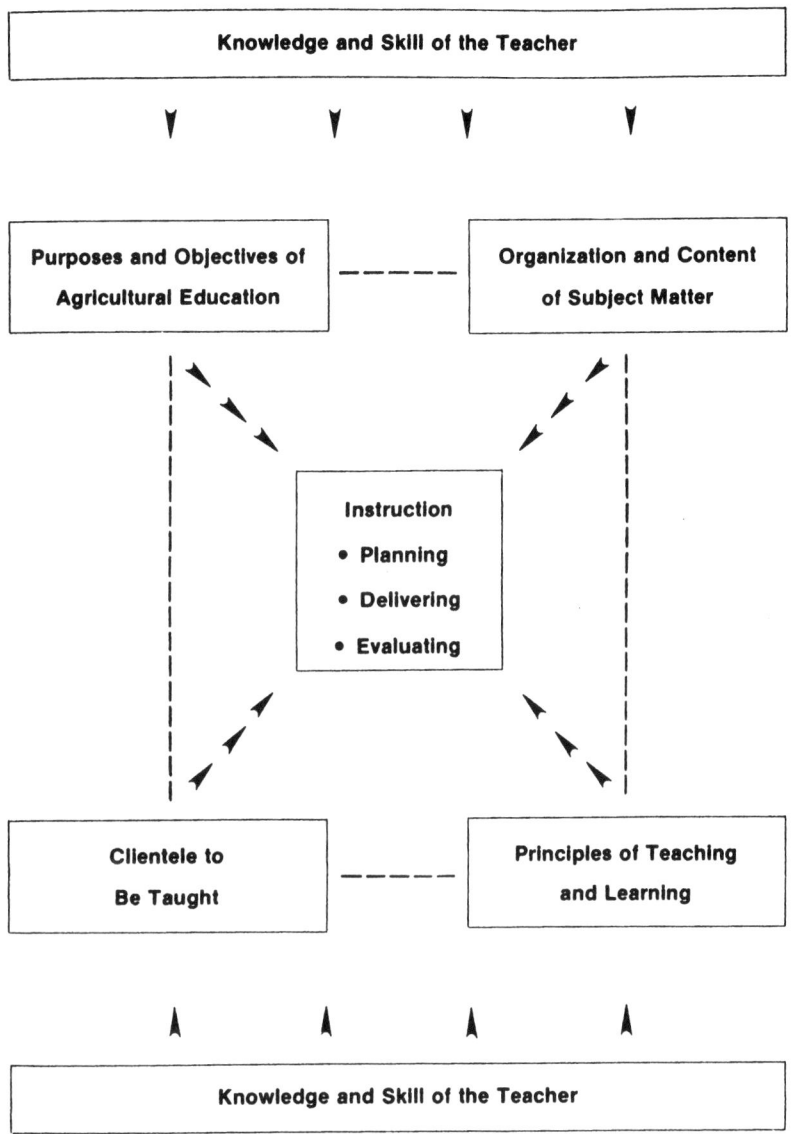

Figure 1-1 Factors influencing decisions about instruction.

ing outcomes. Each chapter is designed to instruct teachers of agriculture and those preparing to teach agriculture in the fundamental theory and the important principles of teaching and learning. Each chapter describes and illustrates how teachers can apply these understandings in planning, delivering, and evaluating instruction. The focus of this chapter is to describe how the objectives of instructional programs, clientele to be taught, subject matter, and the psychology of teaching and learning influence decisions teachers must make about instruction in order for students to achieve high levels of competence.

PUBLIC SCHOOL EDUCATION IN AGRICULTURE

Agricultural subject matter is taught at all levels in the public schools, from kindergarten to the university level. In kindergarten and the elementary grades, agricultural instruction includes animals and plants; nutrition and food; and how people live, work, and play. Outdoor educational activities in elementary schools involve agriculture and conservation of natural resources. Agriculture as a specific course or an identifiable unit of instruction in other courses appears sometimes in middle school, but courses in agriculture are most often offered at the senior high school level; in postsecondary schools, technical institutes, and community colleges; in colleges and universities; and in adult and continuing education programs offered by high schools and postsecondary schools and colleges.

Objectives of Instruction in Agriculture

Instruction pertaining to agricultural topics, as well as specifically identified courses in agriculture, contribute to the attainment of a number of objectives. These objectives vary depending on the level at which instruction is provided and the persons for whom the instruction is offered. It is important for teachers of agriculture to be aware of and understand the range of objectives for educational programs to which agricultural instruction contributes. Few if any agricultural education programs are designed to emphasize equally all of the objectives for agricultural education discussed in this section. Usually, specific agricultural education programs are designed primarily to accomplish one of the objectives described; however, teachers need to be alert to the fact that a particular agriculture course or unit of instruction designed to accomplish a specific objective may, for some students, contribute to the attainment of other equally important, meaningful objectives.

Provide Instruction about Agriculture—Agricultural Literacy. The agricultural industry—the production of food and fiber and the related complex of agribusiness and industry—is an essential and important part of the economic, political, and social concerns of the nation and the world. Persons who are knowledgeable about the community, state, and nation in which they live must have some appreciation for and knowledge and understanding about the role of agriculture in their lives. One purpose of instruction in agriculture is to develop knowledge and skills that contribute to the general education and avocational interests of persons who are not or will not be occupationally engaged in the agricultural industry.

The Committee on Agricultural Education in Secondary Schools appointed by the National Academy of Sciences concluded that instruction about agriculture should be offered to all students, regardless of their career goals or whether they are urban, suburban, or rural.[1] A major goal stated in *The National Strategic Plan and Action Agenda for Agricultural Education*[2] is "All students conversationally literate in agriculture, food, fiber and natural resources systems." Specifically, instruction about agriculture contributes to

- Understanding and appreciation of agriculture for the welfare of all; the interrelationships between agriculture and other aspects of business and commerce; the contribution of agriculture to world trade; and the interrelationships between rural and urban people.
- Understanding and appreciation of the complex processes of food production, processing, and distribution, and the cost of food acquired at each step of the process.
- Awareness of the responsibilities of all citizens in influencing public policies that affect agriculture.
- Familiarization and appreciation of the history of agriculture and rural life, the advances made in agriculture and their impact on all citizens, the values of rural people, and the literature, art, and music that give insight into our rural heritage.
- Knowledge of nutrition that leads to informed choices about diet and health.
- Knowledge and attitudes needed to care for the environment.
- Practical knowledge and skills useful in engaging in avocational interests such as landscaping, conserving natural resources, raising food, performing mechanical skills, and using open spaces in urban areas as well as rural areas for leisure activities.

Instruction about agriculture that has as its major objective the development of avocational and practical arts knowledge and skill is most applicable for students in school and adults who are not engaged in an agricultural occupation. In many cases, subject matter pertaining to agriculture is taught in the elementary grades and in general agriculture and practical arts courses taught in junior high schools and in some senior high schools. Examples of long-standing practical arts programs in agriculture include the school gardening program in the Cleveland Public Schools and the elementary and junior high school agricultural education programs in the Los Angeles School District. Instruction offered by high schools and community colleges as regular or continuing education courses in landscaping, floriculture, lawn care, and mechanical skills for those not engaged in an agricultural occupation are exam-

---

[1] National Academy of Sciences, Committee on Agricultural Education in Secondary Schools. *Understanding Agriculture: New Directions for Education.* Washington, D.C.: National Academy Press, 1988.
[2] *The National Strategic Plan and Action Agenda for Agricultural Education: Reinventing Agricultural Education for the Year 2020.* Washington, D.C.: National Council for Agricultural Education, 2000.

ples of instruction designed primarily to develop avocational and practical arts knowledge and skill in agriculture.

Frequently, instruction about agriculture that is designed to achieve agricultural literacy objectives is offered by elementary school teachers and teachers of other courses in the school rather than by persons whose teaching specialty is agriculture. When this is the case, teachers of agriculture in these schools have an opportunity and a responsibility to provide consultative assistance to other teachers who teach about agriculture and involve students in school activities concerning agriculture and rural life. In some middle schools, and to a lesser extent in senior high schools, teachers who are specialists in agriculture are employed to teach quarter-, semester-, or year-long practical arts courses in agriculture.

Provide Exploration of and Orientation to Occupations Requiring Knowledge and Skill in Agriculture. Specialists in the psychology of occupational and career development indicate that the choice of vocation is one of the major concerns of adolescents, particularly as they proceed from early to late adolescence. Emancipation from parents and the home, attainment of economic self-sufficiency, and recognition as an adult are achieved largely through successfully selecting, preparing for, and becoming established in an occupation.

Occupational decision making is a process, not an event. Many factors, including awareness of and knowledge about occupations, are involved as people make occupational and career decisions. Vocation choice is based on the occupations of which a person is aware. There is evidence that occupational choices are made in terms of what people know about themselves and what they know about the world of work. Emotional needs influence the occupation choice process; however, knowledge about occupational areas and specific jobs is also important. Actual work experience is crucial for the reality testing that is a part of the occupational choice process.

Instruction in and about agriculture, particularly for adolescents and preadolescents, must have as one of its objectives the provision of information and experiences about occupations involving knowledge and skill in agriculture, the type of preparation needed for entry and progress in these occupations, the attributes of those who are successful in these occupations, and the outlook for employment or self-employment and advancement. Agricultural education programs that emphasize occupational exploration and orientation, if this objective is to be achieved, must provide opportunities for students to participate in actual work experiences. For middle school students and some older adolescents this reality testing in the world of work begins with observational experience on farms and in agribusiness firms (e.g., veterinarian offices, landscaping firms, meat processing plants, or farm lending agencies). To be most effective, instruction designed to teach about the world of work must, if at all possible, provide actual work experience in one or more agricultural occupations. Instructional programs designed to emphasize occupational exploration and orientation provide students with information about and experience in a variety of occupations rather than specialized information and extensive experience in one job or occupational area.

Develop Knowledge and Skill for Occupational Competence.
Occupational proficiency—preparation for and advancement in the world of work—is stated frequently as the major objective of agricultural education in the public schools. Much of the agricultural instruction offered in secondary and postsecondary public schools in the United States has as its major objective occupational competence. Agricultural education programs in secondary and postsecondary schools are financed in part by federal and state funds earmarked for occupational education in addition to local funds allocated for the support of public schools.

The Committee on Agricultural Education in Secondary Schools of the National Academy of Sciences[3] recommended that instruction in agriculture designed primarily to develop occupational competence be upgraded to prepare students more effectively for careers in agricultural science, agribusiness, marketing, management, and food production and processing. The *National Strategic Plan and Action Agenda for Agricultural Education*[4] lists the following specific objectives for occupational competency in agriculture.

- Students must be prepared for successful careers in global agriculture, food, fiber, and natural resources systems.
- Every agriculture student must have opportunities for experiential learning and leadership development.

These objectives for public school education in agriculture emphasize the development of occupational competence; however, the objectives also recognize the importance of general education—namely leadership—for occupational competence and success.

Prepare for Advanced Study of Agriculture. An objective of agricultural education programs is to prepare those enrolled for more advanced study of agriculture. For example, instruction in agriculture at the high school level may be designed to prepare graduates for the study of agriculture in postsecondary technical institutes and community colleges or in four-year colleges and universities. Postsecondary agricultural education programs also prepare students for further study of agriculture at the university level.

Agriculture courses of this type offered in community colleges are usually described as transfer courses, indicating that the courses are intended to transfer to a four-year college or university for credit toward a baccalaureate

---

[3] National Academy of Sciences, Committee on Agricultural Education in Secondary Schools, *op. cit.*
[4] *The National Strategic Plan and Action Agenda for Agricultural Education, op. cit.*

degree. Agriculture instruction offered at the secondary and postsecondary levels with the primary objective of occupational competence also allows students to learn about the opportunities and needs for advanced study in agriculture. In these cases, an agricultural education program designed primarily to achieve an occupational competence objective may simultaneously contribute to the preparation of students for advanced study of agriculture.

Agricultural Education as a Part of Public Education

Agricultural education programs in the public schools are designed to accomplish educational objectives that pertain specifically to acquiring appreciation, understanding, knowledge, and skills applicable to the agricultural sciences, agribusiness, and the production and processing of food and fiber. It is also important that agricultural education programs be designed and conducted such that instruction in and about agriculture contributes to the achievement of all purposes of the school. In addition to the career and occupational development of students, public education is concerned with the students' intellectual, social, and cultural development. Instruction in and about agriculture can contribute to other purposes of public education without neglecting a major commitment to achieving specific agricultural education objectives.

Teachers of agriculture must be aware that skills necessary for occupational success include the abilities to read, write, speak, and listen; competence in using numbers; and the ability to work cooperatively and harmoniously with others. These skills are taught in what is usually referred to as general education subjects in the public schools. Teachers of agriculture need to realize that instruction in agriculture also can contribute to the attainment of these skills.

Teachers of agriculture can assist in relating instruction in agriculture more closely with the rest of the school's program in two ways. First, teachers, through their comments and actions, can communicate directly and indirectly to students that what they are studying in English, mathematics, science, and the other core courses is relevant to their interests and goals. Second, teachers of agriculture must make it evident that instruction in agriculture contributes directly to the attainment of general skills relevant to communication, computation, problem solving and decision making, human relations, and leadership. Teachers of agriculture must make a conscious and deliberate effort to maximize this contribution of agricultural education to the overall objectives of the school. Effective teachers of agriculture make important contributions to the understanding and use of the social sciences, mathematics, biological and natural sciences, and English. Agriculture is an applied science, so it is not unreasonable to expect that some of the best teaching of science will be done in courses in agriculture. One of the basic premises of The National FFA Organization is the development of some very important general education attitudes and skills, particularly citizenship and leadership abilities.

Too often agricultural education is regarded primarily as a function of a department of the school, not as a function of the total school system. Teachers of agriculture need to plan and conduct instructional programs such that agricultural education, while achieving specific and unique objectives related to agriculture, is an integral part of the school system. It is also important that teachers use instructional techniques and strategies that make it possible for instruction in agriculture to contribute directly and substantially to all purposes of the school.

Clientele in Agricultural Education

There is a high degree of interdependence among educational objectives for a particular agricultural education program, the content and organization of the educational program, and the clientele who enroll in the instructional program. Not only are agricultural education programs designed to accomplish one or more educational objectives but the programs are also designed for specific clientele. Agricultural education is provided to students enrolled in public schools and adults and youth who have completed or left school. Clientele receiving instruction in and about agriculture range from kindergarten and elementary school students enrolled in agricultural literacy programs to secondary and postsecondary school students enrolled in courses that emphasize occupational competence and preparation for advanced study. Also, out-of-school youth and adults enroll in continuing education programs in agriculture to enhance their knowledge and skill.

Subject Matter for Instruction in and about Agriculture

The subject matter taught in agricultural education programs is broad and diverse. The National Science Foundation's Committee on Agricultural Education defined the agricultural sector as including "supply and service functions involving agricultural inputs; production of agricultural commodities; processing and distribution of agricultural products; use, conservation, development and management of air, land, and water resources; development and maintenance of rural recreational and aesthetic resources; and related economic, sociological, political, environmental, and behavior functions."[5]

The National Academy of Sciences's Committee on Agricultural Education in Secondary Schools[6] defined instruction in and about agriculture as including basic concepts and knowledge about

- Production of agricultural commodities, including food, fiber, wood products, horticultural crops, and other plant and animal products;

---

[5] Committee on Agricultural Education, Commission on Education in Agriculture and Natural Resources, National Science Foundation. *Agricultural Education for the Seventies and Beyond.* Washington, D.C.: American Vocational Association, pp. 6–7.
[6] National Academy of Sciences, Committee on Agricultural Education in Secondary Schools, *op. cit.*, 1988, p. vi.

- Financing, processing, marketing, and distribution of agricultural products;
- Farm production supply and service industries;
- Use and conservation of land and water resources;
- Development and maintenance of recreational resources; and
- Related economic, sociological, political, environmental, and cultural characteristics of the food and fiber system.

The subject matter taught in agricultural education programs in public schools includes the following specialized areas of content:

- Agricultural literacy
- Agricultural science
- Agricultural production
- Agribusiness supplies and services
- Agricultural mechanics and engineering
- Agricultural processing and marketing
- Horticulture
- Aquaculture
- Agricultural and natural resources conservation
- Forestry

Dimensions of a Complete Program of Agricultural Instruction

Classroom Instruction. The core of a successful agricultural education program is formal instruction in the school. Thorough and expert classroom instruction sets standards for all phases of the instructional program and determines to a considerable extent what out-of-classroom or out-of-school activities will be conducted. Instruction in the classroom involves not only group instruction but also individual instruction and supervision. Expert classroom instruction begins with a well-planned, relevant course of study, which is discussed in Chapter 3. Formal instruction that results in high levels of achievement by students requires also that teachers creatively use group and individualized teaching techniques, which are described in Chapters 6 and 7.

Application of Learning. Instruction that results in maximum achievement by students includes techniques and activities that allow students to apply what is taught in the classroom in real or laboratory situations. To be most effective, application of learning must be supervised by the teacher and be accompanied by additional instruction to ensure that what is taught is not only understood but is useful to students. There are three primary systems for providing opportunities for students to combine classroom instruction with the application of what is taught—the school laboratory, supervised agricultural experiences, and the FFA, the student organization. In Part III, Chapters 9, 10, and 11 deal with these three approaches for making what is taught practical and useful.

The School Laboratory. Laboratory instruction is an essential complement to classroom instruction if students are to achieve the highest levels of competence. It is virtually impossible to teach agriculture adequately without laboratories provided by the school. School laboratory facilities most needed for instruction in agriculture include specialized laboratories directly related to the specific areas of instruction. Access to a computer laboratory is essential. Production agriculture programs require a land laboratory; biotechnology instruction must be accompanied by laboratory experience in a properly equipped facility; greenhouses and plant growth areas are required for ornamental horticulture programs; a laboratory adequately equipped for engine analysis, repair, and operation is essential for agricultural mechanics programs. Other school laboratories might include, but are not limited to, forests, nature trails, wetlands, ponds, aquaculture facilities, putting greens and fairways, barns, hydroponics facilities, wildlife habitats, or computer labs.

Supervised Experience. Instruction that emphasizes the development of occupational competence offered in secondary and postsecondary schools places primary emphasis on the development of knowledge and skill needed for entry and advancement in the world of work. Laboratory instruction in programs of this type, is instruction and supervision in the actual careers that students are preparing to enter. For students enrolled in production agriculture programs, a farm or ranch becomes the location for supervised agricultural experience. Such experience in nonfarm agribusiness and industrial firms is essential for students enrolled in other specialized areas such as agricultural mechanics, agricultural business and supplies, ornamental horticulture, and agricultural products. For students enrolled in agricultural science programs, a university, private laboratory, or local research firm becomes the location for supervised agricultural experience.

FFA. The organization for students enrolled in instructional programs in agriculture is a laboratory for acquiring and applying knowledge and skill in leadership, citizenship, and career success. FFA is an integral part of the total agricultural education program and contributes best to the attainment of the objectives of agricultural education and the school when the organization and its activities are developed as laboratory activities that are a part of well-planned and delivered instruction in agriculture.

Use of Community Resources. Well-organized and conducted agricultural education programs are community oriented. Instruction takes place in the community as well as in the school. In-school classroom and laboratory instruction is most realistic and interesting when it reflects the agriculture and agribusiness in the community. Persons as well as the physical facilities in the community are resources useful in teaching. In effect, the community is a laboratory for the instructional program. Involvement of persons and facilities in the community in planning and conducting instructional programs has two major advantages. First, the involvement of resource persons as consultants and teachers as well as the use of the facilities of the community makes instruction relevant and real, and thereby more interesting and meaningful to students. Second, a high degree of involvement by persons in the community is an excellent way for them to get firsthand information about what goes on in school generally as well as specific information about the agricultural education program.

Facilities and Organizations. Some of the most valuable community resources that contribute to successful agricultural education programs are business and industrial firms, farms, parks, recreational areas, governmental agencies serving the agricultural sector, and organizations that relate to agriculture. These community resources in many cases supply the actual subject matter taught. For example, topics such as the organization and operation of agribusiness firms, farm management, career opportunities in agriculture, and qualifications and requirements for employment in agricultural occupations are best taught through actual case examples and, if possible, with actual contact with both the facilities and the persons owning, managing, and working in the various farm and agribusiness firms and organizations. Community facilities are used effectively as sites for field trips and as training stations for students' supervised agricultural experiences. Community facilities and the persons associated with them provide excellent opportunities for independent study activities for students.

People in the Community. People living and working in the community provide a rich resource of specialized knowledge and skills to contribute to an effective instructional program. The use of experts as resource people is a teaching technique that has advantages both to teachers and to students. People in the community who have high levels of expertise in the subjects being taught readily respond to opportunities to assist with classroom and laboratory instruction, to instruct students during field trips, and to consult with students who are conducting independent studies or class assignments. Farmers, extension educators, and employees in agribusiness firms provide on-the-job supervision and instruction to students who are placed on farms and in agribusinesses for supervised agricultural experience. FFA alumni can be of assistance to the teacher. The appropriate use of resource people in teaching agriculture ensures that up-to-date information is being delivered and that technically correct skills are being taught. Resource people, therefore, provide excellent opportunities to update and expand the knowledge and skill level of teachers of agriculture.

Advisory Committees. A practice that has proven effective for making the expertise of people in the community available to teachers and the school is the formation of a school-sponsored advisory committee for agricultural education. Properly organized and appropriately used advisory committees contribute much to planning, conducting, and evaluating instruction in agriculture. It is important that the school administration and the school's governing board approve policies for the formation and use of the advisory committee. Teachers must be skilled in using advisory committees in an advisory and consultative capacity. These committees play an important role in linking the school and the community; they offer advice about the objectives the program should strive to achieve, about the clientele to be served, about the programs that contribute most directly to the accomplishment of objectives, and about the extent to which objectives have been accomplished. Advisory committees are a means whereby accurate information about the school and the agricultural education program is communicated to persons in the community. Teachers who direct agricultural education programs that are community oriented and contribute best to the acquisition of knowledge and skill by those enrolled have an organized and functioning advisory committee as a part of a complete program of agricultural education.

Parents. In addition to school–community cooperative efforts, effective and successful agricultural education programs require cooperative and mutually supportive relationships among teachers, students, and parents or guardians. It is essential that teachers take the initiative in informing parents about the agricultural education program and activities, in communicating to parents what the expectations are for students enrolled, and in involving parents in appropriate activities when that involvement contributes directly to effective teaching and learning. Parents supervise the agricultural experience activities of students when students' places of employment are the home farms or the agribusinesses owned by the parents. Highly competent teachers put high priority on knowing personally the parents of students and having firsthand knowledge of students' home situations. This direct knowledge of students, their parents, and the situations in which they live and work can best be obtained by teachers visiting the homes and workplaces of students. Students' backgrounds, their interests and motivations, and their parents' support and aspirations for them is vital information for teachers who make decisions about instruction that results in students attaining a high level of knowledge and skill in agriculture.

Teachers and Administrators in the School. Earlier in this chapter the point was made that agricultural education programs contribute directly to all purposes and objectives of the school. Also it was proposed that the general education and other specialized courses taken by students enrolled in agriculture contribute to the achievement of specific objectives in agriculture classes. The interrelatedness of instruction in and about agriculture with the total instructional program in the school is real and visible when teachers of agriculture make conscious and deliberate efforts to work cooperatively with administrators, teachers, and counselors in the school. It is imperative that agriculture teachers conduct programs in a manner that makes it evident that school administrators have the same degree of administrative and supervisory responsibilities for the agricultural education program as they do for other programs in the school. Teachers of agriculture must be aware of and follow school policies pertaining to the discipline of students, absence from school for field trips and supervised experience, use of school laboratories to provide products and services to the community, and all other matters that in some way differentiate agricultural education from other programs in the school. Informing administrators about the agricultural education program and the accomplishments of students is necessary. However, teachers of agriculture must ensure that school administrators are actively involved in all aspects of the program so that they consider instruction in agriculture a viable and important part of the school's program and activities.

Successful teachers of agriculture keep other teachers and other professional personnel, such as guidance counselors, informed about all aspects of the agricultural education program. Information about the agricultural education program does not automatically get communicated to other professional personnel in the school. Teachers should use every opportunity not only to provide information about the school's agricultural education program but also to involve teachers, counselors, and administrators in appropriate roles, such as consultants on special topics, advisors for student activities, and guests at special functions such as the annual FFA banquet. Teachers of agriculture, if they expect other teachers and professional personnel in the school to know about and take interest in the agricultural education program, must make deliberate efforts to acquire firsthand information about what is taught in other courses in the school so that the importance and relevance of these courses to agriculture can be emphasized. It is important that teachers of agriculture confer with other professionals in the school about mutual concerns that relate to all facets of a student's educational experiences. Teachers of agriculture who operate as contributing and cooperative members of a school's faculty accomplish much in demonstrating that agricultural education is an important part of the school, not exclusively the function of one department in the school.

ORGANIZATION AND CONTENT OF SUBJECT MATTER

Another factor influencing decisions teachers make about planning, delivering, and evaluating instruction is the organization and content of the subject matter. (See Figure 1-1.) For example, different instructional techniques are required for teaching facts about the agricultural industry in the community, for teaching welding skills, for teaching the application of principles of plant growth, or for teaching affective attributes such as working cooperatively with others. What is to be taught and how that subject matter is structured and organized are important in determining how instruction can be provided in the most realistic, interesting, and effective manner.

A review of the factors influencing decisions about teaching indicated in Figure 1-1 reinforces the interrelatedness among objectives, subject matter, clientele, and the psychology of learning. By now it should be clear that the purposes and objectives of the agricultural education program determine to a considerable extent the subject matter that will be taught. What is known about how people learn—the principles of teaching and learning—has direct and important implications for how the subject matter should be organized and structured to facilitate most advantageously the teaching-learning process. This point is highlighted in Chapter 2, which discusses some of the major principles of teaching and learning that concern the organization and content of subject matter. Some characteristics of those who are to be taught—the clientele—play a decisive role not only in defining the subject matter that must be taught but also in directing the sequence in which units of instruction are taught most beneficially.

Organizing and structuring subject matter is a basic concern of planning the course of study. Chapter 3, "Planning the Course of Study," provides instruction and examples relating to determining course content and to organizing and sequencing subject matter such that instruction is meaningful to students and results in high levels of achievement by those who successfully complete the instructional program.

CHARACTERISTICS AND INTERESTS OF CLIENTELE

The characteristics, interests, and aspirations of those taught, whether high school students or adults, are potent factors influencing decisions about instruction. The interests and aspirations of students and their families, students' backgrounds and previous levels of success in school, prior experiences and instruction related to the subject matter, and academic attributes and study skills must influence the planning, delivery, and evaluation of instruction if maximum levels of competence are to be achieved.

The direct association between the purposes and objectives of a particular agricultural education program and the clientele the program serves has been emphasized previously. Almost without exception, a specific course designed to accomplish certain objectives can be directed toward a specific clientele group possessing distinctive characteristics that have direct and significant implications for instructional strategies and techniques. Expert teachers make special efforts to understand the unique characteristics of their students and to be aware of students' interests, aspirations, and motivating forces. Understanding students and what motivates them in concert with a mastery of the basic principles of teaching and learning equips teachers with the ability to plan for and provide instruction that accomplishes efficiently and effectively the learning outcomes sought.

Characteristics of learners influence directly the selection and use of instructional techniques. Even though a specific course is designed for an identifiable group of persons with similar characteristics, teachers need to be alert to diversity within the group. Groups that appear to be relatively homogeneous for certain characteristics frequently display a great deal of diversity and variation in background, interest, prior instruction and experience, learning style, and academic skills. It is the diversity within a group that must be recognized and addressed if instructional techniques are to reflect a real concern for individual needs and interests.

Interests and Aspirations

Psychologists have identified some major social and personality needs that motivate individuals in addition to the physical needs for food, drink, and sex. The social and personality needs that motivate learners include the needs for status, security, affection, independence, and achievement. The adolescent's needs for status and independence are particularly important. Activities and experiences that are a part of or accompany agricultural education courses and programs will be sought by persons who perceive instruction in agriculture as an effective means for fulfilling these social and personality needs, especially the need for status, independence, and achievement. When high school students enrolled in agricultural education programs are asked to name the most important reasons for enrolling, their responses indicate clearly that they believe instruction in agriculture will contribute to the attainment of a variety of their goals and objectives. Following are the responses of some high school students enrolled in agricultural education courses to the question "Why did you enroll in agricultural education?"[7]

Occupational Orientation and Exploration. "I enrolled in agricultural mechanics because I didn't know exactly what I wanted to do, and now I find I like mechanics pretty well." "So I can think about what I want to do in life—work or go to college."

Preparation for Further Schooling. A student in agricultural mechanics plans "on going to technical school to further my education." Another student studying animal science believes "it would help me in my future career which is becoming a veterinarian."

Preparation for Employment or Self-Employment. "I want to manage a farm so I can stay on the farm for a living."

---

[7] Warmbrod, J. Robert. "Individual Goals and Vocational Education." In Alfred H. Krebs (Ed.), *The Individual and His Education*. Washington, D.C.: American Vocational Association, 1972, pp. 119–128.

Practical Arts Knowledge and Skill. A ninth-grade student elected agricultural education "because I am interested in agriculture." A student in an environmental science course "enrolled to learn all I can about the pollution problem and wildlife. This course helps me understand what I and others will face in the future."

Independence—Status—Achievement. "Need money" was the way an eleventh-grade student expressed the reason for enrolling in an agricultural business program that included on-the-job supervised agricultural experience. "I love flowers. I have a certain feeling of pride in myself when they are in bloom and people buy them. This course is a great opportunity for me to really understand the basics of floriculture."

Highly competent teachers of agriculture are aware of the motivating forces behind students' decisions to enroll in agriculture courses. The students' motivations provide a base for making instruction interesting and useful.

Background and Home Environment

Parents or guardians, as well as students and teachers, are important in shaping the interests and goals of adolescents. Teachers of high school courses in agriculture find it important to know what aspirations parents hold for their daughters and sons. The attitudes displayed in the home about school in general, frequency of attendance, level of achievement expected, and the perceived value of instruction in agriculture play an important role in influencing students' behaviors and attitudes. Parents' level of education, occupational status, and income are important factors that have profound influence on how students value schooling, as well as determining the resources available for assisting students in achieving their goals and participating in activities that are an integral part of instruction. For example, a student from a family with limited financial resources can hardly be expected to provide the materials for constructing an agricultural mechanics project required as a part of laboratory instruction. Likewise, it is difficult for students who must work part time to help support a family to demonstrate a high level of interest and enthusiasm for learning about and participating in FFA leadership activities that take place primarily during after-school hours.

Experience Related to the Subject Matter

Careful attention must be given to the previous experience students have concerning what is being taught and whether students are participating concurrently with instruction in work or supervised agricultural experiences. The extent of students' work experience involving the subject matter is important. Equally important is the type of work or occupational experience students bring to the instructional setting. Students who are working or have worked independently in situations in which they make decisions bring to the classroom or laboratory different insights and skills than do students whose experience is primarily observing and assisting others. Students who already possess specialized knowledge and skills in certain aspects of the subject matter can be used effectively as resource persons for the entire class or as tutors or coaches for other persons in the group. Experience of those enrolled in an adult education program is a major factor in selecting teaching techniques most appropriate for adults in contrast to teaching techniques most appropriate for high school students.

Previous Instruction

Teachers who are in the best position to make instructional decisions that contribute to students achieving high levels of competence also discover and attend to the previous instruction of students in agriculture and other courses that are closely related, particularly science and mathematics. Students who have completed ninth- and tenth-grade courses in agriculture prior to enrolling in an ornamental horticulture course in the eleventh grade can be taught differently from other students in the course who have no prior instruction in agriculture. Teachers of agriculture need to be alert to what is being taught about agriculture to students in other subjects in the school. If agriculture teachers are knowledgeable about what is taught in science and mathematics courses, for example, they can reinforce this knowledge and skill by demonstrating its application and use in their agricultural instruction.

Academic Attributes—Success in School

Learners, whether high school students or adults, vary not only in their interests and motivations for learning but also in intellectual skills and attributes that are associated closely with success in school. Intellectual ability, usually described as level of intelligence, is a major factor that is for most students closely associated with achievement. Other important attributes include reading level, the ability to reason and solve problems, and the determination and desire to stay with a task in order to achieve a high level of competence. It is important that teachers have accurate knowledge of their students' intellectual and academic attributes if appropriate teaching strategies are to be used. For example, students who cannot or will not read can be expected to demonstrate only meager achievement if printed material is the only resource students use to acquire new knowledge.

Students' enthusiasm for school in general and for specific courses is influenced strongly by the success they experience. This factor should be considered by teachers as they plan and teach. Instructional strategies need to be devised that ensure, wherever possible, that all students experience some degree of success in learning and that successful achievement is recognized and rewarded. In some cases it will be necessary for teachers to make special efforts to communicate to students and prospective students that the instructional strategies and techniques will, in fact, be different from what they have experienced in the past. For example, dropouts who experienced limited success in school may show little enthusiasm for a proposed course unless it is made clear to them how the new course will allow them to succeed in acquiring skills and knowledge they believe are important.

Study Skills

Too often teachers overlook the fact that students as well as teachers need to be adept in the teaching strategies and techniques used. Successful teachers make sure that learners are active participants in the teaching-learning process. Students, to be active learners, must have both the desire and the requisite study skills for active participation in the various techniques and activities employed by the teacher. The following illustrations emphasize this point. Instructional techniques requiring students to use books, magazines, and other publications as resource materials are not efficient or effective unless students know how to use an index or table of contents to locate relevant sections of a publication. Students with limited knowledge, skill, or experience in using computers are at a disadvantage in using the Internet and other electronic media. Independent study as a technique of instruction results not only in meager achievement but also in a frustrated learner unless students have been taught skills that are necessary for independent study, such as how to identify and state problems and questions, how to locate and use appropriate resources, how to evaluate information and opinions, how to formulate and test conclusions, and how to report the results of the study orally or in writing.

PRINCIPLES OF TEACHING AND LEARNING

Knowledge and understanding of the psychology of learning are basic to making decisions about and using appropriate instructional strategies and techniques. Some understanding by the teacher of the conditions that stimulate learning and how learning takes place is essential if instruction is to result in a high level of competence achieved by those who are taught. Teachers who are familiar with and understand the basic tenets of how people learn have the capacity to create innovative teaching techniques and use instructional media that are most appropriate to achieve the learning outcomes sought. Teachers who understand how learning takes place possess the ability both to diagnose problems encountered in teaching and to prescribe techniques and strategies that improve the teaching-learning process.

Fundamental to what is being taught through this book is the indispensable and direct connection between the practice of teaching and the psychology of learning. Each chapter is undergirded by the basic premise that teacher behaviors of planning, delivering, and evaluating instruction must be grounded firmly in what is known about how people learn.

Teaching and Learning

Teaching is best described as guiding and directing the learning process such that learners acquire new knowledge, skills, or attitudes; increase their enthusiasm for learning; and develop further their skills. One psychologist describes teaching this way.

> There is no such operation as teaching in and of itself. No one can teach anyone anything; he [sic] can only arrange conditions whereby a learner might learn. Among such conditions are showing and telling, but whether or not learning goes on depends more on the learner than the teacher.[8]

This description of teaching makes it clear that the learner, as well as the teacher, plays a central role in the teaching–learning process. In this context learning is usually described as a process by which persons, through their own activity, become changed in behavior. An essential element of learning is that unless the learner processes the subject matter being studied in a meaningful and understandable manner, little will be learned or retained. Bender and Boucher describe learning as follows.

> When a person learns, s/he feels, thinks, and acts differently because his or her interests, skills, abilities, understandings, attitudes and appreciations have been changed.... It is important to understand that learning is a self-active, personal, choice making process. A person learns what he or she wants to learn.[9]

Principles of Learning

In Chapter 2 learning theory and research substantiate some of the basic principles of learning. The connection between the principles of learning and the practice of teaching is illustrated by describing teacher behaviors that exemplify how the principles of learning are translated into teaching practice. Principles of learning will be presented that pertain to:

- The organization and structure of subject matter
- Motivation
- Reward and reinforcement

---

[8] Bugelski, B. R. *Some Practical Laws of Learning*. Bloomington, IN: The Phi Delta Kappa Educational Foundation, n.d., p. 30.
[9] Bender, Ralph E., and Boucher, Leon W. *Classroom Climate for Effective Learning*. Columbus, OH: Department of Agricultural Education, 1977, p. 4.

- Techniques of teaching
- Transfer of learning

KNOWLEDGE AND SKILL OF THE TEACHER

Throughout this chapter the case has been made that the objectives of the instructional program, the organization and content of subject matter, the interests and characteristics of clientele, and the principles of teaching and learning singly and collectively influence decisions teachers make about planning, delivering, and evaluating instruction. But there is one additional factor that holds the key to the process: the knowledge and skill of the teacher. Teachers must put the pieces together and integrate the system if learners are to acquire new knowledge, skills, and attitudes.

One dimension of teachers' knowledge and skill is expertise in the science and technology of agriculture. Expert teachers are able to relate this knowledge and skill in a meaningful way to the world of work in general and to occupations requiring knowledge and skill in agriculture. Highly competent teachers of agriculture understand and value the science and theory that undergirds what is taught. They also possess the ability to perform practical skills. Actual work experience in agriculture is essential if teachers are to achieve the level of technical competence required for teaching agriculture successfully.

Another essential dimension of teachers' knowledge and skill is their level of teaching competence, including an understanding of and the ability to use the principles of learning. Scientific, technical, and pedagogical competence are required if teachers are to direct the learning process such that students achieve to an optimum level the outcomes sought through instruction and study in and about agriculture.

It is important that persons preparing to teach agriculture and those who are teachers realize that it is essential that current knowledge and skill be continually updated and new knowledge and skill acquired if teaching is to be most effective. The continuing professional development of teachers pertains to their scientific, technical, and pedagogical competence. Teachers who are most successful are career-long learners both of what they teach and of how it is most effectively and efficiently taught.

SUMMARY

Factors influencing decisions about teaching include the purposes and objectives of the instructional program, the clientele being taught, the organization and content of the subject matter, principles of teaching and learning, and the knowledge and skill of the teacher. Objectives of instruction in agriculture relate to avocational and practical arts instruction (agricultural literacy), exploration of and orientation to agricultural occupations, development of occupational competence, and preparation for advanced study. Agricultural education is an integral part of public education. Clientele, subject matter, and instructional strategies describe the nature and functions of agricultural education. Characteristics of clientele, including interests and aspirations, background and home environment, experience related to the subject matter, previous instruction, academic attributes, and study skills influence decisions teachers make about teaching strategies. The knowledge and skill of the teacher are the overriding and integrating factors that result in an effective instructional program.

FOR FURTHER STUDY

Visit local schools offering instructional programs in and about agriculture. Discuss with the teacher purposes and objectives, clientele, subject matter of courses, instructional techniques and strategies, and the knowledge and skill needed by the teacher.

2

Principles of Teaching and Learning

A student teacher in a secondary school was assigned to teach a horticulture class the essentials of grafting. Review her description of the experience.

> Before entering the class I was warned the students have a bad habit of talking during instruction. In addition, they had a reputation for giving student teachers a hard time.
>
> My time to teach arrived. I walked into the room where the students were waiting with that "we're going to be bad" look in their eyes. Before they had a chance to tear the green teacher apart, I asked some follow-up questions on the lesson from the day before. They couldn't help but answer because I made the questions easy and understandable. Their answers pertained to reasons for grafting and examples of grafted material in the area. They really did have an interest in learning how to graft trees!
>
> Consequently, a student asked, "How do we graft trees?" This was the key question for which I had been searching. Immediately, I wrote that question on the board and waited for their responses in determining items to be considered in answering the main question. They came up with meaningful considerations that pertained to the question at hand. Their desire to learn was there! No gossiping or trouble was to be seen or heard the remainder of the period. The students searched the text for the answers and found them. After writing the answers in their notebooks, they were required to graft samples in the classroom. I reminded them that the next day we would go outside and graft some of the fruit trees behind the school.

Why was this a successful teaching experience? What principles of teaching and learning provide the rationale for the teaching strategies the student teacher used? What principles of teaching and learning explain the behavior of students, which differs considerably from what the student teacher expected?

In this chapter principles of teaching and learning are presented and explained. Teachers of agricultural science and those preparing to be teachers need to be aware of two important ideas that are fundamental to effective instruction. First, there is a direct connection between what teachers do in planning, delivering, and evaluating instruction and what students learn, regardless of whether the students are youth or adults. Second, the teacher is the person primarily responsible for the learning activities that take place in the classroom, laboratory, and supervised experience that, in the final analysis, determine the educational value of learning activities.

OBJECTIVES

After studying this chapter, you will be able to

1. Explain the principles of teaching and learning that are basic to optimizing learning outcomes (what students learn).
2. Make operational in the classroom, in the laboratory, and in supervised practice the fundamental principles of teaching and learning.

BASIS FOR EFFECTIVE TEACHING

Principles pertaining to the psychology of learning serve as the basis for effective teaching. Principles that are basic to teaching and learning are only effective if they are applied in practice. Table 2-1, in the chapter appendix, summarizes the principles of teaching and learning and lists the teacher behaviors and practices that can be used to put the principles into practice.

It is essential that teachers possess expert competence in the science, technology, and skills of the specialized areas of agriculture they teach. However, knowledge and skill in subject matter, although essential, is not sufficient if optimum learning outcomes are to be achieved. It is equally important that teachers know about, understand, and be able to use some basic principles of teaching and learning. Knowledge of the basic principles of the psychology of learning enables the teacher to be creative in the development of teaching techniques and instructional media, to understand why certain instructional strategies and techniques do or do not work, and to diagnose teaching situations such that appropriate instructional strategies can be selected to achieve the desired learning outcomes. Teachers who understand the factors basic to effective teaching and learning are able to plan, deliver, and evaluate instruction that results in the acquisition of high levels of competence by those who are taught.

Principles of teaching and learning provide the foundation for all phases of the instructional process. Certain principles provide the rationale for the organization and structure of subject matter; other principles are fundamental to the motivation of students, to the appropriate use of reward and reinforcement, and to the selection of teaching techniques. A major expectation of instruction in agriculture is the application and transfer of learning. Some principles of the psychology of learning are basic to instructional processes that result in the application and transfer of learning from the classroom and laboratory to the real world.

ORGANIZATION AND STRUCTURE OF SUBJECT MATTER
Principle 1

> When the subject matter to be learned possesses meaning, organization, and structure that is clear to students, learning proceeds more rapidly and is retained longer.

Effective teaching involves more than what the teacher does in the classroom and the laboratory. Teaching that is creative, interesting, and challenging to students and results in students achieving a high level of mastery begins with a course of study that makes sense to students. The subject matter should pertain directly to the educational objectives of the course. The content of the course needs to be subdivided into instructional units that indicate clearly the usefulness of what is to be learned, and the course content must be sequenced such that one can see and understand the logic underlying the organization of subject matter.

To learn effectively, students must see clearly how what is being taught contributes directly to the accomplishment of their goals—how what is being taught is and will be useful to them now and in the immediate future. They must be aware of the logic that undergirds the organization and sequence of the instructional units that comprise the course.

It is the responsibility of the teacher to ensure that students understand what learning outcomes are to be achieved, what subject matter will be taught, how the subject matter will be useful, what the planned sequence of instructional units and activities is, and why the course is organized and sequenced as it is. To implement this strategy, at the beginning of the course the teacher should share the course of study with the students. By doing so, students are given a preview of what to expect and why. Openness with students about what is going on and why continues almost daily throughout the course. The effective teacher goes to great lengths to make sure that what is being taught and why it is being taught is clear to students. If students are to achieve and be enthusiastic about instruction, it is important that they not have their energy and attention diverted by their attempts to make sense out of the subject matter, how it is organized, and what they are expected to learn.

Meaning

Students find meaning in subject matter when it is evident how the subject matter can be used. The course of study illustrates and communicates the usefulness of subject matter to students when units of instruction are stated in action terms—tasks and activities to be performed. For example, an instructional unit titled "Landscaping a Residential Site" more directly and clearly denotes the application and use of subject matter than does an instructional unit titled "Principles of Landscape Design" even though the subject matter taught in the two units is the same. "Feeding my Sow and Litter" more clearly conveys the usefulness and application of subject matter to meaningful tasks than does "Swine Nutrition"; "Conducting and Participating in FFA Meetings" is more likely to attract and hold the attention of students than an instructional unit titled "Parliamentary Procedure."

Sequencing

The sequencing of instructional units within a course contributes directly to students' understanding the organization and structure of subject matter. As persons highly knowledgeable in the subject matter being taught, teachers tend to organize and sequence instructional units logically. Certain instructional units precede or follow other topics because of the inherent logic of the subject matter. The arrangement of chapters in textbooks in agriculture illustrates the experts' logical organization of subject matter. Teachers need to be aware that the inherent logic of subject matter is most evident and sensible to those who are experts in the area such that the interrelationships among elements of the subject matter are clear.

Students, who obviously are *not* experts in the subject matter they are about to be taught, do not automatically see or even care about an expert's logical organization of subject matter. Instead, students approach subject matter from a psychological organization and structure. Students are most interested in studying the various phases or topics of a particular subject in an order that corresponds to what is going on in the world around them. For example, high school students learning how to select and buy breeding swine do not first have to study the history and origin of the breeds of swine. In fact, insisting that students memorize the history and origin of the breeds of swine as a prerequisite to the action-oriented instructional unit on selecting and buying breeding animals will almost surely dampen students' interest and enthusiasm for the topic. Likewise, students who are being taught how to participate in FFA meetings do not first have to become experts on the history of FFA and how the organization is organized at the local, state, and national levels. These topics can be taught later, when students see some psychological sense to their acquiring that knowledge.

The contrast of the psychological organization of subject matter (from the perspective of the student) and the logical organization of subject matter (from the perspective of the expert) is aptly illustrated by one author's experience in teaching a production agriculture course to ninth-grade students. The course of study for the first few weeks included a series of instructional units about orientation to the school and to the study of agriculture, requirements for and planning supervised agricultural experience programs, and orientation to and becoming a member of FFA. After a few days of instruction, one student asked this sobering question: "When are we going to study agriculture?" In planning the course of study, teachers must remember that what is taught and the order in which it is taught needs to make psychological sense to students. It is the responsibility of the teacher to ensure that students understand the organization, structure, and sequence of the subject matter they are expected to learn.

Providing Structure. Teachers have numerous opportunities to make clear the meaning, organization, and structure of subject matter during the actual teaching of each instructional unit. The first requirement is that students be aware of and accept enthusiastically rather than passively the learning outcomes to be achieved. It is essential that students know about and understand the instructional objectives being sought if they are to perceive the instruction as meaningful. Teachers can use a number of specific strategies and techniques (what teachers do) to ensure that the organization, structure, and meaning of subject matter is clear to students. These teacher behaviors are labeled "use of structuring comments" and "cognitive clarity of a teacher's presentation."[1]

Teachers provide structure to subject matter when they make it clear to students how the various topics or instructional units of the course fit together. Making it clear when one topic ends and another begins, as well as making evident the interrelatedness of topics, enables students to see and understand the organization and meaning of subject matter. Teachers use structuring comments when they review at the start or the end of a lesson; review, clarify, and emphasize major points during the discussion and study; provide structuring comments before asking questions; and make evident the connections between what is being studied in agriculture and what students are studying in other courses in school. Expert teachers spend considerable time making sure that students understand why certain subject matter is taught and how the subject matter fits with what has already been learned and with

---

[1] Rosenshine, B., and Furst, N. "Research on Teacher Performance Criteria." In B. O. Smith (Ed.), *Research in Teacher Education*. Englewood Cliffs, N.J.: Prentice-Hall, Inc., 1971, pp. 37–72.

what is to be studied in subsequent units of the course. Basically, the strategy of the successful teacher is one of letting students in on what is going on.

Clarity. Clarity of teachers' presentations and of their responses to students' comments and questions contributes to student achievement. Teachers who demonstrate clarity in their teaching explain and demonstrate concepts in a manner that can be understood by students, make points easy to understand, and answer questions in an intelligent and complete manner. Students of teachers who exhibit clarity in their teaching do not have to spend a great deal of time attempting to figure out what is going on. If they do not understand, they know that a question will result in an easy-to-understand explanation of the concepts being taught and how the concepts relate to the world that is real to students.

Principle 2

> Readiness is a prerequisite for learning. Subject matter and learning experiences must be provided that begin where the learner is.

The sequence of courses within an agricultural education program as well as the sequence of instructional units within a particular course are major determinants of the extent to which this principle is put into practice. But regardless of how well sequenced the instruction is in terms of the readiness of students for new learning, individual differences of students make it extremely difficult, if not impossible, for all students to have the desirable prerequisite knowledge and experience for each unit of instruction.

Teachers whose behaviors exhibit that this principle is basic to their teaching go to great lengths to learn about their students—their interests, aspirations, and aptitudes; their previous instruction in and experience related to the subject matter; the expectations of their parents; and their opportunities to participate in laboratory and supervised experiences that accompany classroom instruction. Such a concern for students and their degree of readiness for new learning requires teachers to know students individually and to be familiar with their home situations and workplaces in addition to what they do and how they perform in school.

Students demonstrate their readiness for instruction through the interest and enthusiasm they show when a new instructional unit is introduced, by the questions they ask, their level of achievement, and how they act in the classroom and laboratory. Perceptive teachers attend to the comments and actions of students that indicate they lack the knowledge, skill, or experience essential to achieve the instructional objectives for a particular course or unit of instruction. When it is evident that students lack the prerequisite knowledge, skill, or experience, the teacher must provide or arrange for remedial instruction to make it possible for students to achieve.

MOTIVATION
Principle 3

> Students must be motivated to learn. Learning activities should be provided that reflect the wants, needs, interests, and aspirations of students.

Motivation is "that which gives direction and intensity to behavior."[2] Motivation to learn in school is that which influences students to consider certain subjects and learning activities as interesting and challenging and that which determines the level of intensity and persistence with which students participate in and accomplish learning activities. Motivation to learn in agricultural education is partly a function of the personality and other attributes of students. But motivation to learn is also a function of the quality and variety of learning experiences that are under the direct control of the teacher and the school.

Students demonstrate their motivation for school in general, for particular courses, and for certain educational activities by their interest, by the intensity with which they enter into learning activities, and by their perseverance in completing tasks they undertake. Teachers gauge the motivation of students by the relative degree of interest they display for the course and the activities provided them.

Teachers who design courses and direct educational activities that motivate students in a positive manner must be aware of and sensitive to their interests, wants, and aspirations. When those to be taught are high school students, knowledge of the psychology of adolescence is essential. Regardless of the ages of the persons to be taught, the successful teacher systematically makes an effort to become knowledgeable about the relevant attributes and circumstances of students and their environments that impact directly on the students' motivations for instruction in agriculture and for school in general. Specifically, the teacher will determine and pay attention to students' academic aptitudes and academic achievements in school; their educational and occupational aspirations and plans; the types of nonschool activities in which they participate; the groups with whom they associate; special or unusual talents and experiences they have; the nature of the environment in which they live and work; and the values they, their families, and their friends hold for school in general and, more specifically, for the study of agriculture. Teachers should especially pay attention to the reasons students enroll in courses in agriculture and the types of activities, level of achievement, and degree of recognition that motivates students.

---

[2] Frymier, Jack R. *Motivation and Learning in School.* Bloomington, IN: Phi Delta Kappa Educational Foundation, 1974, p. 2.

Enthusiasm. In addition to skill as a teacher, the enthusiasm of the teacher is an attribute that is considered a major contributing factor to the motivation of students. It is easier for the teacher to learn the subject matter to be taught and teaching skills than it is to learn to be enthusiastic. Complete involvement in teaching, dedication, commitment, and sincere concern for students and their achievements are attributes that students use to describe teachers who model enthusiasm. Teacher enthusiasm is a behavior that research substantiates as being positively associated with the achievement of students.[3] A teacher's movements, gestures, and voice inflections are factors that students associate with a teacher's enthusiasm. Motivation is enhanced by teachers who are stimulating rather than boring, and alert and active rather than apathetic and lackadaisical.

Interest. Students tend to exhibit greater interest when there is activity, love of nature, curiosity, creativeness, gregariousness, desire for approval, altruism, self-advancement, competition, or ownership. Interest of students in a subject tends to be related to their knowledge or skill related to the topic. Interest can be increased in a subject by suggesting links between the new subject and one that the students know and understand. Interest can be created or strengthened by developing suspense, using the novel or unexpected, and using humor. Acceptance by students of group goals, a sense of progress toward goals, and challenging thought about a topic can enhance interest. Individuals can become "caught up" in expressions of interest by the group (class).

Novel learning experiences stimulate students. Talking about their experiences gets students actively involved, which, in turn, motivates them. The use of actual objects and specimens attracts the attention of students, gets them involved, and creates a high level of interest. Learning activities in which students are actors rather than passive observers or participants are stimulating. Students' involvement in the teaching–learning process can be an effective means of motivation. It is the responsibility of the teacher to ensure that student involvement is meaningful and motivating and that it contributes positively both to the instructional process and to the outcomes achieved.

Principle 4

> Students are motivated through their involvement in setting goals and planning learning activities.

When students are actively and appropriately involved in formulating learning goals and in planning, conducting, and evaluating learning activities, they

---

[3] Rosenshine and Furst, *op. cit.*

hold a degree of ownership in the teaching–learning process and, consequently, are motivated to ensure that goals are achieved and activities completed successfully. Students have to be taught how to participate in setting goals and in selecting, carrying out, and evaluating various phases of the teaching–learning process. Active student involvement in goal setting and in planning and evaluating learning activities does not relieve the teacher of major responsibility for ensuring that instruction is adequately planned, delivered, and evaluated. In fact, it is more difficult to teach students about and appropriately involve them in planning for and conducting teaching–learning activities than it is to resort to a teacher-dominated stance for goal setting, planning, and evaluating instruction.

Teacher's "use of student ideas"[4] describes a group of teacher behaviors that correlate positively with student achievement. Teachers use students' ideas when they repeat what students have said, acknowledge the contributions of students during discussion, modify or rephrase the comments of students, apply a student's idea in the analysis and solution to problems, compare one student's idea to the ideas expressed by others, and summarize what has been said by individuals or the group. Students are motivated and in turn achieve at higher levels when teachers indicate by words and actions that they accept students' feelings and ideas when they offer praise and encouragement.

Principle 5

> Success is a strong motivating force.

When students acquire new knowledge and skill they are motivated. Nothing motivates like success. Teachers who demonstrate this principle of teaching and learning ensure that students are engaged in learning activities where success is possible, provide the instruction and supervision needed to enable students to achieve, and make certain that a student's success is made evident and recognized.

Principle 6

> Students are motivated when they attempt tasks that fall in a range of challenge such that success is perceived to be possible but not certain.[5]

---

[4] Rosenshine and Furst, *op. cit.*
[5] Watson, Goodwin. *What Psychology Can We Trust?* New York: Teachers College, Columbia University, 1961.

This principle of teaching and learning, which is closely associated with the preceding principle, emphasizes that challenge is a motivating force when a student sees that success is possible. In implementing this principle, a major concern of the teacher is the realization that what is within the range of challenge for some students is impossible for others to attain and perhaps unchallenging or even boring to other students. It is essential that teachers pay particular attention to individual differences of students when applying this principle. Its application requires the use of subject matter and instructional materials at several levels of difficulty, as well as a variety of learning activities. Effective teachers select and sequence subject matter and develop learning activities that are challenging in that students believe success, although not certain or automatic, is possible and likely to be attained.

REWARD AND REINFORCEMENT
Principle 7

> When students have knowledge of their learning progress, performance will be superior to what it would have been without such knowledge.

If students are to continue to show interest in learning and strive to achieve the learning outcomes sought, they must be given feedback concerning their learning progress. Providing feedback involves both diagnosing the strengths and weaknesses of a student's learning behavior and furnishing additional instruction and supervised practice if needed.

Master teachers use a variety of techniques to provide feedback to students. Tests and examinations are probably the most frequently used techniques. When tests are used in a manner that contributes to learning most effectively, students not only get immediate feedback concerning their performance but also are given the opportunity to participate in additional instructional activities that are designed to develop the knowledge and skill that had not been achieved at a satisfactory level of performance. Using tests as diagnostic devices rather than exclusively for evaluation and grading purposes is an effective way for students to be apprised of their learning progress.

Questioning during classroom discussion affords teachers the opportunity to assess students' knowledge and understanding and, when warranted, can be accompanied immediately by remedial and supplemental instruction. In laboratory situations in which the development of psychomotor skills is the learning outcome sought, the use of this principle requires that teachers observe and diagnose the student's performance of the skill and then demonstrate or model the practices and skills that must be improved. It is important in providing feedback to students that the teacher communicate not only what is incorrect or being performed improperly but also make sure students understand what is correct and what skills and practices are being performed correctly. When incorrect responses and improper performance are detected, reinstruction should be provided as expeditiously as possible.

The gamut of techniques for diagnosing student achievement and performance and simultaneously providing supplemental instruction is determined in large measure by the teacher's innovativeness and creativity. Teachers can provide feedback and supplemental instruction to students through written comments on student notebooks, papers, and other written projects. Some teachers use audio, video, and electronic records to communicate their appraisal of a student's work and to offer suggestions for additional learning activities. Small-group discussions in which the teacher is involved and individual instruction and supervision in the home and at the place of work offer excellent opportunities for teachers to help students become aware of their learning progress.

Successful teachers teach students how to evaluate their own learning progress. In the final analysis, self-evaluation is the goal to be achieved if students are to continue learning throughout life.

Principle 8

> Behaviors that are reinforced (rewarded) are more likely to be learned.

A fundamental principle of the psychology of learning is that behaviors that are rewarded are likely to be learned and retained. If achievement in acquiring facts, deriving principles and applying them to the solution of problems, or analyzing and evaluating information is recognized and rewarded, students are likely to retain this cognitive knowledge and skill to a greater extent than if the learning outcomes achieved are not rewarded. When the behaviors to be learned are psychomotor skills, the behaviors are more likely to recur and be retained if the correct performance of the skills is rewarded. Providing reinforcement when actions students deliberately take demonstrate desirable attitudes is an effective strategy when the learning outcomes desired are attitudes.[6]

In providing reinforcement when desirable learning outcomes are observed, teachers need to take special care to ensure that the reinforcement offered is actually viewed by students as a reward. Again, this illustrates the necessity for teachers to know students individually. This is especially true in the case of high school students because under varying circumstances and at different times students place varying degrees of value on teachers' actions and comments that are designed to reward the acquisition and demonstration of competence. What

---

[6] Gagne, Robert M. "The Learning Basis of Teaching Methods." In N. L. Gage (Ed.), *The Psychology of Teaching Methods*. Chicago: National Society for the Study of Education, 1976, pp. 21–43.

some high school students view as rewarding may not be viewed as rewarding by other students. Generally, recognition by teachers and peers for behavior that demonstrates a high level of mastery is considered by most learners to be rewarding; therefore, reinforcing the learning that has occurred.

Criticism, the opposite of reward, should be used cautiously. The research is consistent in showing a negative relationship between a teacher's use of criticism and student achievement.[7] However, the fact that the use of criticism is frequently accompanied by a lack of achievement should not be used to justify teaching practices that avoid or minimize the importance and necessity for teachers to inform students when their work is inaccurate or incomplete. Effective teaching demands that teachers provide feedback to students not only when students are right but also when they are wrong. When students' responses and performances are right, feedback not only recognizes achievement but also serves to make sure students know and understand what they have accomplished correctly. When students' responses and performances are wrong, feedback not only communicates the fact of inaccuracy or incompleteness but also provides instruction and suggestions such that students may through further study and instruction achieve an acceptable level of mastery.

Sometimes teachers accept or ignore incorrect responses in the classroom or improper performance in the laboratory. This practice should be avoided. Instead, the teacher should make it clear that the observed achievement or performance is not satisfactory, then provide additional instruction to correct the deficiencies observed. In effect, the acceptance of mediocre and inferior achievement reinforces not only an unacceptable level of achievement but also the learning behaviors that resulted in less than acceptable mastery. When achievement and performance are substandard, teachers must accomplish two important tasks immediately. First, students must be made aware that achievement and performance are not acceptable and be informed why their work is so judged. Second, and of even greater importance, students must be retaught or given the opportunity for reinstruction such that acceptable levels of competence can be attained.

Principle 9

> To be most effective, reward (reinforcement) must follow as immediately as possible the desired behavior and be clearly connected with that behavior by the student.[8]

The desirability of providing immediate feedback to students cannot be overemphasized. To have maximum effectiveness, reward should closely fol-

---

[7] Rosenshine and Furst, *op. cit.*

[8] Watson, *op. cit.*

low and be clearly connected to the desired behavior. During classroom and laboratory instruction, reinforcement can be quickly conveyed by a teacher's comments with the teacher making sure that the student understands what behaviors and level of performance are being reinforced. When reinforcing a student's performance, the expert teacher does not assume that the student is automatically aware of the behavior that resulted in the mastery of knowledge or the acquisition of skills. If maximum payoff is to be realized from the use of this principle of teaching and learning, students must not only know that they have demonstrated a satisfactory level of achievement or performance but they must also be aware of what they did right to achieve the level of mastery they have demonstrated.

TECHNIQUES OF TEACHING
Principle 10

> Directed learning is more effective than undirected learning.[9]

If students are to achieve optimum levels of competence, it is essential that the teacher direct the teaching–learning process. Teacher direction means that the teacher assumes responsibility for (1) making sure that students know and understand what is to be taught and what learning outcomes are expected, (2) selecting and using skillfully and efficiently appropriate teaching techniques, (3) supervising the learning activities in which students participate, (4) designing and securing the appropriate instructional media and materials, (5) directing the application of knowledge and skills learned, and (6) evaluating the effectiveness of instruction. Directed learning ensures that both the teacher and the students know what is going on and why.

Teachers who are most expert in directing learning use **task-oriented** and **businesslike** behaviors and provide students **opportunity to learn criterion material**.[10] The extent to which a teacher is task oriented, achievement oriented, and businesslike determines the extent to which teaching–learning activities are structured and directed. The teacher's use of task-oriented and businesslike behaviors is correlated positively with student achievement. Teachers who are task and achievement oriented place high priority on students learning. Task-oriented teachers encourage students to work hard, provide opportunities for independent study, and supervise the learning activities of students.

---

[9] Hatch, Winslow R., and Bennet, Ann. *Effectiveness in Teaching.* Washington, D.C.: U S. Department of Health, Education, and Welfare, 1963.

[10] Rosenshine and Furst, *op. cit.*

Investigations of the relationship between teacher behavior and student achievement consistently reveal that higher levels of competence are achieved when students are provided ample opportunity to learn the knowledge and skills that are specified in the instructional objectives. Time spent on learning tasks that pertain specifically and directly to the learning outcomes sought is an important factor contributing to student achievement.

Principle 11

> To maximize learning, students should *inquire into* rather than *be instructed in* the subject matter. Problem-oriented approaches to teaching improve learning.[11]

Learning occurs when students through their own activity have changes in behavior. That which is to be learned must be processed by students such that it is meaningful and understandable. The clear and simple message of this principle is that "students learn" and "teachers teach." If students are to learn, they must have a clear notion of what it is they are supposed to know and be able to do. They must make serious efforts to achieve the expected outcomes. Through their own activity, study, and practice that is directed and supervised by the teacher, they acquire, retain, and use knowledge, skills, and attitudes. The job of the teacher is to structure and direct the teaching–learning process (principle 10) such that students are provided opportunities for inquiring into that which is to be learned. Teachers have an array of techniques and strategies from which to choose in facilitating students becoming actively involved such that their behavior—whether thinking, doing, or feeling—is changed. For both the learner and the teacher, learning is an active rather than passive process.

Variability. The "teacher's use of variety or variability"[12] is positively correlated with student achievement. Research has consistently demonstrated that the achievement of students is higher when teachers use a variety of instructional materials and teaching procedures. A wide variety of written, electronic, audio, and video instructional materials are needed. It is essential that instructional materials be available that can be used by students with varying interests and aptitudes. If opportunities are provided students to participate in an inquiry mode of learning, teachers must have available and use appropriate audio, visual, and electronic media.

Variety in teaching requires that teachers use several diagnostic and evaluative devices for providing feedback to students. Written tests can be

---

[11] Hatch and Bennet, *op. cit.*

[12] Rosenshine and Furst, *op. cit.*

used to assess competence in the analysis of problem situations, problem solving, and the generation and application of general principles in addition to the acquisition of facts and the recall of information. Performance tests, oral tests, and the assessment of actual performance in the laboratory and in supervised practice provide variety in assessment techniques, which allows students with different interests, aptitudes, and levels of competence to demonstrate achievement.

Using variety in teaching requires that a broad range of activities and tasks be provided so students can be involved actively in the acquisition and application of knowledge and in the development, practice, and use of skills. Students vary greatly not only in their interests and motivation for learning but also in the prerequisite knowledge and skill they possess and in their learning styles. Some students learn best when they work independently, whereas others learn most effectively in group activities. Teachers who demonstrate variability in teaching provide opportunities for both individual and group learning activities, supervise students' learning tasks to the extent necessary to ensure that students learn, allow students to participate in a broad range of learning tasks, and select teaching techniques that are best suited to accomplish effectively and efficiently the specific learning outcomes sought for each instructional unit.

When selecting teaching techniques for a particular lesson or unit of instruction, the teacher needs to consider two important factors. First is the nature of the learning outcomes to be achieved. For example, consideration has to be given to whether students are to acquire facts and information, problem-solving skills, psychomotor skills, or personal development skills such as leading or participating in group discussions. Second, the teacher must consider whether students possess the necessary study and inquiry skills to enable them to participate actively and productively in the particular learning tasks that will be used.

The specific learning outcomes to be attained must be considered when decisions are made concerning what teaching techniques to use. If a primary focus of the lesson is the acquisition of certain facts, teaching techniques are selected that either present the facts in as clear and understandable fashion as possible or assist students in locating and recording the information expeditiously. If the primary learning outcomes are to develop problem-solving abilities, to create a positive attitude toward inquiry, or to make reading more discriminating, group discussion techniques are most likely to be effective. Research indicates that group discussion is effective in achieving mastery of subject matter content; developing positive attitudes toward concepts taught; solving problems that require group commitment for implementation; and developing skills related to listening, speaking, and leadership.[13] If the development of psychomotor skills

---

[13] Gall, Meredith D., and Gall, Joyce P. "The Discussion Method." In N. L. Gage (Ed.), *The Psychology of Teaching Methods*. Chicago: National Society for the Study of Education, 1976, pp. 166–216.

is the learning outcome sought, teaching techniques are required that emphasize demonstrations and modeling accompanied by supervised practice.

Problem-Oriented Teaching. Problem-oriented teaching places high priority on students inquiring into subject matter. With problem-oriented teaching strategies, teachers pay close attention to the study and learning skills possessed by students that are essential to their involvement in learning tasks. Students have to be taught how to participate in learning activities that are basic to a problem-oriented approach to teaching. To participate successfully in learning tasks that require students to inquire into subject matter, students need competence in setting goals, defining problems and questions, locating pertinent knowledge and information, developing alternative answers and solutions, evaluating the validity of data and information, and assessing the consequences of alternative solutions. Teachers who use problem-oriented teaching techniques successfully find it necessary to teach skills of systematic inquiry if students are to be actively and productively involved in learning tasks.

The teacher's use of questions is important with a problem-oriented approach to teaching. Research indicates that teachers should use both lower-cognitive-level "what" and "where" questions and higher-cognitive-level "why" and "how" questions. There is some evidence to indicate that probing—encouraging students to elaborate on their answers—contributes positively to higher achievement.[14] The use of questions is one strategy teachers use to structure and direct the learning process. Research indicates that in teacher-directed instruction learning is organized around questions posed by the teacher, the teacher asking direct and narrow questions that have definite answers, the teacher immediately providing feedback indicating whether answers are right or wrong, the teacher asking additional questions after correct answers are given, and the teacher making sure that the correct answer is given after incorrect answers are given.[15]

Principle 12

> Students learn what they practice.

Often practice is associated only with psychomotor skills. Obviously, practice is of prime importance when the learning outcome to be achieved is the development of new psychomotor skills or the further refinement of present skills. However, practice is also applicable to cognitive and attitudinal

---

[14] Rosenshine and Furst, *op. cit.*

[15] Rosenshine, Barak. "Classroom Instruction." In N. L. Gage (Ed.), *The Psychology of Teaching Methods.* Chicago: National Society for the Study of Education, 1976, pp. 335–371.

skills as well. If the teaching techniques a teacher uses emphasize the memorization of facts, students practice memorizing and many become highly proficient at it. If the teaching techniques a teacher uses emphasize a problem-oriented approach to teaching and learning, students, through practice, develop skills that enable them to participate actively in learning tasks that encourage, if not demand, that they inquire into, rather than be instructed in, the subject matter. The importance of this principle to effective teaching and learning cannot be emphasized too strongly. The teaching techniques a teacher uses invariably are accompanied by students practicing certain learning behaviors, cognitive skills, psychomotor skills, and attitudes. It is the teacher's responsibility to ensure that the behaviors and skills students practice contribute to high levels of achievement and performance, to an intensified motivation to continue learning, to positive attitudes toward learning, and to the ability to apply and use constructively what is being learned.

Practice in itself is not enough. Sheer repetition does not ensure that the hoped for learning outcomes will be achieved. Students can practice error as well as success. It is imperative that practice be supervised by the teacher such that reward and reinforcement (principles 7, 8, and 9), including modeling and reinstruction, can be used to ensure that practice leads to desirable learning outcomes. Practice without appropriate feedback, whether reinforcement or reinstruction, is a poor way to attempt to learn.

Principle 13

> Supervised practice that is most effective occurs in a functional educational experience.

Supervised practice contributes optimally to the attainment of learning outcomes when students practice in situations that resemble as closely as possible the actual situations in which they are to apply and use the knowledge, skills, and attitudes that are being learned. The educational experiences provided students for practicing cognitive, psychomotor, and attitudinal skills must be perceived by students as sensible, meaningful, and functional.

If supervised practice is to make sense to students, teachers must deal with real situations, problems, and current and accurate data and information in the classroom. For certain learning outcomes it is essential that supervised practice be provided in laboratories that are equipped with facilities that are up-to-date and operate properly. Supervised practice in the community and the world of work is basic to agricultural education programs. This principle of teaching and learning provides the rationale for supervised practice in the classroom, the laboratory, and the FFA and through supervised experience.

TRANSFER OF LEARNING
Principle 14

> Learning is most likely to be used (transferred) if it is learned in a situation as much like that in which it is to be used as possible and immediately preceding the time when it is needed.

Principle 15

> Transfer of learning is more likely to take place when what is to be transferred is a generalization, a general rule, or a formula.

Principle 16

> Students can learn to transfer what they have learned; teachers must teach students how to transfer learning to laboratory and real-life situations.

The desired outcomes of effective teaching involve more than students answering questions correctly and keeping legible notes with correctly spelled words and grammatically correct sentences. The major learning outcomes sought are that students acquire new knowledge and skill; that they understand the concepts and principles that are basic to the knowledge and skill they acquire; and that they use this knowledge, skill, and understanding not only to facilitate further learning but also in achieving useful and productive occupational and personal lives.

Teaching for transfer of learning requires teachers to emphasize the usefulness of what is being taught to the immediate and long-term needs and interests of students. Descriptions of the principles previously presented in this chapter pertaining to organization and structure of subject matter, motivation, reward and reinforcement, and techniques of teaching provide the rationale for teaching strategies and techniques that result in students being equipped to apply and use the knowledge, skill, and understanding that has been acquired.

In the final analysis, transfer of learning comes down to students understanding what has been taught, preferably through experiences that provide opportunities for supervised practice and application. Teachers who effectively teach for transfer make it clear to students what new knowledge, concepts, principles, and skills have been learned and point out to students through examples and supervised practice how, when, and where the new learning can be applied and used. Teaching that explicitly makes direct connections between classroom and laboratory instruction and real-world situations is essential for teaching students to transfer learning.

An important element in teaching for transfer is stated in the principle that transfer is "more likely to take place when what is to be transferred is a generalization, a general rule, or a formula." This requires that students' answers to questions and solutions to problems go beyond recording facts, figures, and statements from a reference. It is the responsibility of the teacher to guide students in understanding "why" certain answers or solutions are correct. Understanding "why" means that underlying concepts and principles are made explicit and their meaning understood. This is a major ingredient of "directed learning" (principle 10). In addition to directing learning activities that lead to students deriving or becoming aware of concepts and principles, the effective teacher provides examples and situations that demonstrate to students how and when the concepts and principles apply to other settings and situations.

While observing an experienced high school teacher of agriculture, one of the authors noted the following example of transfer of learning. During the class period the teacher administered a test. As students worked on the test, the observer noted that one student seated nearby, before writing an answer to a single question, proceeded to read each question and write brief notes in the margin beside each question. The observer, curious about what the student had written, approached the student's desk and quietly asked how he was doing. In doing so, the observer noted that in the margin next to one question the student had written "candy bar rule." Later in discussion with the teacher, the observer asked why the student would write "candy bar rule" by one of the questions on the test. "Oh, that's easy to explain," the teacher replied, "for problems of that type we learned that the calculations involved follow the general rule 'if five candy bars cost _____, then one candy bar costs _____'." The student recognized a situation where the general candy bar rule applied, thus immediate transfer of learning.

SUMMARY

Principles of teaching and learning provide the rationale for strategies and techniques teachers use to plan, deliver, and evaluate instruction in and about agriculture. The principles described in this chapter are substantiated by research in the psychology of learning. Using these principles as the basis for formulating and selecting strategies and techniques for planning, delivering, and evaluating instruction contributes to students acquiring and using appropriate knowledge, skill, and understanding—learning outcomes sought from effective teaching.

FOR FURTHER STUDY

1. Analyze the teaching episode presented at the beginning of this chapter.
    a. What techniques did the teacher use to motivate students?
    b. In what learning activities did students engage?
    c. What techniques did the teacher use to elicit responses from students?
    d. What type of questions did the teacher ask?
    e. What learning outcomes were achieved?
    f. What principles of teaching and learning were demonstrated?
    g. How do you explain why the actual behavior of students during the class session was substantially different from what the student teacher expected?
2. Examine a lesson plan or a unit of instruction. Identify the principles of teaching and learning used in the lesson.
3. Observe a videotape of a high school class, an elementary class, or a college class. Identify the techniques used by the teacher to apply principles of teaching and learning. Tie specific techniques to specific principles.

Appendix

Table 2-1 How Teachers Put into Practice the Principles of Teaching and Learning

Principles of Teaching and Learning That Contribute to the Achievement of Learning Outcomes (What Students Learn)	Teaching Practices That Make Operational the Principles of Teaching and Learning (What Teachers Do)
Organization and Structure of Subject Matter	
Principle 1 When the subject matter to be learned possesses meaning, organization, and structure that is clear to students, learning proceeds more rapidly and is retained longer.	Students are informed about the content and organization of the course of study: • Units of instruction are titled to indicate application and use of subject matter. • Sequence of instruction makes psychological sense to students. Use of structuring comments: • Make the interrelationships of topics clear to students. • Make clear when one topic ends and another begins. • Review at start or end of a lesson. • Emphasize major points during study and discussion. • Provide structuring comments before asking questions. • Make evident the connection between what is studied in agriculture and what is studied in other courses in school. Cognitive clarity of teacher's comments: • Explain and demonstrate concepts in a manner that students understand. • Make points easy to understand. • Answer questions in an intelligent and complete manner.
Principle 2 Readiness is a prerequisite for learning. Subject matter and learning experiences must be provided that begin where the learner is.	Information about students is obtained: • Interests and aspirations. • Aptitude for learning. • Previous instruction and experience. • Expectations of parents. • Opportunities and facilities for supervised practice. Remedial instruction is provided when needed.

(continued)

Table 2-1 Continued

Motivation

Principle 3
Students must be motivated to learn. Learning activities should be provided that reflect the wants, needs, interests, and aspirations of students.

Awareness of and sensitivity to the wants, needs, interests, and aspirations of students is required.

Information is obtained about reasons students enroll in the course.

Teacher's enthusiasm:
- Complete involvement, dedication, and commitment.
- Sincere concern for students.
- Movements, gestures, and voice inflections.

Principle 4
Students are motivated through their involvement in setting goals and planning learning activities.

Teacher has major responsibility for planning, delivering, and evaluating instruction. Students are taught how to participate in setting goals and in selecting, carrying out, and evaluating learning activities.

Teacher's use of student ideas:
- Acknowledge the contributions of students during discussion.
- Repeat, modify, or rephrase student comments.
- Apply student ideas to solutions of problems.
- Compare and contrast student comments.
- Summarize student comments.

Ask students to describe experiences.

Use actual objects and specimens.

Principle 5
Success is a strong motivating force.

Subject matter is provided at several levels of difficulty.

Principle 6
Students are motivated when they attempt tasks that fall in a range of challenge such that success is perceived to be possible but not certain.

A variety of learning activities is provided.

Reward and Reinforcement

Principle 7
When students have knowledge of their learning progress, performance will be superior to what it would have been without such knowledge.

Feedback is provided to students that diagnose strengths and weaknesses.
A variety of techniques is used to provide feedback:
- Tests and examinations.
- Questioning during discussion.
- Observation of student performance.
- Written comments on notebooks, papers, and written reports.

Chap. 2 Principles of Teaching and Learning

Table 2-1 Continued

	Additional instruction and supervised practice is provided when needed.
Principle 8 Behaviors that are reinforced (rewarded) are more likely to be learned.	Reinforcement follows achievement and performance immediately.
Principle 9 To be most effective, reward (reinforcement) must follow as immediately as possible the desired behavior and be clearly connected with that behavior by the student.	Reinforcement offered must be perceived as a reward by students. Criticism is avoided. Feedback indicating inaccuracy and incompleteness is accompanied by reinstruction.

Techniques of Teaching

Principle 10 Directed learning is more effective than undirected learning.	Teacher provides organization and structure to the teaching–learning process. Teacher is responsible for • Making sure students know what is to be taught and what learning outcomes are expected. • Selecting and using appropriate teaching techniques. • Supervising learning activities. • Providing instructional materials and media. • Directing the application of knowledge and skills learned. • Evaluating the effectiveness of instruction. Use of task-oriented and businesslike behaviors: • Encourage students to work hard. • Provide opportunity for independent study. • Emphasize student achievement. • Supervise closely learning activities. Provide opportunity to learn criterion material: • Content covered pertains directly to learning outcomes. • Opportunity given to attain the knowledge and skills specified. • High proportion of time spent on learning tasks.
Principle 11 To maximize learning, students should *inquire into* rather than *be instructed in* the subject matter. Problem-oriented approaches to teaching improve learning.	Teacher's use of variety or variability: • Teaching techniques. • Instructional materials and media. • Diagnostic and evaluative devices. • Individual and group learning tasks. Students are taught how to work independently. Demonstration and modeling techniques are used.

Table 2-1 Continued

	Teachers emphasize in questioning: • Both "what" and "where" questions and "why" and "how" questions. • Probing that encourages students to elaborate on answers. • Direct questions that have definite answers. • Feedback indicating whether answers are right or wrong. • Correct answer when incorrect answers are given.
Principle 12 Students learn what they practice.	• Supervised practice is provided in the classroom, laboratory, community, and world of work.
Principle 13 Supervised practice that is most effective occurs in a functional educational experience.	• Teachers design projects, laboratory exercises, and experiments pertaining directly to what is taught in the classroom. • Teachers supervise activities students engage in during their supervised agricultural experience activities. • Teachers instruct students to plan and carry out activities that require the application and use of subject matter and principles taught during classroom instruction.

Transfer of Learning

Principle 14 Learning is more likely to be used (transferred) if it is learned in a situation as much like that in which it is to be used as possible and immediately preceding the time when it is needed	Teachers use learning activities that lead students to discover and make explicit the concepts and principles that undergird what has been learned.
Principle 15 Transfer of learning is more likely to take place when what is to be transferred is a generalization, a general rule, or a formula.	Teachers, through questions and examples, ensure that students know and understand the underlying concepts and principles.
Principle 16 Students can learn to transfer what they have learned; teachers must teach students how to transfer learning to laboratory and real-life situations.	Teachers use examples from previous instruction where the concepts and principles are applicable. Teachers alert students to future instruction where the concepts and principles will be applicable. Teachers, through experiments, laboratory exercises, and supervised practice, demonstrate the application and use of concepts and principles.

3
Planning the Course of Study

> Ms. Roberts had just accepted an agriculture teaching position in a progressive school district. The principal asked her to stop by the office. During the visit, he requested that she prepare a yearly video/dvd rental order within two weeks, stressing the importance of using media that would relate to her course of study. He also asked that she requisition her laboratory supplies, indicating the necessary delivery dates. She found she was also expected to place on the school calendar the FFA activities for the year. Ms. Roberts knew she would need to plan a course of study so she could appropriately fulfill her principal's expectations.

A course of study is a comprehensive plan that presents the scope and teaching sequence of all the learning activities provided for a particular course. It indicates the content and when the various topics will be taught. A course of study is the blueprint, or design, for instruction. A teacher of agriculture needs such a plan for each course he or she teaches. A course of study is prepared sequentially by assessing community needs, identifying objectives or competencies, specifying course content, and arranging or structuring the content in an appropriate sequence.

OBJECTIVES

After studying this chapter, you will be able to

1. Explain the necessity of developing and using a course of study.
2. Describe the responsibility of the teacher in organizing instruction.

3. Develop a course of study.
4. Explain the need to continually update and adapt the course of study.

REASONS FOR A COURSE OF STUDY

What are some reasons for writing a course of study? Is it worth the time required to develop it? These are common thoughts when approaching the task of writing a course of study, and there are several reasons for developing a well-planned course of study.

Rationale for Content

The process of preparing a course of study requires the teacher to think through what is to be taught. This thinking process begins by assessing community needs that are relative to preparing students for jobs in the community. Community needs, for your purposes, are assessed by examining the types of occupations available to your students and your graduates in the community. Part of your content is then planned to prepare competent persons to enter those occupations. Another part of your content, however, is planned not as occupational content but as agricultural literacy content to enrich your community's knowledge of agriculture.

Communication with Others

Another reason for developing a course of study is to communicate information about the course to others. Administrators, supervisors, community members, agribusiness owners, parents, and students often desire information about a program. The course of study can provide much of that information.

Lesson Planning

A teacher needs to know what subjects to teach in order to plan lessons. Also, the teacher should encourage supervised agricultural experience programs, laboratory work, and FFA activities that relate to instructional content. If students are to study animal science in a course, animal supervised experience programs should be encouraged. If students are to study bedding plant production, bedding plants might be produced in the school greenhouse. If students are learning about meat processing, participation in the FFA meats career development events would be an appropriate activity. Lesson planning, supervised experience programs, laboratory work, and FFA activities will be more interrelated if the course of study is used as their foundation.

Course Structure and Student Learning

As discussed in Chapter 2, when a course is planned and organized, students achieve at a higher level; students have a greater understanding of what is expected of them and teachers interrelate concepts and plan for later reemphasis of principles. Structure of content is essential in order to develop units of instruction that package learning activities in a logical and sequential manner. Students of learning theory suggest that the sequence or order in which subject matter is taught may relate to the amount of knowledge and skills learned and retained for long-term benefit.

The teacher developing a course of study can choose a sequence that teaches basic skills prior to more complex skills, such as teaching record keeping prior to records analysis. He or she may also choose to teach certain problem areas when they are seasonally relevant, such as combine adjustment during the harvest season or wildlife habitats during the fall. The order of content and the amount of time devoted to each topic should be decided by the teacher with the goal of facilitating student learning.

Planning for Resources

The course of study will be helpful to a teacher needing to secure instructional resources. Videotapes and DVDs may be ordered so they will arrive for use when they are relevant to the topic being studied. Valuable Internet web sites can be located prior to introducing a new unit of instruction. Instructional supplies, such as plant cuttings and test tubes, can be ordered so that the plants arrive at the time they need to be cultured for applying instruction. Field trips can be arranged in advance, with some assurance that the field trip will relate to instruction. Resource people can be invited into your classroom with enough advanced notice for them to adequately prepare for instruction.

TEACHER RESPONSIBILITY

The primary responsibility for planning a course of study rests with the teachers because they are expected to be experts on subject matter. He or she should be knowledgeable concerning the community as well. Also, the needs of students are often best understood by the teacher. Therefore, although others can and will be involved in recommending what should be taught, the teacher is best qualified to plan the course of study.

Expert on Subject Matter

Teachers are to be students of their teaching specialty. Therefore, they must stay abreast of the literature relating to the subject matter being taught. Through supervised agricultural experience (SAE) and home visits, teachers will be aware of how current knowledge is being applied. Although beginning teachers will not be experts in all phases of instruction, they should strive to develop expertise in both theory and practice in relevant specializations. Because teachers are the subject matter experts, decisions about course content will be more correct and instruction will be more relevant when teachers concisely strive to be knowledgeable about their content.

Knowledgeable of Community

Agricultural organizations provide educational opportunities in their communities by bringing knowledgeable resource people into the community to present lectures on current information. Teachers of agriculture have a unique opportunity to give local clientele a more rich experience than these visiting lecturers because they can become thoroughly acquainted with the specific type of agriculture and the special problems of students in the local area. Agriculture teachers, therefore, can use their expertise in the subject matter and in education to develop a sound program of instruction to meet community needs.

In some communities students may have a rich tradition in sheep production, or maybe they operate a livestock cooperative. In other communities students may have experience with a pheasant raise-n-release program, or a rich history of ag-science fair winners. The teacher can use these interests in selecting what to teach. A tradition of trout catch-n-release in a community can serve as the basis for excellent instruction in stream water management. The teacher should be willing to adjust the curriculum in response to real and expressed interests of students. The teacher should also provide for individualized, supervised study to allow students to explore and prepare themselves in areas of interests.

Adapting Instruction for Needs

Teachers in the local schools are the ones who must adapt instruction to meet the needs of the students in their classes. A state curriculum guide may be available to provide a framework to plan a local course, however, the specific agricultural situation of the community and other factors influencing decisions about course content suggest the course of study should be a responsibility of the local school and its teachers. A course of study should never be prepared by an individual or by a committee that is removed from the local situation. Instead, a local course of study must be tailor-made by the teacher to suit the individual teaching situation.

FACTORS INFLUENCING DECISIONS ABOUT CONTENT

How does one decide what will be taught in a course? The answer to this question reveals that there are many factors that must be considered in arriving at a course content.

Educational Philosophy

Every teacher has an educational philosophy. In many cases, it has not been set forth in written form. A philosophy reflects the principles that govern the way humans behave. A philosophy furnishes the direction for a type of education. Most schools have developed and adopted a statement of philosophy. In general these statements relate to the transmission of culture, improvement of the social structure and environment, and provision for individual needs.

PHILOSOPHY OF A LOCAL AGRICULTURE DEPARTMENT

As agriculture has become more scientific, more complex, and more extensive, it has required many changes in procedures and practices, both on farms and in related agricultural businesses. This trend is certain to continue; thus, those persons who manage farms and businesses must become even more proficient. It is the responsibility of agricultural education programs to provide training for those who will be and are now engaged in agricultural pursuits. This training must be flexible in order to meet the needs of the individual students and their community. Our modes of living and our communities are also changing, and we must share the responsibility of developing community leaders and citizens.

Teachers of agriculture should develop a philosophy by which their local programs will be guided. The statement might be brief or detailed. An example of a brief but comprehensive statement adopted by a local school is given in the Principles for Agriculture Programs. Philosophy may also be reflected in statements of principles. An example statement of principles is shown on pages 53–54.

The philosophy by which one selects instructional content may have some influence on whether skills are viewed as an end or a means to an end, whether one believes students should prepare intensively for an occupation or explore a number of related occupations, whether it is preparation for postsecondary education, whether one believes agricultural education is preparation for an entry-level skilled or semiskilled occupation, or whether it is preparation for careers in agriculture regardless of level. The educational philosophy, stated or unstated, of the school and agricultural education department will have a major influence on the determination of objectives and selection of course content.

PRINCIPLES FOR AGRICULTURE PROGRAMS

- Agricultural education is an integral part of the total educational program.
- Some formal occupational preparation should be a part of every educational experience.
- A gradually expanding articulated program of career education should be implemented for elementary, secondary, postsecondary, and adult education.

- Agricultural education should provide extensive education for individuals in need of retraining and additional skill development.
- Agricultural education programs should be as dynamic, flexible, and sensitive to change as the economic sector they hope to serve.
- Schools should be organized to provide the breadth of career and technical offerings needed by their students.
- Lay advisory committees should be used in studying occupational needs, program planning and promotion, setting of standards, course of study development, program evaluation, and placement of students.
- Local initiative is essential for success of agricultural education programs.
- Funding should be sufficient to accomplish the objectives of the program.
- Agricultural education should be a year-round program of instruction.
- Technical instruction should be based on the occupational needs, interests, and aspirations of the students and the demands of employers.
- Supervised agricultural experience related to desirable career objectives should be provided as an integral part of the agricultural education curriculum.
- A community agricultural education program should be designed to provide career education at various levels for students of differing ability.
- FFA should be provided as intracurricular activity.
- Standards and conditions under which vocational instruction is given should compare favorably with what is desirable in the occupations concerned.
- A well-organized, supportive guidance service should be provided for students.
- Agricultural education personnel should be occupationally and professionally competent.
- Public relations activities should create and maintain active support for agricultural education.
- Periodic critical evaluation should be conducted to provide for program efficiency and improvement.

Program Objectives

Each agricultural education program should have stated objectives that relate to or flow out of the program philosophy. These, like statements of philosophy, provide direction in planning course content. Objectives are more specific than philosophy. They should indicate what students are expected to be able to know or do on completion of a course or program. They are much more general than instructional objectives. Instruction should be planned so as to accomplish the program objectives. An example set of objectives appears on page 55.

OBJECTIVES FOR HORTICULTURE PROGRAMS

Upon completion of the horticulture instructional program, students will be able to

- Demonstrate abilities in turf, floral production, floral retail, landscaping, nursery production, garden center operation, and fruit and vegetable production at a level necessary for entry into horticulture or continued education in related occupations.
- Demonstrate a proficiency in human and social relationships essential for employment in agriculture.
- Demonstrate a knowledge of the opportunities and qualifications for employment in horticulture.
- Demonstrate the ability to make decisions, manage money, and apply knowledge and skills in supervised agricultural experience programs.
- Demonstrate desirable attitudes toward school, community, and country through the practice of citizenship and by exhibiting desirable behavior patterns and effective leadership.

Teacher Expertise

There are those who question the extent the expertise of the teacher should influence the content of the curriculum. Some say that the content should be based on the knowledge and skills needed by employees in agribusinesses in the community. If teachers lack expertise in certain areas, they should use resource people or other methods to ensure the essential content is learned.

Others would argue that the special expertise of a teacher is important. Teachers are often more enthusiastic and thorough in teaching units in which they are confident of the material. Therefore, they would recommend more emphasis on these units.

The beginning teacher should logically be expected to give additional emphasis to areas of special expertise. This teacher should, however, through study and practice, develop the necessary knowledge and skills to teach the entire range of subject matter that should be included in the course of study. Teachers should not use "student interests" as an excuse for giving either extraordinary emphasis or no emphasis to an area of instruction. Often student interest is provided as the rationale for a curriculum decision when teacher expertise may be the more appropriate reason. For example, teachers who are excellent livestock judges may devote too much time to this topic, whereas teachers who are not capable in this area may choose not to teach it at all. Neither student interest nor teacher expertise should be overemphasized in selecting course content.

Educational Facilities and Equipment

Agricultural knowledge and skills must be taught, regardless of the adequacy of facilities and equipment. When inadequate facilities and equipment exist, the teacher should still instruct students in these areas by using community resources, field trips, simulations, distance education, and other methods.

Teachers should attempt to provide students with needed knowledge and skills. They can do this by making maximum use of existing facilities and equipment and of community resources. The emphasis given to the various content areas, however, may be partially determined by the adequacy of resources that are available to instruct the units in those areas.

Community Resources

The availability of an agricultural business or industry or a person with special expertise in a community may suggest that advantage should be taken of the opportunity for them to instruct students in unique areas. A beekeeper may teach students much about the interrelationship of plant and animal life or about the opportunity for a supervised agricultural experience program in beekeeping. The existence of food science research firms may enable students to learn much about nutrition and research. Teachers should use the resources available in their communities.

Articulation

Articulation is a term used to describe how the content of a course relates to other courses and programs. A teacher should build on rather than replicate content previously experienced by students. For example, if ninth-grade students have taken industrial technical education, the teacher will find that instruction relating to tool usage can begin at a higher level than if students had not taken industrial technical education.

The curriculum should also be articulated with other courses students take. For example, in the biology course taken by agriculture students, plant growth and reproduction topics were taught. The agriculture teacher can, therefore, plan instruction with the assumption that these concepts have been learned in biology class and can now be applied in agriculture class. This critical application of the science principles and concepts from biology can be made in agricultural education programs.

Articulation should also be planned with postsecondary agricultural education programs. Some content of a more advanced nature can be left for later instruction after students graduate from high school.

Two curriculum patterns for agricultural education programs are shown in Figure 3-1. The patterns were designed for a one-teacher department. With additional teachers, additional options can be added for offering courses in agricultural mechanics, horticulture, forestry, or other instructional program areas. A teacher planning for the agribusiness option for grade eleven would need to be aware of what was taught in grades nine and ten. A teacher planning for grade nine would need to be aware of the content of the preagricultural instruction.

Chap. 3 Planning the Course of Study

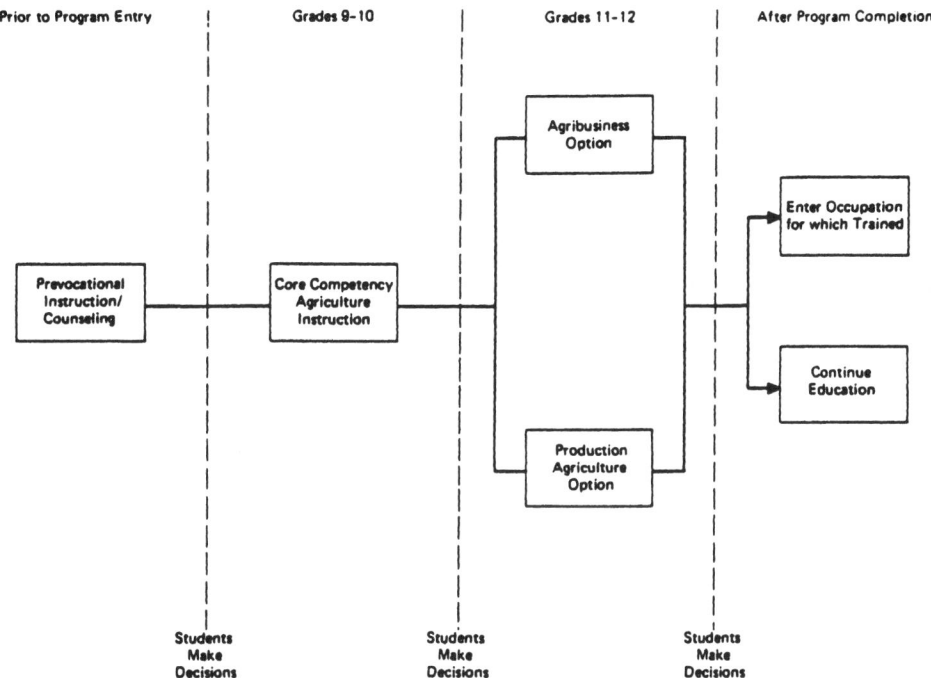

Figure 3–1A Curriculum patterns for agricultural/agribusiness programs (*designed for a one-teacher department*).

Source: From Lee, Jasper S., *Agribusiness Education*. New York: McGraw-Hill, p. 6 (A); and from Neil A. Knobloch, Agricultural Education program, Mid-Prairie High School, Iowa (B).

INFORMATION INFLUENCING CONTENT

Is there information that might assist the teacher in deciding what to teach? Information about the community, about occupations available to students, and about curriculum guides is useful in making content decisions.

Community Surveys

Community surveys can be planned and conducted by teachers or others to gather more information to help in program development. The results of such surveys may provide information about the types of employment opportunities that might be available to graduates. A quick survey of the yellow pages in the telephone directory will often provide useful information about the number and types of agricultural businesses in a community. Students can assist teachers in identifying and obtaining information about agriculturally related businesses. Useful information would include the number of employees in each job classification, the number of workers hired annually, educational requirements

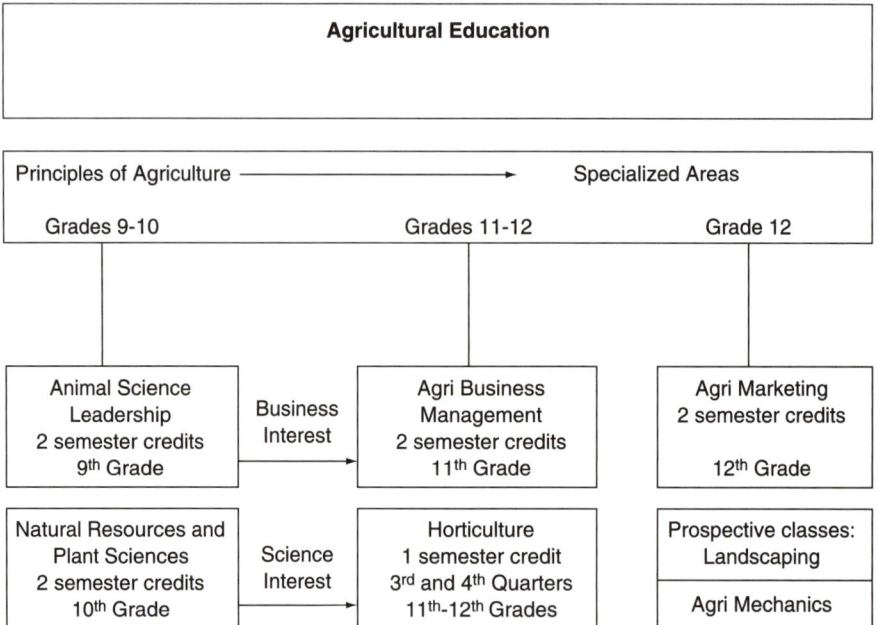

Figure 3–1B Curriculum patterns for agricultural/agribusiness programs (*designed for a one-teacher department*). Continued

by job classification, and desired qualifications of entry-level employees. This information will help the teacher structure course content so students are prepared for career opportunities.

Teachers of agriculture may use the results of a survey that has already been conducted in their communities. The USDA National Ag Statistics Service has data available for each county. These data can be found on the Internet or in published forms at the local county extension office. Data include number of farms, land use, farm income and sales, marketed value of products sold, total cash receipts, livestock cash receipts by type of crop, livestock numbers, crop acreage, and number of full- and part-time farmers. Such information can be used to plan instruction in keeping with the agriculture of a community. For example, if it was found that the community planted large soybean acreage, this information would suggest that soybean production is an important topic in the course of study.

Science Underlying Problem Areas

Basic principles of science underlie many of the problem areas within the broad field of agriculture. For example, an understanding of the principles of plant growth and reproduction is essential for a student who wants to develop expertise in horticulture, production agriculture, agricultural science, natural resources, and so on. Understanding the principles of animal nutrition enables students to plan better livestock feeding programs for efficient and rapid rates of gain. Understanding cellular biology is required for students who wish to learn about genetic engineering in an agricultural science course. In addition, many physics principles underlie instruction in agricultural mechanics. As mentioned in earlier chapters, the teacher should plan to teach, emphasize, or review these principles at the moment in the unit of instruction when understanding is essential for students to achieve competence. To teach the principles alone without applying them will result in instruction perceived by students as lacking meaning. Therefore, teachers should plan to work back and forth between basic principles of science and applied subject matter content.

Competencies Needed

Once the job classifications have been identified for which agricultural employees are to be prepared by the agricultural education program, the essential entry-level competencies should be identified. Many agricultural educators claim the major decisions about what is to be taught should be determined by agribusiness skill requirements. The rationale is that one should find out what these competencies are and then prepare students to perform them.

Many studies have been conducted to identify the essential competencies performed in agricultural occupations. The USDA web site and its links are resources for teachers who are seeking agricultural skills information. Figure 3–2 provides a sample of the kind of information available in a competency study. This is a valuable source of information in planning what knowledge and skills might be taught in a course.

Teachers can conduct their own competency studies by observing workers and recording the things they do and the things they must know. This record, when developed for many of the agricultural occupations in a community, will provide a basis for selecting course content that will serve to prepare competent employees.

Input by Advisory Committees

An advisory committee is a group of citizens from the community who are interested in the local school's agriculture department. Representatives are usually selected for three-year terms on a rotating basis so some of the members' terms expire each year. The committee is often made up of members who are farmers or ranchers, representatives of agricultural business, representatives from county agencies such as the fish and game commission, parents, and former and current students.

Teachers should obtain the assistance of the advisory committee in planning the content of the course of study. Advisory committees should be knowledgeable about agricultural occupations and interested in the educational

Conservation Assistant

Other Titles: Game Biologist, Game Propagator II, Private Shooting Ground Operator and Manager

Job Description

Assists in the operation of the game hatchery and general operation of a game farm. Maintains simple roads, trails, and ditches on state or federal game lands. Aids wildlife manager or conservation officer in the disposing of nuisance stream obstructions, trapping game animals for experiments, and trapping and banding game animals. Assists in maintaining a traveling wildlife exhibit. Plants farm-game habitat improvement plots. Aids in the operation of game animal checking stations. Serves as a production assistant on state-produced wildlife movies. Performs general building construction, and maintenance and repair of buildings. Aids in fire prevention programs and fights fire on game forest lands and game areas. Assists wildlife managers and wildlife biologists in game survey. Installs boundary fences and signs on game refuge lands. Supervises temporary employees as assigned. Performs preventive maintenance on conservation equipment.

Competencies Identified and Validated
(n = 48)

Competencies

1. Operate game farm and wildlife equipment.
   - Use safety standards relating to operation of a particular piece of machinery.
   - Service machinery and equipment according to operator's manual.
   - Recognize malfunctions in equipment.
   - Operate machinery and equipment under a variety of field conditions.
   - Adjust equipment under field conditions for maximum efficiency.
   - Become familiar with operator's manual for each piece of equipment.
   - Attach accessory equipment to basic farm power unit.
   - Prepare machines and equipment for storage.
   - Maintain records of maintenance and repair on machinery and equipment.
   - Maintain a daily log of number of hours each piece of equipment is used.

2. Measure and compute water flow.
   - Use stream flow meter.
   - Use conversion tables.

3. Remove obstructions from ditches and streams.
   - Use general earth-moving hand tools.
   - Operate earth-moving equipment.

4. Identify wildlife species.
   - Identify species of animals.
   - Identify species of birds.
   - Identify feed plants for animals and birds.

Figure 3–2 Information provided by a competency study.

Chap. 3 Planning the Course of Study

5. Assist in maintaining a mobile wildlife exhibit.
   - Hold the required license for type of equipment operated.
   - Drive responsibly and defensively at all times.
   - Protect and secure load.
   - Perform minor repairs.
   - Maintain a favorable condition for live exhibit.
   - Erect or place wildlife exhibit.
6. Aid in operation of game animal checking stations during hunting seasons.
   - Record location from which wildlife was taken.
   - Inquire about numbers, sizes, and species of game animals, game birds, fish or shellfish taken.
   - Measure sizes and make visual checks for other biological data.
   - Interview hunters at assigned checkpoints.
7. Serve as a production assistant on state wildlife movies.
   - Perform duties as assigned.
   - Maintain wildlife animals.
8. Perform general construction maintenance and repair of wildlife facilities.
   - Perform semiskilled-level carpentry work.
   - Clean and sanitize buildings used in wildlife areas.
   - Perform semiskilled-level electricity work.
   - Perform semiskilled-level plumbing work.
9. Aid in fighting game land forest fires.
   - Clear fire lanes by removing brush, trees, and ground litter.
   - Patrol burned area after fire to watch for hot sparks.
   - Suppress fire by felling trees, digging trenches, and extinguishing flames and embers, using ax, chain saw, shovel, and hand- or engine-driven pumps.
   - Ignite backfires to hasten burnout of major fire.
   - Operate firefighting equipment such as bulldozers and fire plows.
10. Aid in the gathering and recording of wildlife data.
    - Inventory number and species of wildlife.
    - Inventory available feed for wildlife.
    - Use rain gauge.
11. Install boundary fences and signs on state game lands.
    - Use hand tools where necessary to erect fences and signs.
    - Use a prepared map to determine locations.
    - Stretch wire fencing.
    - Operate mechanical equipment to drive posts.
12. Supervise temporary employees as assigned.
    - Explain assignments to be done.
    - Assign specific jobs.
    - Evaluate jobs accomplished.

Figure 3-2 Information provided by a competency study. Continued

program of the school. Many teachers annually meet with an advisory committee to review the course of study. The committee members may review competencies they desire students to be able to perform and the units of instruction to make recommendations about the relevancy of subject matter. Teachers could prepare a master list of competencies representing the agricultural opportunities of the community. They then have their advisory committee rate each competency concerning its importance in the curriculum. Others have the advisory committee discuss the relative importance of preparing students in each area of competence. Involving an advisory committee results in a more relevant course of study and a spirit of community involvement and interest in the instructional program.

Curriculum Guides

In deciding what should be taught, the teacher should consider the content that has been recommended by experts in the field of agriculture. These recommendations are sometimes in the form of curriculum guides or state core curricula. Teachers must also consider how the experts have recommended the subject matter be organized into courses. For example, should teachers plan semester courses on each subject or should they integrate various subjects into the course of study for each semester?

Numerous curriculum guides have been prepared by committees of teachers that outline suggested content for various courses in agriculture. These guides can be a useful reference for the teacher in planning what to teach. The teacher will, however, want to add and subtract material to make the content more relevant in a local community. Some guides have been prepared for teachers in a particular state to use in preparing their local courses of study. Other guides have been prepared for use in a specific local agriculture program. A source of information about available curriculum guides is the ERIC Clearinghouse, Ohio State University, 1900 Kenny Road, Columbus, Ohio 43210. Many state departments of education also provide this information through their Internet addresses.

THE COURSE OF STUDY

A course of study generally contains an introductory statement or situation specifying some background information about the community, school, and department and also introducing the main concepts of the course. This is where statements concerning philosophy, purpose, and objectives are placed. The intended grade level of students is usually specified. Often the teacher outlines expectations of students who participate in the agricultural education program. Some teachers place policies in the introduction to the course of study relating to the laboratory, supervised agricultural experience program, student discipline, FFA participation, and grading. The introductory material serves as an orientation to the course.

In a single-teacher agriculture department, the teacher is responsible for developing the content of all courses. She or he can independently determine the content of each course. However, in a multiple-teacher department, the teachers need to articulate jointly the content across courses. The total content for the course of study in a local school should be a joint decision of all involved instructors. Decisions need to be made about the instructional areas that will be offered and in which course(s) the content will be taught. The courses to be offered should also be periodically reviewed.

The expertise of the teacher can help in deciding in which course certain material is to be taught. Grouping items that fit a particular teacher's expertise will help in assigning the most technically competent instructor to each course.

Problem Areas

The major determinant of problem areas for a course is the competence needed by students to enter and advance in agricultural careers. Traditionally, a great deal of emphasis has also been placed on the leadership and personal development of students. The FFA organization serves as an intracurricular laboratory for this area of instruction. Another problem area often included relates to the need for informing students of career opportunities in agriculture and related fields.

Some problem areas require emphasis in every course offered. The use of records in decision making is an example of a problem area needing continual emphasis. Other problem areas may be taught in only one course.

A list of problem areas for each course should be prepared. The problem areas should be organized by instructional area. Figure 3-3 shows a partial list of problem areas for a general course in production agriculture. Figure 3-4 shows a partial list for horticulture.

The relevancy of the list of problem areas should be evaluated against the occupational needs of the community and the student potential to apply instruction in the area. Instruction may be applied in the laboratory, in student supervised agricultural experience programs, or in FFA activities. A problem area on orientation and one or more relating to supervised agricultural experience programs should be included in each course. Other problem areas should relate to the planned content for the particular course. Most problem areas should require one to four days of instructional time.

Time Allocation

An essential part of a course of study is the estimate of time for each unit and problem area. The time distribution is important because it gives the teacher a guide for preparing lesson plans. Without the allotment of time, teachers often approach the end of a school year with essential units not taught. Sometimes an inexperienced teacher may find material is being taught too quickly. The time allotment simply provides a guide to follow. The teacher should later feel free to make adjustments, within reason, as needed. There should be sufficient flexibility to consider unanticipated problems that need attention.

SEQUENCED COURSE OUTLINE

School: _____ Teacher: _____ Course/Year: _____

Number of hours in each month for each task

Instructional Areas—Problem Areas	A	S	O	N	D	J	F	M	A	M	J	Total
Personal Development												
• Leading groups		5										5
• Participating in the FFA		12	1	1	1	1	1	1	1	1		20
• Developing communications		7										7
• Participating on committees			3									3
• Participating in social events							3					3
Farm Management												
• Developing records		5	1	1	1	1	1	1	1	1		13
• Purchasing supplies									8			8
• Marketing products									10			10
Agricultural Mechanics												
• Using and maintaining hand tools					8							8
• Using and maintaining power tools						9						9
• Operating tractors and related equipment						9						9
• Operating arc and gas welders						10	5			15		15
• Constructing wood projects						10	5					15
• Assembling equipment									5	13		18
• Following safety procedures				8			3					11
Livestock Production												
• Handling and caring for livestock							12					12
• Feeding livestock							5					5
Subtotal		29	5	10	10	30	30	2	25	30		171

Figure 3-3 Partial sequenced course outline for agricultural instruction.

SEQUENCED COURSE OUTLINE

School: _____ Teacher: _____ Course/Year: _____

Number of hours in each month for each task

Instructional Areas—Problem Areas	A	S	O	N	D	J	F	M	A	M	J	Total
Personal Development												
• Developing awareness of policies		7										7
Plant Production												
• Understanding soil				20								20
• Understanding plant growth			23	2								25
Personal Development												
• Developing the individual		11					9					20
• Developing leadership		4			1		5					10
• Developing citizenship		1			2		6					9
Plant Production												
• Selecting greenhouse crops		1	1	1	5	1	1	1	1	1		13
• Preparing for planting							1					1
• Planting crops					10			4	3			17
• Managing crops					7				1			8
• Controlling pests					1							1
• Harvesting, grading, storing			1		1			1				3
• Marketing					2							2
• Operating the greenhouse								2				2
Retail Floriculture												
• Constructing basic designs		5	5									10
• Constructing related products							3					3
Subtotal		29	30	23	29	1	25	8	5	1		151

Figure 3–4 Partial sequenced course outline for horticulture instruction.

The time should be assigned for each problem area within a course. The instructor should ensure that the total time approximates that which is available. Some allowance needs to be made for unanticipated events during the school year, which reduce the actual time available for teaching and learning of the planned problem areas. The plan can be developed with days or hours as the unit of measure. Figures 3-3 and 3-4 show partial time allocation (see the "Total" column) for production agriculture and horticulture examples, respectively.

Sequencing of Instruction

How does one choose an appropriate order in which to teach the problem areas that have been selected for a course? What are some principles or concepts that influence sequencing decisions? The order for the course content is quite important. This order is influenced by the following factors.

Seasonal Sequencing. Problem solving in teaching and learning is best used when problems are realistic and meaningful. In most areas of agriculture the problems are seasonal in nature. Crop harvesting methods are best taught just prior to and during the harvesting season. Dendrology is best taught prior to leaf drop in the autumn. Controlling weeds in lawns is best taught when a weedy lawn can be used as a laboratory, possibly early spring. Because the school year generally extends from August/September through May/June, autumn and spring often are excellent times to schedule seasonal events. Although some seasonal activities occur in winter months, especially with livestock and in the southern regions of the country, this time of year may have more time available for problem areas that are less seasonal in nature such as cooperative ways of doing business, MIG/TIG welding, or leadership education.

Logical Sequencing. Instruction should be logically sequenced. More basic or prerequisite skills should be taught prior to the more complex skills. Teachers should develop an order that is as educationally sound as possible. For example, the planning of a cropping system for a farm should follow a study of soil surveys and land capability classifications. The budgeting of an aquaculture enterprise should follow a study of fish and aquatic life nutrition. The spraying of a greenhouse crop for insects should follow a study of the safe use of chemicals.

Calendar of Activities. A teaching calendar that lists major academic and intracurricular activities by date should be developed. Included on the calendar might be such events as fairs, career development events (CDEs), skills competitions, FFA meetings, the recognition banquet, FFA fund-raising activities, schoolwide career shadowing, dates reports are due, and FFA conventions. In horticulture, important holidays related to floral retail work should be listed. The purpose of developing the calendar is to plan instruction that is relevant at the time when related activities provide opportunity for real-life application, thus student (and teacher) motivation.

Laboratory Schedule. In today's agricultural instruction programs, school laboratories include, but are not limited to, greenhouses, farms, agribusinesses, school wetlands, nature trails, agriculture mechanics labs, floral retail shops, computer labs, aquaculture labs, biotechnology labs, and research labs. It is wise to develop a laboratory use schedule to prevent underutilization of the laboratory during part of the year and overutilization at other times. Different teachers in multiple-teacher departments may have access to the laboratory at different times of the year. In some schools students in different classes share the laboratory. Horticulture teachers with greenhouse facilities should develop a cropping schedule or plan. This plan will assist in deciding when certain problem areas should be taught. For example, classes on germinating seeds should be scheduled for February so that bedding plants are ready for Mother's Day sales. Teachers who use this type of facility should also develop land laboratory schedules. For example, school farms and nature trails can have target dates and deadlines that guide effective instruction. The laboratory schedule will partially determine when certain problem areas are taught. It will also dictate that some problem areas not needing a laboratory be taught at times other than when the laboratory is scheduled. The laboratory schedule should be flexible enough to allow for some adjustments as necessary to take advantage of "teachable moments."

Planning for Variability. Teachers should plan for variability, that is, using the rich and varied content in agriculture to stimulate and motivate students and their interests. The content or subject matter being studied should be varied. It is usually recommended that a specific topic of study should not last for more than one week. If students are in the laboratory, then they may desire up to three weeks on a single topic.

There are several reasons for planning for variability of instructional content. One important reason is that the interest span of students often does not expand over a long period of instruction. Their interest may be greater if the same instructional content is taught in several shorter time periods.

Another feature of a course of study related to the variability issue is that topics of high interest to students should be interspersed throughout the year. If all items of high interest are taught together and all items that are less interesting are taught together, students are not likely to respond as well as when there is variety.

Scope and Sequence Charts. Scope and sequence charts are useful as worksheets when a teacher plans the instructional sequence of a course. One example of such a chart for production agriculture is shown in Figure 3-3. The horticulture example is Figure 3-4. An instructor would use several such sheets for one course. The problem areas are listed at the left side of the sheet. The hours for each problem area are listed at the right in the Total column. As the decision is made to teach a certain problem area, it is recorded in the month column(s). Be careful not to plan for more hours than are available in any one month because school calendar hours fluctuate with holidays and teacher in-service hours.

In using the form, the instructor should first fill in the hours needed each month for problem areas having to do with supervised agricultural experience program records (possibly plan for one hour at the end of each month), FFA committee work (possibly one hour each month close to an FFA meeting date), and other continuing routine items. The teaching calendar, laboratory schedule, and list of seasonal topics should then be used to schedule problem areas by month. Principles of logical sequencing and variability should be applied in scheduling remaining problem areas. Teachers should expect some trial and error in organizing and scheduling topics by month.

Sequenced Course Outline

The sequence chart can serve as a sequenced course outline, providing the teacher with a list of problem areas, the approximate time of the school year in which each will be taught, and the time allocated to each problem area.

Another format for a sequenced course outline is shown in Figure 3-5. It is for a senior course in agricultural business. The date for each week is listed at the left side of the outline. For each week, special events from the teaching calendar are recorded. Note, for example, that the National FFA Convention in October is a field trip that must be taken into consideration when planning the sequenced course outline. Problem areas in this example are listed by week. The sequenced course outline that you create may be placed on a bulletin board in the classroom or in the office, so that teachers, students, parents, and others who are interested in the course of study can view it.

After preparing the sequenced course outline, the teacher has a usable "road map" or organization of instructional content. The major remaining planning activity is the development of units of instruction, daily plans, and the continual updating of the course of study. The development of daily plans is discussed in Chapter 5.

UPDATING THE COURSE OF STUDY

The course of study, which, again, organizes the entire agricultural education program for a given school district, must be viewed as a plan that will be continually updated. The student population changes from year to year, community conditions sometimes change, and information becomes available that an instructor will incorporate into the course of study.

A Dynamic Document

Instructors should plan for at least an annual review and revision of the course of study for each class. This review should be an agenda item for the advisory committee to the agricultural education program. Although drastic revisions are usually not necessary, some adjustment is often desirable.

Most agricultural education programs face formal evaluations at least once every five years. These reviews may be for accreditation or for the purpose of approving the program for state and federal technical education funds.

Chap. 3 Planning the Course of Study

The course of study is usually reviewed as a part of these evaluations. The agriculture teachers should implement recommendations resulting from these processes.

Revision

As the course is taught, notes concerning needed changes should be made directly on the course outline. Often a teacher will find a different time allocation is needed, or that the time of year is wrong for a topic. Perhaps a reordering is needed to enhance the learning sequence. These ideas, if not noted at the time, are often lost until the subject matter is again taught the next year.

Time should be scheduled during the summer months for a period of concentrated work on revising the course of study. Development and revision is best done when other matters are not allowed to interrupt the process. During the time that has been scheduled, therefore, the teacher should make this work the number one priority.

SUMMARY

A course of study is necessary for the teacher. It provides the outline of content that serves as a basis for further planning of instruction. The process of preparing a course of study enables teachers to think through many important questions related to the process of teaching and learning. Teachers should commit themselves to course of study development as an essential function of their work.

FOR FURTHER STUDY

1. Obtain the course of study for a local agriculture program. Analyze it with the following in mind:
    a. Key concepts present in the philosophy
    b. Three examples of logical sequencing
    c. Three examples of seasonal sequencing
    d. The time allocation for problem areas
    e. The laboratory schedule
    f. The teaching calendar
    g. Variability of content
2. Prepare U.S. Census of Agriculture information that is relevant for course planning for the county in which you want to teach.
3. Using the yellow pages of the telephone directory, prepare a list of agriculturally related businesses for a local community and then list the entry-level skills a student graduating from your program might need in order to be hired by one of the agribusinesses selected.

SEQUENCED COURSE OUTLINE

Date	Special Events	Class
		Problem Areas
	Teachers' meetings	Orienting students, developing record books
	Labor Day parade	Participating in the FFA
	Planning citrus sales	Developing a sales campaign
		Organizing an FFA fruit sale
		Handling and processing production agricultural products
	FFA meeting	Managing business money
		Transporting agricultural products (Speaker topic)
	Farm science review	Selecting building materials
		Following safety precautions
		Updating record books
	Pumpkin show	Exploring Value-Added agriculture
	Begin citrus sales	Handling and applying fertilizer
	Soil CDE	Handling chemical applications
		Storing agricultural products
	FFA meeting	Participating in business meetings
	National FFA Convention	Performing office procedures
	End grading period	Speculating on futures
		Updating record books
	Parliamentary Procedure CDE	Developing good work habits
	FFA meeting	Developing abilities to work with others
	Parent conferences	Developing employer/employee relations

Figure 3-5 Sequenced course outline for agribusiness.

Date	Special Events	Class
	Deliver citrus	Handling production agriculture products
		Servicing agricultural supplies
	FFA meeting	Exhibiting proper dress
	Public speaking CDE	Developing an advertising plan
		Updating record books
		(Additional weeks of the school year would follow)

Figure 3–5 Sequenced course outline for agribusiness. *(Continued)*

4

Learning as Problem Solving

During the first class session of experienced teachers of agriculture enrolled in a graduate course on teaching methods, the professor called to their attention a phenomenon observed from the classroom windows. Some 200 yards from the building was a field used as a pasture for the university's dairy herd. Within the field, beginning at an entry gate, was a clearly demarked strip of weeds with bright yellow blossoms, twenty to twenty-five feet wide. The very visible and identifiable strip of blooming weeds extended from the entry gate in a defined pattern to the center of the field. Weeds with the characteristics observed in the distinct strip were not visible at other locations in the field. The university's football stadium is located another 200 yards beyond the pasture.

The professor asked the teachers, "How do you explain the unique pattern of blooming weeds we see in the pasture?" Immediately, the teachers identified the weed. They were aware that the field was a pasture for the dairy herd. Soon speculation began about possible explanations for the observed pattern of flowering weeds. Consensus was reached quickly that the pattern observed was probably not a random occurrence. Possible explanations proposed included (1) weed seeds were in loads of manure that had been spread in a particular pattern; (2) weed seeds were in residue from a prior use of the land, perhaps from previous experimental work; and (3) weed seeds could have been scattered in the observed pattern by a prankster. After further discussion, including some joking banter, the professor concluded the discussion with the comment, "Why don't you think about this some more; if you come up with any other ideas, how about sharing your thinking with us during the next class session?"

> At the beginning of the next class session, the phenomenon of the distinct pattern of weeds in the dairy cattle pasture was revisited by the teachers. Two teachers reported they had solved the mystery. First, they had inspected the site and confirmed that the weed had been identified correctly during the previous class session. Second, they reported the observation that there was a residue of straw in the area corresponding to the distinct pattern of weeds. Because they knew that the field was used for parking during the football season, their inquiry with traffic and parking officials confirmed that the previous autumn on a rainy football Saturday straw had been scattered in the field in the pattern observed to facilitate fans walking from their vehicles to the exit gate, then to the stadium. Hence, the distinct pattern of brightly blooming weeds is apparently accounted for by weed seeds in the straw that was spread on the field in the systematic pattern observed.

OBJECTIVES

After studying this chapter, you will be able to

1. List and explain the steps in the learning process.
2. List and explain the steps in the problem-solving approach to teaching.
3. Compare and contrast the learning process and the problem-solving approach to teaching.
4. Explain what is to be accomplished and techniques to use, and identify the principles of learning used for each step in the problem-solving approach to teaching.

Every day people learn on their own without the presence of teachers. How is it that people learn without experiencing formal instruction? What process do people follow as they encounter problems, questions, or obstacles that require them to think and study in order to solve problems confronting them?

By identifying the process that people use rather automatically—and use successfully—then could one not teach students in a classroom or laboratory following the same process? Would it not make sense for teachers to teach people by following the same process that people generally follow in learning on their own?

John Dewey describes this process in his book, *How We Think*, as steps in reflective thinking. The process can also be described as a chain of reasoning, the method of science, the scientific method, or the learning process. The steps comprising the process are listed in the accompanying learning process chart. Analyze Case 4-1 in terms of the learning process outlined in the following chart.

THE LEARNING PROCESS

1. Experiencing a provocative situation
2. Defining the problem—clarifying questions to be answered
3. Seeking data and information
4. Formulating possible solutions
5. Testing proposed solutions
6. Evaluating the results

Case 4-1. *The Leaking Copper Pipe.*

> You discover that a copper water pipe in your basement has a leak around one of the soldered joints. It gets worse daily and has the potential of ruining the basement and escalating the water bill. You decide to "resweat" the joint yourself.
>
> You turn off the water at the main valve, loosen the soldered joint, clean it, apply flux, heat it, and apply solder. The solder drips to the floor rather than fusing with the copper. You are sure you know how to solder, but after you repeatedly go through the procedure, water flows out more than ever.

In Case 4-1, how do the steps of the learning process apply? Not only would you experience a provocative situation when you found the leak, which was the first problem, but you would also experience a provocative situation when you applied what you thought was the solution (soldering) and it didn't work. When the soldering did not work, that was yet another problem. The problem of the soldering not working cannot be solved until you move to step 2 of the learning process, that is, defining the problem.

So often, people try to omit this important step. We try to jump ahead to formulating solutions and testing them. In fact, that is what happened in Case 4-1.

Being provoked is not enough. Being perplexed is not enough. Being interested and wanting to know the answer is not enough. We should figure out what is wrong and decide on what questions to ask before rushing on by trial and error.

Once we have defined the problem or determined questions that must be answered, then we can productively gather data or information and arrive at possible solutions. In the case of the soldering problem, one needs to ask questions such as, "Why isn't the solder bonding with the copper?" "What conditions are essential for soldering to work?" and "How does soldering work?" Then we can read; ask others for advice; search the Internet; study pictures, diagrams, or videos; and use our own reasoning. In doing all of this, we arrive at answers and possible solutions that can then be tested and evaluated.

In the case of the soldering, you would learn that, in order for soldering to work, the surface must be thoroughly clean, flux must be used for further chemical cleaning, there must be a good fit between the pipes, and the temperature of the copper pipes must reach about 400° Fahrenheit. Then you ask yourself if your earlier attempts at soldering met all of those conditions. Through this process of reasoning, you would realize that the only condition you were not sure of was the temperature. As you read and discussed the matter with others, you would find the caution: Be sure *all* water is drained from the pipes. If this is not done, the metal will not reach the proper temperature because the water conducts away too much of the heat.

This would lead you to deduce that the pipes were not drained well enough. You would then drain them further (notice that this could be yet another problem to solve) and determine if the joint would sweat satisfactorily. In so doing, you would be evaluating your proposed solution.

Notice that in the process of solving one problem, other problems are often discovered. It should also be pointed out that people do not always solve a problem by following the steps of the learning process in consecutive order. A person moves back and forth among the steps in an interactive fashion.

If following such a process allows people to learn on their own, then teachers can use this process to guide students in their learning. Teachers who use a problem-solving approach in teaching can build on the learning process presented in the earlier chart. The problem-solving framework that corresponds to the steps in the learning process is presented in the next chart.

THE PROBLEM-SOLVING APPROACH TO TEACHING

1. Interest approach
2. Objectives to be achieved
3. Problems to be solved—questions to be answered
4. Problem solution
5. Testing solutions through application
6. Evaluation of solutions

An explanation of each step follows. The teacher begins the unit of instruction with an interest approach that is designed to present a provocative situation. Students are then led in a discussion of why they need to know this information and formulate a list of objectives to be achieved by studying the unit. They are then asked to develop a list of questions they need to answer. By developing a list of objectives to be achieved by studying this unit, and deriving a list of questions to be answered, the students will have rather carefully and clearly defined the problem. In order to solve the problem (or answer the questions), the students will need to gather data and information, and formulate possible solutions. Specific suggestions for how to accomplish this problem–solution phase are found in Chapter 5, and specific techniques that may be used to provide the needed data and information are presented in Chapters 6 and 7.

Students then test the solutions or answers in class, lab, FFA activities, or through their SAE programs. Specific techniques teachers may use to guide this testing of solutions are found in Chapters 9, 10, and 11. Solutions are then evaluated by the students and by the teacher to determine how much has been learned. The next chart compares the steps in the learning process with the steps in the problem-solving approach to teaching.

Learning Process	Problem-Solving Approach to Teaching
1. Experiencing a provocative situation	1. Interest approach
2. Defining the problem	2. Objectives to be achieved
	3. Questions to be answered
3. Seeking data and information	4. Problem solution
4. Formulating possible solutions	• Develop possible solutions
	• Acquire new knowledge, skill, and experience
	• Formulate conclusions and general principles
5. Testing proposed solutions	5. Testing solutions through application
6. Evaluating the results	6. Evaluation of solutions

A PRINCIPLE OF TEACHING AND LEARNING APPLIED TO PROBLEM-SOLVING APPROACH TO TEACHING

For each step of the problem-solving approach to teaching, the principles of learning that support the step are presented, what the step is designed to accomplish is explained, and techniques that can be used to accomplish the step are suggested. A full discussion of teaching techniques is presented in Chapters 6 and 7.

Interest Approach

The interest approach corresponds to the first step of the learning process, that is, "experiencing a provocative situation."

Principles of Learning That Are Applicable

1. Students must be motivated to learn. Learning activities should be provided that consider the wants, needs, interests, and aspirations of students.

2. Readiness is a prerequisite for learning. Subject matter and learning experiences must be provided that begin where the learner is.
3. Motivation (interest) is strongest when students perceive that learning can be useful.

What to Accomplish. In the next step, the teacher seeks to gain the attention of the students. They have been studying another unit of agricultural subject matter, and now they need to focus in on something else. Prior to coming to the agriculture class, they have been in other classes, and during the change of classes, they've been involved in numerous and varied activities. The interest approach is designed to draw the attention of the students to a common point. But it must accomplish more–it must also create a felt need on their part for studying the unit that is being introduced.

This felt need to know must make students realize they do not know enough about the subject to be successful in their area of study. It should be so urgent that students personally feel they need to learn more.

The interest approach should also make students aware of their situation and experiences that relate to the unit of instruction. Finally, it should set the stage for the establishment of the objectives to be achieved through the unit of instruction.

Techniques to Use. Teachers can create the necessary provocative situation that encourages interest in numerous ways. Some possibilities are

- Raising perplexing questions.
- Showing specimens or samples, such as an injured owl or a broken flywheel.
- Presenting a case study in which a job to be done is outlined and then students are asked to explain how to do it.
- Showing an Internet connection of a veterinarian performing an operation.
- Giving a skillful demonstration.
- Showing pictures of success and failure–good and bad.
- Giving a project assignment.
- Conducting a provocative skit or role play.
- Providing a computer spreadsheet for completion.
- Asking questions that encourage students to describe their experiences about the subject matter to be taught.

When using each of these techniques, students have to be led to realize that they do not know enough or do not have the skills required for solving the problem, answering the questions, or otherwise handling the subject matter being presented. Yet they must also end up wanting to be able to do what they have been challenged to do and led to see that it is important for them to move from where they currently are to where they would like to be.

Objectives to Be Achieved

The second phase of the problem-solving approach to teaching partially accomplishes the "defining the problem" step of the learning process.

Principles of Learning That Are Applicable

1. Students are motivated through their involvement in setting goals and planning learning activities.
2. Students progress in acquiring new knowledge and skills only as far as needed to accomplish their purposes.

What to Accomplish. At this point in the problem-solving approach, the teacher's goal is to lead students to realize that the subject before them is worth studying and admit that they need to study it. The teacher asks lead questions (presented in Chapter 5) that cause the students to discover reasons why they need to study the unit. Furthermore, the teacher has the students identify goals they hope to achieve by studying the unit. In the process of thinking through their goals or objectives for studying the unit and identifying reasons why this subject matter is worth knowing, students are able to begin to more completely define the problem area. They will complete the definition of the problem step when they develop a list of questions to be answered (problems to be solved) in order to meet the objectives they have set for the unit of instruction.

Techniques to Use. The teacher develops the objectives to be achieved with the class by raising carefully selected lead questions (Chapter 5) and skillfully conducting a discussion (Chapter 6) in order to develop (usually on the chalkboard) a list of the students' objectives to be achieved from studying the unit of instruction.

Questions to Be Answered

The third step of the problem-solving approach to teaching completes the "defining the problem" step of the learning process.

Principles of Learning That Are Applicable

1. When the subject matter to be learned possesses meaning, organization, and structure that is clear to students, learning proceeds rapidly and is retained longer.
2. Students are motivated when they attempt tasks that fall in a range of challenge such that success is perceived to be possible but not certain.

What to Accomplish. Here the teacher seeks to lead the students to think about the questions they need to answer so they can achieve the objectives that have been established. This step is designed to completely define the problem area by identifying all the questions with which it is associated.

Hence, it is here that the subject matter to be learned is presented. Although students may not think of all the questions they need to answer, it is important for them to learn how to go about identifying the questions to arrive at the best possible conclusions and solutions.

Techniques to Use. Just as with the "objectives to be achieved" step, for the "questions to be answered" step, the teacher relies on lead questions (Chapter 5) and discussion (Chapter 6) to develop a list of questions the class will study in order to complete the problem area. Whenever students fail to think of all the questions for which they need answers, the teacher completes the list using their anticipated lists from the unit of instruction. In the case of subjects where the students have too little experience to be able to deduce the list of questions, the teacher should provide the list for them.

Problem Solution

During the fourth phase of the problem-solving approach to teaching, the following steps in the learning process are completed: "seeking data and information," and "formulating possible solutions."

Principles of Learning That Are Applicable

1. Students think (formulate and test possible solutions to problems) when they encounter an obstacle, a difficulty, or a challenge in a situation that interests them.
2. Learning is an active rather than passive process.
3. Directed learning is more effective than undirected learning.
4. To maximize learning, students should "inquire into" rather than be "instructed in" the subject matter.
5. Problem-oriented approaches to teaching improve learning.
6. Students learn what they practice.
7. Behaviors that are reinforced (rewarded) are more likely to be learned.
8. To be most effective, reward (reinforcement) must follow as immediately as possible the desired behavior and be clearly connected with that behavior by the student.
9. Transfer of learning is more likely to take place when what is to be transferred is a generalization, a general rule, or a formula.

What to Accomplish. During this phase of problem solving, the teacher seeks to assist the class in obtaining, studying, and evaluating facts, concepts, and skills necessary to answer the questions, solve the problems, and develop the conclusions needed to master the subject matter. In essence, the teacher and the students cooperatively decide on the content to be studied.

The preceding list of principles of learning for this phase of problem solving are not the only principles of learning that apply to this phase of problem solving. For example, as each problem is solved, interest must be re-created. As the teacher prepares the class for supervised study, the class needs to further define the problem. So, although certain principles of learning are associated with one phase of problem solving more than another, they are, nevertheless, often used in more than one place in the problem-solving approach to teaching.

Techniques to Use.   It is at this step of the problem-solving approach that the greatest amount of time is spent by teachers and students. It is here where subject matter and skills are learned. Hence, it is here on which all the techniques of teaching have to be drawn. These techniques of teaching are presented in Chapters 6 and 7. They are

- Lecture
- Discussion
- Demonstrations
- Field trips
- Role-playing
- Resource people
- Cooperative learning
- Supervised study
- Independent study
- Experiments
- Use of student notebooks
- Use of instructional sheets, job sheets, and skill sheets

Testing Solutions through Application

When students apply what they have learned in the classroom and laboratory, they are completing the "testing proposed solutions" step of the learning process.

Principles of Learning That Are Applicable

1. Supervised practice that is most effective occurs in a functional educational experience.
2. Learning is most likely to be used if it is learned in a situation as much like that in which it is to be used as possible and immediately preceding the time when it is needed.

What to Accomplish.   This is a very important phase of problem-solving teaching. This is the "learning to do" stage of the process. Granted, there is room for and need for "doing" in the problem-solution stage, but doing is essential in the application stage. At this point, the teacher wants the students to put into practice what they have learned.

Chap. 4 Learning as Problem Solving **81**

This is the opportunity to have students try out the conclusions and approved practices arrived at during classroom study to see if and how they work. In the process, they can prove for themselves the relevance and usefulness of their course of study.

Application also allows students to practice new knowledge and skills. This practice develops the proficiency they will need after they leave the program.

Another important thing this step accomplishes is that it allows and even forces individualization. The teacher has to work with each student in accordance with his or her situation, learning style, interests, and aptitudes.

Techniques to Use. Every conceivable opportunity needs to be used to provide practice. Application should occur in the classroom whenever possible. It will also occur in the school laboratory, through FFA activities, during field trips, and as a part of students' supervised agricultural experience programs. More specific details for guiding students in the application of their learning are offered in Chapters 9, 10, and 11.

Evaluation of Solutions

The sixth step of problem solving is the same as the final step of the learning process: "evaluating the results."

Principle of Learning That Is Applicable. When students have knowledge of their learning progress, performance will be superior to what it would have been without such knowledge.

What to Accomplish. Students try what they have learned and decide if it works. They use their knowledge and new skill and determine whether it produces the results they had come to expect it would.

Of course, they also are able to evaluate their own progress. Their understanding of what was studied and their levels of skill affect outcomes. Thus, the teacher needs to help them learn to evaluate their performances as well as the validity of what they have studied.

Likewise, the results obtained during the process of gathering information, formulating solutions, and applying what has been learned also provide evaluative information for teachers. This information allows teachers to determine not only how well students have learned but also how well the teachers have taught.

Techniques to Use. Many forms of evaluation are used and not just at the conclusion of the problem area. Students can complete paper and pencil tests to assess cognitive outcomes. Students can also use checklists to determine if their products reach the level of quality desired. They can further use checklists to see if they have followed the proper procedures. Of course, the teacher can use rubrics to evaluate the students' performances. Evaluation is also accomplished with project work and through the analysis of financial records. Projects may be further evaluated through competition in the FFA. Teachers should also use appropriate instruments to have students evaluate their teaching. Detailed information regarding evaluation appears in Chapter 14.

SUMMARY

Much of the learning in life comes through some form of problem solving. This is true for the homeowner who is trying to start the lawn mower. It is just as true for the researcher who is seeking an answer to a perplexing scientific problem. Likewise, it is true for students as they move from where they are to where they need to be to become successful individuals.

The learning process is a problem-solving process. Hence, agriculture teachers need to teach their students using the learning process as the foundation for classroom and laboratory activities. The use of the problem-solving approach to teaching makes operational a systematic and effective process for teaching and learning.

FOR FURTHER STUDY

1. Analyze the episode described at the beginning of this chapter.
   a. What incidents or activities described in the episode correspond to each step in the learning process and the problem-solving approach to teaching?
   b. What principles of teaching and learning are demonstrated during the episode?
   c. What was the role of the teacher in leading the problem-solving exercise?
   d. What sources of information did the teachers use to identify the blooming weeds? In explaining the unique pattern of blooming weeds in the pasture?
   e. What new information did the teachers have during the second class session that made possible the solution of the problem? What is the source of this new information?
2. Observe an agriculture teacher who is teaching students a new process. During the occasion, write down each step of the learning process that is used. Also, note the principles of learning that are used with each step.
3. Recall a recent problem with which you were confronted. Analyze it according to the steps in the learning process. See if you omitted any steps. Ask yourself if you could have solved the problem more efficiently by following the problem-solving procedure more closely. If so, how and why?
4. Explain how teaching, learning, and research are similar by using the steps in the learning process to formulate your answer.
5. Select a topic in agriculture or natural resources and outline how you could teach it using the problem-solving approach.

PART II Methods for Teaching and Learning

Teachers are expected to be competent in the technical subject areas they teach. They are also expected to be skilled in the area of helping students to learn. Knowledge of the science and art of teaching will assist a teacher in the effective communication of subject matter knowledge and skills to students. Teachers need to be competent in planning for instruction, in using group and individual teaching techniques, and in managing student behavior. Part II is designed to assist current and prospective teachers in developing these abilities.

Whereas the course of study provides the blueprint or design for what is to be taught for an entire agricultural instruction program, and the sequenced course outline describes when it is to be taught and for how long, the written plan for the unit of instruction provides the organization, structure, and the procedure for how each class session is to be taught. The written plan for the unit of instruction is an essential ingredient for effective teaching. The reader will, by studying Chapter 5, be able to plan for problem-solving teaching. Examples are provided for each step of the process.

Group teaching techniques are appropriate for providing instruction to a group of students in the same setting. Chapter 6 relates each technique to problem-solving teaching, explains how each technique might be used, and indicates how to select the most appropriate techniques. The major techniques discussed include lectures, discussions, demonstrations, field trips, role-plays, cooperative learning, and use of resource people. The necessity of learning to effectively use instructional media is also highlighted.

There are differences among learners that can be addressed by using individualized teaching techniques. Students have varying levels of aptitude, interest, energy, previous experience, opportunity to use information, and long-term goals. These techniques help students learn how to learn. The techniques discussed in Chapter 7 include supervised study, experiments, and independent study. Tools to aid in learning through individualized methods are discussed.

Without good class control, teachers cannot be effective teachers. Managing student behavior is the topic for Chapter 8. The teacher is regarded as the key for an acceptable classroom atmosphere. One of the public's chief concerns with education is discipline in the schools. The chapter suggests that teaching performance and student behavior are interrelated. Teacher technical competence, student–teacher rapport, student interest, and organization and clarity of course content all set the stage for managing the classroom. Specific strategies are forwarded to establish and maintain an appropriate climate for learning. Parental involvement and promotion of individual responsibility is regarded as essential.

5

Planning for Instruction

> You are a newly hired teacher. You have been asked to examine the sequenced course outline contained in the course of study for the local agriculture program, and select a problem area for which you will develop the written plan for teaching the unit of instruction. You are to include instructional objectives, reasons for studying the unit, questions to be asked, answers to the questions, and approved practices.
>
> The preceding assignment is the first task for all teachers of agriculture. Classroom teaching is the reason for the existence of agriculture teachers. To those of you who read the assignment and felt unprepared to proceed, once this chapter has been read, you will have achieved the ability to complete the assignment.

OBJECTIVES

After studying this chapter, you will be able to

1. Develop your rationale for written plans.
2. Select a topic and write a plan for teaching it.
3. Explain the relationship between planning for instruction, the problem-solving approach to teaching, and techniques of instruction.
4. Explain each component of a written plan.
5. Complete each component of a written plan.
6. Develop daily plans that accomplish the objectives of the unit of instruction.

SELECTING THE UNIT OF INSTRUCTION

Prior to planning lessons, teachers complete the course of study. Included in the course of study are the course offerings for the entire agricultural instruction program (e.g., environmental science, animal science, landscape design and maintenance, and others). For each of these course offerings there is a sequenced course outline. The sequenced course outline contains a list of instructional areas (see Chapter 3) sequenced to the school calendar. Each instructional area contains a list of problem areas for which the teacher will write a unit of instruction. Each unit of instruction contains a list of questions to be answered. Day-by-day learning activities for students are organized around these questions to be answered.

For example (see the feature below), a teacher's course of study might include the following course offering: Landscape Design and Maintenance. The sequenced course outline for that particular course will include an instructional area called Turfgrass. The teacher would need to develop a written unit of instruction for each problem area (for example, Establishing My Home Lawn). The unit of instruction would then be written to address a number of questions to be answered such as, "How do I test the soil?" Later in the year the teacher will write another unit of instruction to teach the problem area, "Maintaining My Home Lawn."

Course title: (from course of study)	Landscape Design and Maintenance
Instructional area: (from sequenced course outline)	Turfgrass
Problem area: (for which the unit of instruction is planned)	Establishing My Home Lawn
Questions to be answered (from written unit of instruction)	How do I test the soil?
	How do I prepare the seedbed?
	What equipment do I need to sow grass seed?
	How do I sow the grass seed?
	How do I mulch and water to establish the grass?
Problem area: (for which the unit of instruction is planned)	Maintaining My Home Lawn
Questions to be answered (from written unit of instruction)	How do I properly mow the lawn?
	What fertilizer should I use?
	What are the recommended watering rates?
	How do I control for diseases?
	What insects and pests are common?
	How are insects and pests controlled?
	How do I control weeds in my lawn?

When planning the course of study from which the preceding example was taken, the teacher included a problem area or unit of instruction titled, "Establishing My Home Lawn." At the time the course of study, including the sequenced course outline for each course offering, was developed the teacher had little more in mind than the basic instructional areas and problem areas that were listed. Now the teacher needs to take a problem area and decide on the list of questions that must be answered before students fully grasp the content. The teacher develops a plan for helping students find answers to the questions. In doing this, the teacher begins the necessary preparation for successful classroom teaching.

The teacher begins with the title of the unit of instruction, which restates the problem area, "Establishing My Home Lawn." This title, this basic idea or general notion, then has to be developed. A content outline must be brainstormed. This structuring, or filling-out of details, is included in the planning of the unit of instruction, and is often referred to as lesson planning.

Not only do teachers need to decide on specific content to be taught as a part of each unit of instruction but they also must plan for ways to teach so that students master the subject matter in the problem area. This requires serious contemplation. Subject matter as well as pedagogy must be selected and sequenced. Without prior thought (planning), optimum teaching effectiveness is seldom achieved. Yet some teachers resist the disciplined effort such planning requires and quite often ask if plans need to be in writing.

RATIONALE AND NEED FOR WRITTEN PLANS

Remember the principle of learning, "When the subject matter to be learned possesses meaning, organization, and structure that is clear to students, learning proceeds rapidly and is retained longer." This is the heart of developing written plans. Through written plans, the teacher is bound to develop sensible and complete organization and structure that can be made clear to students. Otherwise, organization is fragmented and the structure of the subject matter and learning experiences is much more imperfect.

Efficiency is gained by recording the planning before teaching a problem area. Less time is spent rethinking, and the plan can be used for subsequent classes.

The act of writing out a plan forces the teacher to process the subject matter. When teachers think through a concept thoroughly enough to figure out how best to make it clear to others, they understand it better and are able to teach it with greater authority and clarity.

By taking time to write down the plan for teaching, the teacher is more likely to be able to draw on and make use of more of the principles of learning. A teacher is also apt to do a better job of properly selecting and using a variety of techniques of teaching.

RELATIONSHIP BETWEEN PLANNING FOR INSTRUCTION AND THE PROBLEM-SOLVING APPROACH TO TEACHING

When teachers encounter a need to plan a new unit of instruction, they encounter problems in two ways. First is the problem of planning a unit of instruction. Second is the need to solve the problem or problems they intend to teach their students. In managing both of these kinds of problems, teachers can make use of the very kind of problem solving the authors are advocating.

When you as the teacher determine the solution to the problem to be taught, and when you develop a plan for teaching the solution, you encounter the same felt need to know that you are always trying to impress on your students. The next step is to clearly define the problem; how can you best plan this unit of instruction, and what is the solution to this "question to be answered?" After clearly defining the problem, you gather the information that is pertinent to the "question to be answered" that is being taught. You then need to develop the best possible answers and apply them in planning instruction. As the unit is taught, as well as after it is completed, you need to evaluate the effectiveness of the plan. In following the steps that were just identified, you have followed the same steps of problem solving through which students will be guided in their instruction.

RELATIONSHIP BETWEEN PLANNING FOR INSTRUCTION AND TECHNIQUES OF TEACHING

Techniques of teaching are presented in Chapters 6 and 7. They represent the tools that teachers use either to present information or to guide students in their discovery of knowledge. Techniques of teaching are used in problem-solving teaching during the data-gathering step. They are also used to create interest and to develop the discussion needed to lead the class in constructing lists of reasons for studying the unit and questions to be answered.

First, you need to understand the overall scheme of the planning phase. Then you need to study and learn how to use each of the teaching techniques presented in Chapters 6 and 7. Not until you have mastered both the planning process and the techniques of teaching will you be able to effectively plan and present instruction in the classroom and laboratory.

Consider the following assignment. If it was actually your assignment to complete, would you be able to complete it satisfactorily?

> Prepare to instruct students on "Establishing My Home Lawn." Outline the parts that must be included, and complete each part so that you can teach the problem area in a masterful fashion.

In order to meet this challenge, what must you know and be able to do?

This is the task that teachers face when they develop written plans for the unit of instruction. The written plans for the unit of instruction include the following parts:

Chap. 5 Planning for Instruction

Outline of Unit of Instruction

1. Title
2. Situation
3. Instructional objectives
4. Interest approach
5. Reasons for studying the unit
6. Questions to be answered
7. Answering questions, acquiring knowledge, and developing skills
8. Application of learning
9. References and teaching aids
10. Evaluation procedures

In order to understand how to plan using this approach, each step in the plan is explained and illustrated with an example. The sample unit of instruction will be "Establishing My Home Lawn."

TITLE

The title for the unit of instruction should be stated in action terms to convey to the students that this is a unit in which they will be involved. Action can be denoted in the title by using the *ing* ending, such as calibrating, timing, constructing, selecting, or designing. Another way of putting action into the title for the unit of instruction is to use a "how to" title such as "how to castrate pigs," "how to provide low-voltage outdoor lighting," or "how to feed cattle."

The title of the unit of instruction should accurately describe what the unit is about and should provide at least a preliminary idea of the scope of the unit. It should also follow the filing system the teacher uses to file plans as well as the pattern of organization used in the students' agricultural education notebooks.

Consider the example being used for illustrative purposes in this chapter, "Establishing My Home Lawn." This title conveys a sense of "we'll be establishing a lawn," which implies activity, involvement, and decision making on the part of the learner. Such a title enables the teacher to file the plan under the teacher's notes that deal with turfgrass, and the students can be told to place their notes and handouts for this unit of study under whichever section of their notebook the teacher selects, such as agronomy.

SITUATION

Before teachers of agriculture begin planning a unit of instruction, they need to think about the local situation in which the unit will be taught. Teachers do not plan to teach information for information's sake; rather, they plan to teach what is relevant to the students' specific situations and needs. In the case of the unit "Establishing My Home Lawn," it is important for the teacher to reflect on the specifics of the community, the students in the class, and the specific home lawn(s) that will be established as a part of the laboratory activities that are included in the unit of instruction (see Case 5-1) and as part of a student's home improvement SAE or entrepreneurship SAE. That which is taught will be learned with reference to specific situations in which students find themselves.

The situation forms a frame of reference within which the unit is taught. By writing down the situation, instructors force themselves to keep their instruction relevant and specifically focused on local needs. After teaching for awhile, teachers are so familiar with the local situation they do not have to write it down, but they still start their planning process by considering it. This situational information becomes the foundation of the plan. It provides the basis for identifying content via a list of problems and concerns. The situation should describe pertinent agricultural practices in this community as they relate to this particular unit of instruction. One would include information on the scope of the problem area, practices that are currently being used, problems that have been observed in the community, experiences each student has had in this area of instruction, and any other community-related information that needs to be considered as a teacher plans instruction.

It is also appropriate in the situation to include all state proficiency skills or competencies that are required to be taught in agricultural education, and all math and science standards, and state career or technical education standards that need to be met through the agricultural education curriculum. Case 5-1 provides an example of that which should be addressed in the situation section of a unit of instruction.

Case 5-1. *Situation.*

- The class has not studied establishing home lawns previously.
- Three of the twenty-one students have helped establish a home lawn previously.
- The class is to establish, as a laboratory project, a home lawn for Mr. Dobson.
- Most homeowners in the area fail to get their seedbed fine enough to cover the seed properly or to use the best variety of grass for their conditions and intended use.
- Five of the students will have jobs with lawn care services this summer.
- Three of the students wish to start their own lawn establishment, renovation, and care businesses.
- Several new housing developers are in need of people to establish new lawns from seed.
- Three of the students will have SAE improvement projects on improving the home lawn.
- One of the students in the class is working toward an FFA Turfgrass Management Proficiency Award.

Chap. 5 Planning for Instruction

In order to effectively delineate the pertinent information for the situation, teachers have to consider a number of sources, such as home or supervised agricultural education visits, awareness of current practices used and problems encountered, agricultural census information, and data from students' record books (such as average yields, profits, hours worked, and rates of gain).

Once the situation is developed, it not only serves as a backdrop against which planning is conducted but it can also serve other purposes. Specific problems observed can become the basis for creating a desire to study the unit. Production data can be used to create interest also. For example, a teacher could point out to the class that the average corn yield in the state is 147 bushels per acre; in the county it is 121 bushels per acre. The best young farmers average 137 bushels per acre, and the class average last year was 100 bushels per acre. This situation data is provocative and stimulates interest in the problem area. The process of developing the situation also reveals specific problems to be indicated under the "questions tob e answered" section. With the situation clearly in mind, the teacher is ready to develop specific instructional objectives for the unit of instruction.

INSTRUCTIONAL OBJECTIVES

Prior to teaching a unit of instruction or even planning for teaching a unit, it is essential that teachers decide what they want their students to know and be able to do once the instruction has been provided. The goal has to be defined and developed before the teacher can achieve it or even determine how to achieve it. The teacher's statements of intended learning outcomes are called instructional objectives. It is important to remember that these are the teacher's goals for the unit of instruction. This is what the teacher wants to achieve from the unit.

Why Are Objectives Needed?

In writing instructional objectives, teachers make definite decisions about the content of the problem area. They force themselves to establish parameters that in turn help define and limit the scope and content of the problem area. Objectives also help teachers decide what is truly relevant and worthy of students' learning versus that which is "nice to know."

Instructional objectives also help teachers begin to make decisions about the sequence of instruction. As objectives are developed, teachers begin to realize what knowledge and skills need to be learned and in what order. A final reason for writing instructional objectives is that they provide a basis for evaluation. If teachers decide in advance that they want students to know and be able to do certain things, share those expectations with students, and plan their instruction so as to accomplish those objectives, then it only makes sense to evaluate success (of student and teacher) based on the instructional objectives with which they began.

Case 5-2. *What is the Objective?*

> The following is your assignment. Leave where you are now and without using any map or talking to anyone, drive to Chase City. When you get there, call the following number, and you will receive further directions: 777-888-9999. Some questions:
>
> 1. How likely will you be to reach your assigned destination?
> 2. How can you plan to get there with the most efficiency?
> 3. How will you know when you have reached your intended goal?

Analysis of Case 5-2. Read Case 5-2. Chances are a person would have very little success completing the assignment. It's very difficult to plan a journey if you don't know where you are headed and have no mileposts to guide you; and so it is with planning instruction. If teachers do not know where they wish to go or how to get there, then the resulting instruction is apt to be haphazard, confusing, and unproductive. Yet, many teachers try to begin planning instruction before they take time to adequately decide what they hope to accomplish.

Domains of Learning

Objectives or desired learning outcomes fall into three domains of learning: cognitive, psychomotor, and affective.

Cognitive behavior or outcomes deal with the acquisition of facts, knowledge, information, or concepts. Psychomotor behaviors are in the realm of manipulative skills—using the mind in combination with motor skills. These are prominent outcomes of agricultural instruction. The psychomotor domain is the whole area of "hands-on"—actual performance of skills. Psychomotor learning is not accomplished without appropriate cognitive understanding. Affective behaviors have to do with attitudes, values, aesthetics, and appreciation. This is the most difficult area to have reflected in the list of teacher objectives. However, the affective area is a crucial domain of learning that agriculture teachers wish to stress.

In the list of objectives presented in Case 5-3, which objectives do you believe are addressing the cognitive domain, which the psychomotor, and which the affective? Objectives 1, 2, 3, 4, 5, 8, and 9 address the cognitive domain. Objectives 6 and 7 address the psychomotor domain. None of the objectives are written to address the affective domain. An example of an objective written for the affective domain might be, "To appreciate the beauty of a well established home lawn."

Writing Instructional Objectives

The most important characteristic of an instructional objective is that it clearly and completely communicates to the teacher and students what it is the students should know and be able to do on completion of the unit of instruction. The major focus of instructional objectives should be on specifying observable (measurable) behaviors that students need to exhibit once they have studied the problem area. These behaviors are called the performance.

When writing the objectives, teachers should use verbs that are action oriented. Examples include *explain, describe, select, compare and contrast, recommend, identify*, and *prescribe*. The key idea is that the verb should specify a behavior of the student that can be observed and measured so that the teacher can determine whether the instructional objective has been met and to what extent it has been met. Keep in mind that these action, measurable verbs also fulfill cognitive requirements. For example, "to explain" requires a higher level of cognitive activity for the student than "to list." See Appendix B for guidance on selecting verbs that challenge students to think across all cognitive levels.

Many writers insist that instructional objectives also specify the conditions under which the student's behavior will be measured and the criteria that the student's performance must meet. Certainly, teachers must decide on these things if they are to be able to adequately determine whether an objective has been met. Thus, a completely written objective contains three parts: (a) the behavior, (b) the conditions under which the behavior is to occur, and (c) the criteria by which the performance is judged. Consider the following objective as an example.

> Given an E6014 electrode, the learner is to weld a stringer bead such that the width of the bead is $2\frac{1}{2}$ times the diameter of the bare end of the electrode and the height is $1\frac{1}{2}$ times the diameter of the bare end of the electrode.

The measurable behavior (performance) is indicated by "weld a stringer bead," with "weld" being the action verb. The condition under which the behavior will be measured is indicated by the words "Given an E6014 electrode." The criteria are indicated by the words "such that the width of the bead is $2\frac{1}{2}$ times the diameter of the bare end of the electrode and the height is $1\frac{1}{2}$ times the diameter of the bare end of the electrode." Hence, the sample objective contains all three parts.

The minimum that teachers should include in their written instructional objectives is the performance or observable behavior. For the sample plan on "Establishing My Home Lawn," the instructional objectives in Case 5–3 are appropriate.

Relationship of Objectives to Content to Be Taught

It should be apparent that the list of instructional objectives is parallel to the subject matter content of the unit of instruction being planned. In essence, the objectives identify the major knowledge, skills, and values that are to be taught. It is from this list of major ideas to be addressed that the rest of planning flows. In fact, when teachers plan their list of questions to be answered, the questions to be answered will parallel the list of instructional objectives.

Case 5-3. *Instructional Objectives**

> The learner will be able to
>
> 1. Interpret soil test data and apply the correct amount of fertilizer or lime to a given lawn site.
> 2. List the steps to follow in properly preparing a seedbed for a home lawn.
> 3. Select the appropriate variety of seed for various home lawn situations.
> 4. Develop a list of tools (and their costs) that are needed to sow seed.
> 5. Compare the advantages and disadvantages of the methods of seeding lawns that were taught in class.
> 6. Demonstrate ability to use the seeding techniques taught in class.
> 7. Apply the appropriate mulch to cover the seed.
> 8. Explain why mulches are needed.
> 9. Solve problems for giving estimates on establishing home lawns.
>
> *Notice that these objectives specify only the performance. It is assumed that in delineating the performance the teacher has in mind, the conditions under which the performance will occur and the criteria it must meet will have to be considered.

Finally, then, it is the list of instructional objectives that indicates to the teacher what the evaluation should include and should stress as major points in the content being taught.

INTEREST APPROACH

Teachers need to have the basic aims of the interest approach, as presented in Chapter 4, in mind when they plan the interest approach for their problem area. They must also remember the principles of learning on which motivation and interest rest (see Chapter 2).

Sources of Interest Approaches

Teachers need to constantly look for good ideas for interest approaches–ideas that will create a desire in students to know more about a topic. Lessons should not simply be funny, sensational, or dramatic; rather, the study should begin in a way that causes learners to be interested in overcoming inabilities and solving real problems that affect them as individuals.

True felt needs and provocative situations grow out of personal situations students face. Hence, teachers can develop many good interest approaches based on the students' supervised agricultural experiences. These include production projects, improvement projects, research projects, entrepreneurship projects, projects in business or on another persons' farm, and school laboratory projects. By clearly portraying the specifics related to students' experiences and programs, genuine desires for knowledge can be generated. For example, if a group of students who are enrolled in small animal care knew they had three dogs to groom in lab and they didn't know how to proceed, then they would easily recognize a "need to know." Likewise, if several students have sows about to farrow and, when engaged in a class discussion, they are not sure how to best care for the sow at farrowing, then they are apt to "want to know" more. The teacher needs to become committed to learning how to apply the unit of instruction to students' situations.

Another good source of interest approaches is the use of case studies or case problems. Teachers can develop fictitious letters from agriculturalists that ask for advice or request that the members of the class complete a project the agriculturalists do not know how to complete—but should be able to complete—or give descriptions of common dilemmas faced by people in the occupational area and ask students what course of action needs to be taken.

Experiments are an excellent source of interest approaches. The teacher could bring in two containers, one with corn seed planted one and a half inches deep, another with corn planted three and a half inches deep, and raise thought-provoking questions as to which would do better and why. The students can then write hypothesis statements about the potential outcome of the experiment. When studying nutrition, students could write hypothesis statements, then begin feeding two groups of baby chicks two different feeds to test diets. The class would then study the problem area in class to find out how and why diets matter.

Demonstrations are also a good way to generate interest. The teacher could begin a unit of study on fish hatcheries by expertly removing the eggs from the female and then turning to several students to see if they can do the same. If they lacked the necessary knowledge and skill, they would immediately realize it and admit that they needed to know more.

Another way of creating interest is to plan a series of puzzling questions. Get students to commit themselves to an answer and then pursue the question—why? When students conclude they are not sure, then the teacher is ready to move them to the point where they realize it is important that they learn more.

Whatever the source or manner of creating interest, it is essential that the teacher plan the scenario in writing.

What to Write Down

For the interest approach and all other parts of the unit of instruction, there are two categories of information that teachers need to write down. One is a set of directions that the teacher writes to himself or herself, the other is the key content (see page 103–105).

Directions to self, at a glance, tell the teacher what to do and the order in which to do it. These are thought-out directives that are purposely ordered to achieve the desired effect. They include statements such as, "Show the transparency on average yields," "Distribute Mr. Leuellen's letter," "Have the class suggest specific reasons why an engine will not start," or "Ask the following questions: (a) When do we need to start lilies if we want them to be ready for Easter? (b) How should they be planted? (c) What fertilization program is needed?"

Examine Case 5-4, which is an example written plan for an interest approach. The written plan is divided into two columns with directions to self written in the left column and key content written in the right column. Directions to self include "Distribute," "Show Transparency 1. How much should we charge for the job?" "Ask," and "Probe." Key content includes Dr. Dobson's letter, and the lead question. Separating the directions to self from the key content allows the teacher to think through and write the key content in a logical flow in the right column without the intrusion of the directions for how it will be done. The teacher then, likewise, thinks through and plans for the logical flow of the classroom processes by examining the directions to self in the left column. Also, by separating the directions to self from the key content, the teacher can very quickly glance at the plan, even from a distance (i.e., from the blackboard to the podium that is holding the plan) and see that the next direction to self indicates, **"Transparency 1"** in the left column. The teacher sees where he or she is headed with the lesson, and subsequently begins, mentally and physically, to move in that direction.

Case 5-4. *Interest Approach*

Directions to Self	Key Content
Distribute the attached letter to the class.	Letter from Mr. Dobson: Dear _____:
Show transparency 1, "How much should we charge for the job?"	I would like to request your class to put in a new lawn at my home which is currently being built. The area to be seeded is 1/2 acre. I will pay you for the job. Please advise me of your decision and give me an estimate of cost if you are willing to accept the job.
Ask students to develop a list of tools and supplies we would need if we took the job.	
Probe: When the class as a whole begins to experience problems answering the questions (because they need to know more before they can respond), probe as to whether they want to take on the job (anticipated answer is "Yes").	
Ask students	"Why is it important for us to know how to establish a new lawn?" (This question provides the transition to the next phase of the problem-solving approach.)

The other category of information that must be written down as a part of the plan is key content. Key content may include facts and figures. It may be key notes

Chap. 5 Planning for Instruction

presented on the board or on an overhead, or a handout. It may be the key ideas of a case situation or a real situation faced by a student in the class. Whatever it is, enough must be recorded in writing such that at a glance it is clear and complete enough for the teacher so that he or she can fluently develop a high level of interest in the students for studying the unit. Consider Case 5-4.

Analysis of Case 5-4. First, look at what was recorded in the teacher's plan. The key content for this interest approach is the letter from Mr. Dobson (written in the right column). The directions to the teacher (written in the left column) include comments given before and after the letter. These directions are sequenced so as to gain commitment from the learners and then give the class specific assignments that they will want to complete but cannot without additional information and skill. These directions lead the class to a closing point where they conclude that they need to know more and agree it is worth learning.

The techniques the teacher uses deal primarily with offering the students an experience as a class that they would like to have, and then helping them understand that their lack of knowledge and skill prevents them from having the experience. This is an actual group situation that personally affects all class members and thus provides an actual felt need for the entire class. Individual student situations (creating home lawns) could have also been used to create this felt need if they were available. The teacher then makes use of probing questions to help the students discover and admit their need for further study.

The final phase of the interest approach involves planning for closure and transition to the next step—establishing reasons for studying the unit. Teachers must remember that the interest approach should be designed as a logical way of beginning study in an area—it is not a free-standing part of a unit of instruction. In addition, it is not a feature that is unrelated or unconnected to the next step in the learning process. The interest approach must be planned and carried out such that it makes its point clearly and leads students to conclude that they do not have all of the answers they need to have.

Once planning for closure is accomplished, the teacher needs to plan a transition into reasons for studying the unit. Remember that the purpose of the interest approach is to get the students to decide they need to know what is about to be taught; the purpose of the "reasons for studying the unit" is to get the students to discover why they need to know what is being studied. So, the interest approach ends with this transition comment, "But, why do we need to learn this?"

The teacher then needs to plan some bridging comments that link the content of the interest approach with a class discussion that reveals why students want to study the unit of instruction. It can be considered to be a connecting part of the lesson plan. In essence, the teacher has to deliver the following message, "OK, you say it's important to learn how to . . . (whatever the problem area that is being introduced). Help me understand why this is important."

REASONS FOR STUDYNG THE UNIT

Two facets of planning are needed for establishing the reasons for studying a unit at the beginning of the unit of instruction. Because establishing reasons for studying the unit reveals students' objectives and goals, the teacher has to lead students into discovering and verbalizing why they believe it is important. The best way to accomplish this is to develop several leading questions that, when asked, will cause students to think of their reasons for wanting to learn that which is about to be taught. The following lead questions will serve this purpose well.

Some Lead Questions for Establishing Reasons for Studying This Unit

1. Why do you need to learn how to (whatever the problem area is)?
2. What are some goals you wish to achieve by learning how to (whatever the problem area is)?
3. What are some objectives you hope to accomplish through studying (whatever the problem area is)?
4. Why is it important to learn how to (whatever the problem area is)?

Next, the teacher should record in the plan some possible responses students are likely to give. The primary reason for recording several likely responses is to have some "hints" written on paper in front of you in case students have difficulty thinking of reasons for studying the unit of instruction. It is important to remember that these possible responses are simply extra ammunition for the teacher that can be used to generate discussion if and when needed. Students do not need to give the same responses or any of the possible responses that are recorded on the teacher's plan. To force them to give the teacher's anticipated responses would remove the psychological leverage this portion of the plan is meant to provide. Examine Case 5–5, Reasons for Studying This Unit.

Case 5–5. *Reasons for Studying This Unit.*

Lead Questions

1. What are some reasons you believe you need to learn how to establish a home lawn?
2. What are some goals you should strive to accomplish in learning how to establish a home lawn?

Possible Reasons Students Want to Study This Problem Area

1. To do a good job on Mr. Dobson's lawn for the laboratory project.
2. To learn enough to start a personal lawn service.
3. To get out of the classroom and work outdoors.
4. To be able to re-establish my own lawn for my supervised experience improvement project.

Analysis of Case 5-5. Notice that two lead questions were recorded. Their purpose is to prompt students to think of reasons they believe they need to learn how to establish a home lawn. Always give students the chance to individually write on paper at least one reason of their own before opening the verbal exchange in the class. This way students are given time to think, and especially to think independently, before the class discussion proceeds. In case no one offers a reason, the plan contains some plausible reasons students might suggest. If no student suggests a reason for studying the unit, the teacher could ask if one of the reasons listed on the teacher's plan was a viable reason. If so, it would be recorded on the board, and then the teacher would encourage class members to add one, two, or three other reasons that they would think of on their own.

QUESTIONS TO BE ANSWERED (PROBLEMS TO BE SOLVED)

Before examining the specific aspects of planning involved in this section of the unit of instruction, the reader needs to distinguish between problems and questions. Real problems are "questions that must be answered" in order to remove a felt need that personally affects the learner. When the data gathering is related to a personal concern, then one is dealing with a problem. If a student's cow has bloat and he or she wants to know how to treat the cow for bloat, that is a real problem. If a student is interested in cattle and would like to know how to treat cows for bloat, this is a question. Thus, what is a problem for some students is a question for others. The more often your teaching can use real problems that actually affect one or more members of the class, the more exciting the study will be.

When students study questions, they could be seeking information for information's sake. With problems, students have a different frame of reference. They are seeking information in order to remove a problem that personally affects one or more of them.

Questions to be answered are the core of the problem area in the problem-solving approach to teaching. They lead to the specific content and final practice that the problem area is designed to produce.

By thinking of the questions to be answered (problems to be solved) before addressing specific content to be taught, the teacher, when planning, is forced to consider the entire problem area and think it completely through. This section of the plan provides a framework for reasoning. Then when the students are led through this step of the plan as the problem area is presented, they too are forced to think through (or reason through) the entire area of study.

The questions to be answered also provide the framework for the problem area being studied. The questions define the subject matter to be taught. They provide a way of organizing teaching and hence learning. Additionally, the list of questions to be answered provides guidance, direction, and structure.

Identifying Questions to Be Answered

In order to clearly and completely identify the best questions to be answered for a unit of instruction, teachers of agriculture should begin by focusing on the title of the problem area and deciding what questions logically need answering. Then the teacher should reason through all the questions related to the topic. Sometimes it is helpful to mentally review the process to which the unit of instruction refers, the task to be completed, or the information to be gained in order to be sure the list of questions generated by the students is complete. In doing this, the teacher may reflect on personal experiences and consider his or her own expertise and skill in the area. Also, by comparing the list of questions to the list of teaching objectives, the teacher can usually detect any omissions.

Additionally, the list of questions to be answered grows out of the situation for which the problem area is being studied and will reveal the important problems and questions that must be studied. Be careful not to try to pull your words from the students, or to try to pull your complete list from them. Instead, use the students' words, knowing that what they mean parallels what you anticipated they would say. You also will add to their list the items that you had planned to study that they have not verbalized.

Once the total list of questions to be answered has been identified, the teacher then needs to carefully sequence the list. The list of questions needs to be arranged in a sensible order, with questions of a prerequisite nature addressed before those for which they are prerequisite. Thus, by answering the first question (or solving the first problem), the class is better prepared to study the next one. The learning from studying each question builds cumulatively.

Planning for Developing the List of Questions

The same approach is used for planning this section of the unit of instruction as was used to plan the reasons for studying this unit section. First the teacher identifies and records a list of lead questions designed to help the learners think of the "questions to be answered" that require study before they can solve the problem with which the unit of instruction deals, and before they can address the reasons for studying this unit that they have identified.

Some Lead Questions for Developing a List of Questions to Be Answered

1. What are some questions you will need to answer before you can (whatever the problem area is about) or reach your objectives?

2. What are some things you need to know and be able to do in order to (whatever the problem area is about) or reach your objectives?

3. What are some problems you have to solve in order to be able to (whatever the problem area is about)?

Chap. 5 Planning for Instruction 101

Case 5-6. *Questions to Be Answered.*

Lead Questions

1. What are some questions you must answer before you can satisfactorily establish a home lawn?
2. Give me some problems you think we need to solve before we start on Mr. Dobson's lawn or as we work on it.
3. What are some things we must be able to do if we are to do a good job of establishing lawns?

Anticipated Questions the Class Should Identify

1. When should we establish a home lawn?
2. How do we get the ground ready for planting?
3. What kind of grass should we plant and at what rate?
4. What fertilizer should we use and how much should we use?
5. How do we plant the seeds?
6. What do we do after we have sown the seeds?
7. How do we know how much to charge for the job?

Analysis of Case 5-6. Teachers can be certain that when they lead the class in developing this list of questions to be answered, it will not be a perfect match to the list the teacher has written in the plan. Thus, a teacher must be sure the list in the plan is complete. The teacher, as the expert, must ensure that all appropriate questions are included in the list to be studied for the unit. If the students are not able to think of all the questions, the teacher must finish the list for the class. The list generated in class should then be reviewed with the class to establish a clear order, combine questions that are too fragmented, and summarize this important overview to the unit of instruction.

By leading the class in reviewing and ordering the list of questions to be answered developed in class, the teacher helps the students gain practice at revising their first thinking. They learn to group similar concerns to gain greater efficiency. They also learn to logically order their inquiry to facilitate greater understanding.

The important benefit to the teacher is that the class list of problems and questions then is more nearly matched with the order the teacher has followed in developing the plan for instruction.

ANSWERING QUESTIONS (SOLVING PROBLEMS), ACQUIRING KNOWLEDGE, AND DEVELOPING SKILLS

The planning for how to help students solve problems, acquire needed knowledge, and develop new skills is the very heart of the process of planning instruction. As teachers determine how best to teach students in any problem area, they have to simultaneously consider the specific knowledge, skill, practices, and principles that will be taught. A crucial part of planning instruction is identifying and selecting content.

Teachers need to assess their own knowledge, complete additional background study, and visit with practitioners to be sure they have identified what students need to learn in order to solve the problems being studied. To become fully acquainted with subject matter to be presented, teachers must review sources of information and teaching materials, such as books, pamphlets, Internet sites, CD-ROMs, DVDs, visual aids, experiment station bulletins, and extension publications.

As teachers identify and select content, subject matter, key points, practices, and principles, they concomitantly begin to make at least preliminary decisions about how they might best be able to help students learn what they need to know in order to solve current and future problems. As content and potential methods are brought into focus, teachers then move on to planning specific solutions to each of the questions to be answered in the unit of instruction.

Planning for Solving Problems, Acquiring Knowledge, and Developing Skills

The first planning activity for each question to be answered is a trial discussion. This requires developing, on paper, a strategy for completing a trial discussion. It may consist of jotting down several questions that will be used to re-create interest, create a division of opinion, determine existing levels of student knowledge about the question, or helping students realize that a more thorough period of study is needed. This part of the plan must also indicate to the teacher how to smoothly move into presenting the final solution of the problem.

As teachers plan trial solutions as well as actual solutions to each problem, they will need to select techniques of teaching that are appropriate for the given situation. These techniques are thoroughly explained in Chapters 6 and 7.

In planning for the detailed teaching (and learning) that must follow the trial discussion, two fundamental elements must be included in the written plan: directions to self and key content.

1. Directions to self present instructions or guidelines in sequential order (this is the information recorded in the left column of the written plan).
2. Key content (concepts, principles, practices, and facts) that must be learned in the process of studying the problem as presented (this is the information recorded in the right column of the written plan).

Each element is discussed in detail.

Directions to Self. Directions to self are not simply a matter of teachers writing scripts for themselves. Rather, teachers need to list some

Chap. 5 Planning for Instruction   103

catchy phrases that can be read and comprehended at a glance (noted in the left column of the written plan). This portion of the teacher's plan is a guide to the orchestration of the teaching episode. Consider the accompanying chart of directions to self and key content.

Consider the following examples of the kinds of directions that are needed.

Directions to Self	Key Content
1. **Show** transparency 7.	(This key content is printed on the transparency.)
2. **Discuss** the following points:	1. Remove all dead and diseased wood 2. Remove all crossed branches 3. Create desired form
3. **Demonstrate** the procedure to follow:	see attached notes (notes regarding the procedure are found in the supporting materials attached at the end of the unit of instruction). Steps: 1. Carefully pull back branches 2. Identify by shoots growing perpendicular to main branches 3. Use pruners to cut the shoots close to the main branches 4. Treat the exposed wood for disease control

Notice that in direction 3 ("Demonstrate . . ."), additional notes are referenced. It would be in these notes that the details of the key content (facts, figures, key points) are recorded. Likewise, in direction 2 ("Discuss . . ."), the list of points (noted in the main body of the teacher's directions to self) contains the specifics to be presented or otherwise learned. As was discussed earlier, most teachers are successful if they keep their subject matter notes (key content) separated from their pedagogical directions (directions to self).

Key Content. In the case of the key content of the planned solution, it is crucial that the subject matter recorded is clear, precise, and thorough. It may be recorded in the form of notes to be put on the writing surface or transparencies to be presented, slides, a detailed demonstration, study questions, conclusions for a supervised study period, or any number of techniques discussed in Chapters 6 and 7. The crucial point is that if instruction is to be clear and thorough, the key content must be written in the plan.

Following are examples of ways in which key content is recorded in the left column of the written plan:

Example 1: Using a transparency to record key content.

Directions to Self	Key Content
Show Transparency 2: Guidelines to Follow in Transplanting Seedlings	(This key content to be taught is printed on the transparency, "Guidelines to Follow in Transplanting Seedlings.")
	1. Transplant as soon as the seedlings show the first true leaves.
	2. Transplant into containers in which they will be sold.
	3. Make holes in soil to receive one seedling per hole:
	a. Dibbles
	i. Single
	ii. Board with several
Show Slides 12–16	(The key content to be taught is printed on the slides.)
	b. Mechanized methods (slides 12–16 illustrate each type)
	(*and so forth*)

Example 2: Using notes to yourself of technical details to record key content.

Directions to Self	Key Content
Blackboard: Soilless mixes	Notes for question 3 in unit of instruction
	Types of soilless mixes—
	1. Jiffy-mix
	2. Redi-earth
	3. Metro-mix
	4. Pro-mix
	5. Ball-mix
Show: Samples of each as each type is written on the board	

Example 3: Using attached copies from specific resources to record key content

Directions to Self	Key Content
Blackboard: List the parts of the ignition system.	(Full list on page 10 of Briggs and Stratton; copied and attached.)
Describe the function of each part.	(Chapter 2 in Briggs and Stratton; copied and attached.)

Case 5-7. *Problem 5: How to Plant the Seeds.*

Directions to Self	Key Content
Bring in small box of seed that can be shaken out of top for spot seeding. **Bring in** large bag of seed. (Trial discussion) **Ask** a student to come forward and show the class how to spread the seed for best results–we want 3 lbs per 1,000 sq ft. (Note: If there is clear division of opinion and lack of understanding, the trial discussion ends, so move on into remainder of solution.) **Show Transparency 2**: What is the recommended way to apply seed?	Notes for question 5 in unit of instruction "What is the recommended way to apply seed?" a. Use drop-type spreader. b. Follow length of lawn. c. First apply two header strips. d. Then go back and forth. e. Shut off spreader gradually as you approach header strips. f. Overlap wheel tracks for even coverage g. May prefer to sow twice at lowest recommended rate—with second seeding done at a 90-degree angle to first.
Blackboard: Draw sketch of seeding method 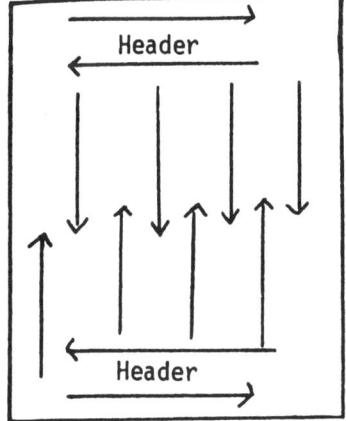	

(continued)

Discuss seeding	How deep should seed be covered? Answer: 0.1–0.3 inch with soil
Discuss tools	What tools are needed? Tools: bamboo or leaf rake
Demonstrate tool use	How are the tools used? Use back side, apply light pressure
Supervised study: Students use the publications	Lawn Care (Scotts page 255) and Lawn Establishment (Extension Publication # H-1011).
Blackboard (remind students to get notes into their notebooks)	After study, develop conclusions with the aid of the class (see attached notes for key content to be taught).
Field trip to lab site	
Demonstrate the steps developed in the conclusion	

Not only do teachers need to write down key content but they also need to jot down ideas on transmitting the key content to the class. They may make a note to show a slide or slides, jot down an illustration number, make a note to show a sample, work through a case problem, show an Internet site, give students a web address to view, or note other techniques designed to reinforce and apply the key content. Without these additional notes and directions, teaching becomes very superficial.

Analysis of Case 5–7. The teacher began the trial discussion to see if the class needed thorough instruction in this area or simply a review. This was accomplished by presenting students with a concrete task. The next step was to set up a division of opinion and help the students realize that they need to be more certain about how to perform this phase of the operation. This trial discussion is a mini-interest approach. It is a way to re-create interest throughout the unit of instruction.

The class begins data collection by studying in two different references under the teacher's direction and supervision. This technique allows the students to be creative in their learning and also makes use of the principle of learning that students learn more when they inquire into subject matter than when they are instructed in it. Their task was clearly set with the use of study questions. By using the transparency, the teacher was able to specifically prepare prior to class. Notice also that the teacher was sure to develop final conclusions with the class.

The two column format for recording plans for teaching solutions to problems (or addressing questions to be answered) were clearly evident. The direction to the teacher (in the left column) consisted of phrases such as "Ask student to come forward ...," "Demonstrate tool use," and "Blackboard." The key content was provided (in the right column) by attached notes, transparency content, and notes in the column that the teacher could use in securing closure.

Each problem identified in the list of questions to be answered for the unit of instruction must be planned in the manner just described. It is essential that at the conclusion of each question to be answered the class be led to develop a conclusion or agreed-on solution. In cases where several students have the problem and must solve it for their own situations there may be several different individual solutions. Then, after all the questions have been answered by the class, it is essential that the entire unit of instruction be summarized. Certainly, some of this may be done as various questions are answered (problems are solved), but often a major portion of this final pulling together is completed after all the individual questions have been answered. Students should always be directed to open their notebooks to the beginning of the unit of instruction, examine the list of questions to be answered that were developed earlier by the class, and consider the status of the problem situation (i.e., which questions have been answered and which will be answered in the coming days). A major strategy that is used in agricultural education to summarize a problem area involves providing for the application of learning.

APPLICATION OF LEARNING

The application of learning is key in agriculture instruction. By applying learning, students see its relevance, more clearly understand it, and retain it more permanently.

On concluding the solution of all the problems, a good strategy for gaining closure and moving toward planned application is for the teacher to develop with the class a list of approved practices to use for success.

Approved Practices for Success

Approved practices are those ways of doing things that are currently accepted by the industry as the best way to do them. They have been tried, through history or research, and have proved to be superior.

There are approved practices in growing corn, soybeans, fish, chrysanthemums, beef cattle, white pines, gerbils, and anything else that is produced in agriculture. Likewise, there are approved practices to follow in doing mechanical work, such as tuning an engine or sharpening a lawn mower blade. There are also approved practices to follow in completing landscape designs, grooming poodles, and making corsages.

Case 5-8. *Approved Practices for "Establishing My Home Lawn."*

> 1. Grade the soil such that there is a slight slope away from the house in all directions.
> 2. Spread high-quality (debris-free) topsoil. Apply water with a sprinkler to check for low spots.
> 3. Till the soil to a depth of two inches, incorporating fertilizer and lime, according to soil test recommendations. Add any needed amendments. Soil should be tilled until there are no chunks larger than the size of a pea.
> 4. Wet the soil to firm it down and wait several days before seeding.
> 5. Spread seed evenly according to recommendations on the package. Seed twice at 90-degree angles to each pass.
> 6. Rake seed in lightly with back side of bamboo or leaf rake.
> 7. Fertilize at rate recommended for analysis of fertilizer used and in accordance with soil test results.
> 8. Apply clean wheat or oat straw such that you can see 50 percent of the ground.
> 9. Water two or three times daily until seedlings are established.

Once the unit of instruction has been studied, the teacher needs to lead the class in developing a list of approved practices. Of course, the teacher needs to have written these down in his or her plan in advance. For the sample plan on "Establishing My Home Lawn," the list of approved practices in Case 5-8 would be written in the teacher's unit of instruction.

Plans of Practice

Once approved practices for success have been discovered and agreed on, the teacher then guides the students in developing plans for applying what has been learned. In the case of the agriculture laboratory at the high school, it may be a production plan (drawing, bill of materials, equipment needed, procedures to follow).

In a nonfarm laboratory, it may be a list of steps to follow to complete application of learning by doing a specific job. For students who will apply learning at home via production or improvement projects, it may be a "plan of practice" such as the one in Case 5-9.

Case 5-9. *Plan of Practice.*

> *Establishing My Home Lawn*
>
> I will sow Scotts Family mixture at the rate of 13 oz per 1,000 sq ft on August 20 using a drop spreader set at the 5¼ setting. I will sow north–south and then repeat going east–west. After sowing the seeds, I will rake it very lightly and then cover with clean oat straw such that about ½ of the ground can still be seen. Then I will water thoroughly and keep moist until the seeds have germinated.

General Principles

As students develop answers to questions and lists of approved practices for problem areas, it is important that the teacher point out or help students realize general principles that can be used in other areas. General principles are broad generalizations that can be applied to other topics. They represent basic truths that are used widely in a discipline. They bring out the "why" behind fact and practice. As students learn to discover and use basic generalizations in agricultural subject matter, they have a new freedom. They are no longer locked into the specifics of facts as they apply to one situation.

In the unit of instruction on Establishing My Home Lawn, students would learn specific facts, such as when and how to sow seed, how deep to cover seeds, methods of covering them, and with what to mulch them. They also need to learn the general principles behind these facts and skills that can be transferred to other areas of turf or plant science in general. General principles such as the one that follows should be pointed out and explained.

> Lawns should be established in the fall. They will develop fewer weeds because soon after the weed seedlings germinate they are killed by frost and cease to offer competition. There will also be a stronger root system because there is more moisture and evenings are cooler as contrasted with planting in the spring when each passing week brings the prospect of less moisture, and hotter days and warmer nights.

Such general principles need to be emphasized by the teacher so that students begin to realize the broad utility of their learning. When a list of general principles is developed with the class following the development of a list of approved practices, it helps unify the problem area and facilitates transfer of learning.

REFERENCES AND TEACHING AIDS

One might readily ask, "Why write down a list of references and teaching materials that are needed to teach a problem area?" The basic reason is to provide a summary checklist in order to be fully prepared for embarking on the problem area. This checklist can be reviewed prior to beginning the unit of instruction. By including such information in summary form, the teacher need not review the entire plan to determine what is needed. The best time to make such a list is immediately on completion of planning while what is needed is fresh in the teacher's mind. The list includes, but is not limited to, books, media, equipment, web sites, materials, events, resource people, computer programs, and anything else needed to teach the unit.

EVALUATION PROCEDURE

An important part of planning for instruction involves planning how to evaluate learning and teaching. Teachers should write down their quiz questions, outlines for laboratory practical exams, and any other form of evaluation that is needed. The specifics of planning for evaluation are presented in Chapter 14.

REVISING AND UPDATING PLANS FOR INSTRUCTION

Once a complete unit of instruction is written, it will have to be revised and updated periodically. As new technology develops, key points and approved practices will have to be updated to include the current best way of doing things.

As teachers teach from a plan, they also encounter problems with students' abilities to understand. Sometimes well-planned learning activities do not proceed as the teacher intended, hence the need to try other approaches. Then there is the gaining of new knowledge of how to teach and the development of new materials for use in teaching. Teachers will want to revise their plans to include such developments.

Some revising is done daily based on feedback from the previous day's class. Often major revisions are made prior to the time when the unit of instruction is next taught.

Teachers need to make notes in the margins of plans and to add new pages of notes, transparencies, spreadsheets, and other aids to reflect thoughtful revisions.

DAILY PLANNING FOR INSTRUCTION

When planning for an entire unit of instruction that may last from several days to three weeks, it will be necessary to do additional planning on a daily basis. For example, the teacher may have to conclude class midway through a problem solution. The next day it is necessary to think of (plan for) a clear and logical way of reentering the problem being studied. A daily plan accomplishes just that. It serves to connect what has occurred the previous day with what is to occur in the current day's class. A daily plan template can be developed on the computer and either printed in hard copy or used directly from the computer for the teacher to prepare for the next day's lesson.

This daily plan is good for one time only. It is a temporary, nonreusable aid. The accompanying sample daily plan shown illustrates the concept (see Appendix A).

Notice that the daily plan includes routine notes such as announcements, reminders, and assignments, as well as connecting comments for the teacher to bridge previous learning with today's lesson and future lessons. It also has a brief overview of the current day's class to provide a clear mental set. There is also a new interest approach for the day's class.

SUMMARY

Planning for instruction is the key to becoming an effective teacher. It forces the teacher to be clear about what is to be accomplished, the content to be included, and how the content is to be delivered. Thorough planning allows for

Sample Daily Plan

Directions to Self	Key Content
Announce FFA meeting today 5th period **Reminder** of field trip tomorrow	Yesterday we discussed how to plant the seed. (Interest approach) We left off with our conclusions, which were:
Show Transparency 5: "Conclusions from Supervised Study" (question 5).	(Content is listed on overhead)
Write objective on the board	Today's objective: Practice the planting procedure on one of the plots after the demonstration. *Supplies needed*
Field trip to demonstration plot **Opportunity to learn**: Students will show me they can properly use the equipment to plant the seed according to the process we studied.	Take: Rakes–5 Wheelbarrows–3 Shovels–4 Carton seeder Drop seeder Cyclone seeder
	Today we had the chance to apply the seed at the proper rate following the procedure we learned in class. Tomorrow we will take the field trip to the lab.

flexibility and spontaneity because the goal or end product are always in sight and serve to refocus the teacher and learners.

In planning the unit of instruction, the teacher makes predetermined use of the principles of learning and the techniques of teaching. This is where each individual's creativity and worth as a teacher is displayed. Once teachers master the ability to plan well, they are free to enjoy teaching.

FOR FURTHER STUDY

1. Select a problem area and develop a unit of instruction around a list of questions to be answered. Check it against the suggestions offered in this chapter.
2. Ask your teacher educator for some sample units of instruction that you can compare to the suggestions made in this chapter.

6

Group Teaching Techniques

> "OK, class, listen up. I am going to tell you how to make a splice graft. Then we'll go to the lab, and you can make some.
>
> "First, you make a 45-degree cut on the stock wood. Then you make a 45-degree cut of the same length on the scion wood. Then you perform a reverse cut on each piece of wood. Next you slip the scion wood onto the stock wood such that the 'tongues' of each cut overlap. As you slip the two pieces together, you must be sure to line up the cambium layer of the two pieces. Then you wrap the graft and you're done.
>
> "OK. Any questions? Good. Now let's go to the lab and do it."

What are the chances of this teaching technique being successful? The chances are quite slim, but why is this so? It is because the teacher is trying to tell the group how to perform a meticulous psychomotor operation. The teacher's technical knowledge is accurate and the teacher may, in fact, be a very articulate spokesperson. Nevertheless, the technique is wrong for the situation.

As teaching situations vary, teachers must vary their teaching techniques. In this book, two broad classifications of teaching techniques are presented: group teaching techniques in this chapter and individualized teaching techniques in Chapter 7.

Chap. 6 Group Teaching Techniques

OBJECTIVES

After studying this chapter, you will be able to

1. Explain the relationship of group teaching techniques to problem-solving teaching.
2. List and explain the use of the basic group teaching techniques.
3. Write a daily plan that incorporates the use of each of the group teaching techniques presented in this chapter.
4. Use each of the group teaching techniques presented in this chapter.
5. Select the appropriate group teaching technique for a given teaching situation.

THE ROLE OF GROUP TEACHING TECHNIQUES

Group teaching techniques are those techniques that are appropriate for providing instruction to a group of students in the same setting. For example, a class can learn much about controlling plant diseases by being engaged in a well-guided class discussion where notes are taken from information developed on the board or on overhead transparencies. Students can also learn very effectively from taking a well-planned and conducted class field trip. Group teaching occurs in the classroom, laboratory, field, and workplace. For example, many demonstrations given to the whole group are essential to successful performance in the agricultural instruction laboratory.

Much of agricultural instruction is presented to groups of students all studying the same topic with the same or similar learning outcomes being sought. Such instruction and the appropriate teaching techniques for group settings are essential.

THE NEED FOR GROUP TECHNIQUES

One of the justifications for group instruction and the teaching techniques appropriate for such instruction is that the time available to the teacher and the students is limited. Therefore teachers must teach students in groups whenever the students have common needs and the knowledge, skills, or attitudes being taught are teachable in a group setting. Take the case of learning how to establish a record keeping system. Every student will need this information, and it is easily taught by presenting a good case example on the overhead projector, chalkboard, or computer-projected image. Students follow the example illustrated and develop the system on their own work sheets.

Furthermore, all students in the class usually need some common factual information. If you were teaching a unit of instruction on estimating timber yields, and none of the students had ever cruised timber, they would all need to learn the basic procedure to be followed.

Another factor that makes the use of group teaching techniques not only desirable but also necessary is the lack of sufficient learning resources. When teachers lack sufficient student references, self-paced modules, or computer stations, for example, they find it necessary to use group teaching techniques to help students learn efficiently the information they need to know.

OPERATIONALIZING THE DATA-GATHERING STEP IN PROBLEM SOLVING

Another key role of group teaching techniques and individualized techniques is to provide students with the factual information needed to solve the problem presented to the class. Once a teacher has decided that group instruction is best for the situation, the particular group teaching technique that is most appropriate is used to help students gain the information and understanding needed to solve the problem at hand.

Once students have gained the basic information, teachers often find it helpful to use a group setting to aid the class in developing tentative solutions based on the factual information that has been presented. Group techniques are also helpful to gather tentative solutions and draw final conclusions to problems that students are studying, using the individual techniques discussed in Chapter 7.

USING INSTRUCTIONAL MEDIA IN TEACHING

Before presenting specific techniques of teaching, it is important to point out that teachers need to develop the ability to select and use instructional media effectively with all techniques of instruction. Instructional media consists of any type of visual or audio material that can be used to enhance learning.

People use all of their senses to learn, and they generally learn more when they make use of more than one sense at a time. Thus, for example, students should be encouraged to listen as well as see material.

Another reason for using media to supplement any technique of teaching is that such media adds variety, and a variety of approaches to learning is beneficial. The use of media also provides learners with more vivid frames of reference against which to apply the instruction being received. Table 6–1 presents some examples of media and their possible uses. Media can supplement any of the teaching techniques.

THE BASIC GROUP TEACHING TECHNIQUES USED IN TEACHING AGRICULTURE

There are seven basic group teaching techniques that work well in helping students develop the competencies they need in order to successfully enter the labor market. These group techniques are

1. Lecture
2. Discussion

Table 6–1 Examples of Media That Can Be Used with Various Teaching Techniques

Media	Possible Uses
Chalkboard/dry-erase board	To record key points, steps, procedures, illustrations, and diagrams, and to develop conclusions.
Overhead projector	To present key information, charts, tables of numbers, diagrams, sketches, and steps to follow, or to summarize remarks.
Slides	To show pictures that illustrate the point being made. Can be used as a basis for discussion.
Audio recordings	To "bring" experts to the class to stimulate discussion, to use to elaborate on a point, and to use as a basis for group analysis through discussion, such as evaluating the effectiveness of a speaker.
Video recordings or movies	To present a case that prompts discussion, to explain a crucial concept, to show a process to follow, to show a close-up view, and to use as a way of concluding a discussion.
Hook and loop display	To show or illustrate key points as they are presented by either the teacher or the class.
Charts	To illustrate technical details or present a problem for analysis.
Handouts	To provide basic information, explain steps and key points, reveal drawings on parts, and provide structure for following the steps of a demonstration.
Models	To point out key features and illustrate working parts.
Cutaways	To illustrate a process, show hidden views, and explain interrelated concepts.
Printed reference materials (books, booklets, activity guides, and job sheets)	To supplement teacher talk as a basis for supervised study, to use for study prior to seeing a demonstration or holding a discussion, to use to follow along during the demonstration, to use as guides in practicing skills, and to use as the basis for a class report or independent study.
PowerPoint	To enhance teaching through visual stimulation; to portray tables, graphs, charts, or digital photos.
Internet	To show interworking parts as live examples. Gather pictures and dates from libraries around the world. Project to the entire group or use on an individual basis.

3. Demonstrations
4. Field Trips
5. Role-play
6. Resource People
7. Cooperative Learning

Each technique of teaching in a group setting is presented. Once the reader has studied each technique, he or she will be able to use the technique effectively in group instruction.

LECTURE

Recall the teaching situation with which the chapter started: the teacher telling the class show to make a splice graft. For that particular teaching assignment, lecturing was not the appropriate teaching technique to use. However, there are teaching situations for which the lecture works very well.

Lecturing is a good group teaching technique for disseminating factual information. It is also useful when the material being taught is not readily available in a form other than through telling the class. For example, the history of the local FFA chapter may not be written in a form that can be distributed to the class, so a lecture format is chosen to deliver the content.

Sometimes teachers choose to lecture because they need to organize and present a segment of content in a particular way; oftentimes this method is chosen because the teacher has discovered that it makes the most sense to their students. For example, prior to earning the Greenhand degree, or preliminary FFA degree, students often must pass a "greenhand" examination that tests students' knowledge of national, state, and local FFA history and events. The teacher might choose the lecture technique to sequentially present much of the important content of the FFA unit of instruction, before offering students the exam. Lecturing is acceptable when that which is taught is only needed for short-term retention. It is also used for providing oral directions, suggestions, and introductory comments.

In teaching agriculture, lecturing has acquired a bad name. This is probably because it has been overused and often abused. Under such circumstances any teaching technique would be undesirable.

Nevertheless, there are several potential problems with lecturing. Teachers need to be familiar with these problems and learn to resolve them.

Problems with Lecturing

So often teachers think of lecturing as talking and seem to conclude that they do not need to plan for that. Nothing could be more false. Teachers will use "lecturing" to *fill time* as opposed to "planning a lecture." No teaching technique can be expected to work very well if the teacher is inadequately prepared.

When teachers do plan a lecture, it is often disorganized. There may not be a logical sequence to the organization. Sometimes there is no good flow between the parts of the lecture. It is also not uncommon to find that in planning a lecture teachers fail to plan for interim summaries and fail to develop clear conclusions.

Another difficulty with lecturing as a teaching technique is that lecturers often lack animation. They allow their stage presence to be lifeless, stale, and boring. Teachers also fail to use clear illustrations—illustrations that help learners create mental images that provide important frames of reference. Not only are there too few verbal illustrations but teachers also fail to incorporate sufficient visual, as well as other, media to help make their instruction as clear as possible.

Perhaps the most deadly problem with lecturing is the monotonous monologue. Nothing is quite so boring as an uninterrupted monologue.

Each of these problems can and must be overcome through good planning and good delivery.

Planning a Lecture

When planning a lecture to be used to solve a problem or part thereof, it is necessary to develop a clear and complete outline of the content, key points, or concepts that need to be taught. This means that teachers must have a thorough knowledge of areas in which lecturing is the teaching technique to be used.

This outline should be in written form and have a definite pattern of organization and flow that enhances clarity. There are a number of possible approaches to the organization of the outline. Some examples are

- Chronological
- Spatial
- Cause to effect/vice versa
- Problems and solutions (problems, possible solutions to each, conclusions for each)
- Topical

Whatever the pattern, the segments of the outline must fit together with clarity and fluency.

The plan must include marginal notes (directions to self) that instruct the teacher on how to best get a key point across to the learners. Perhaps the note reminds the teacher to show a transparency, put key notes on the writing surface, or relate an example.

Finally, the teacher needs to plan a summary to the lecture. This should be "drawing together," and concluding with a restatement of the major points that have been made. Consider Case 6-1, a sample daily plan of a brief lecture on the history of the FFA.

Case 6-1. *A Sample Daily Plan of a Brief Lecture on the History of the FFA.*

DIRECTIONS TO SELF	KEY CONTENT
Interest Approach	
Tell story of four FFA founders.	I. "On a hot day in September 1925, four men were sitting around an oak table in Blacksburg, Virginia."
Show PowerPoint of founders at table.	A. They were Newman, Groseclose, Saunders, and McGill.
	B. Newman was concerned about the students studying agriculture in Virginia.
	1. He felt they were sneered at by their city students.
	2. They did not belong to any school clubs.
	3. They seemed to feel they were inferior.
	(continued)

Tell story of "Jamie" and how such feelings affected him.	

	C. Newman believed the agriculture students were outstanding individuals.
	  1. He felt they were equal to their city students.
	  2. They had chores to do and lived in the more remote areas of the school district, which prohibited them from remaining after school for certain activities.
	  3. They had plenty of academic, social, and occupational ability, but seemed to be their own worst enemy.
	II. "The group tried to come up with ways and means of solving this problem."
	  A. The most popular suggestion was to form an organization for these students.
	  B. It was agreed to begin work immediately.
	  C. The result was the FFV (Future Farmers of Virginia), the forerunner of the FFA.

Write on blackboard **Hand out** a half sheet of colored paper. **Students write** (opportunity to show they have learned):	*Objective*: "Given this short scenario, write one reason that you believe the FFA was started." "I want you to show me that you were processing this story by giving you an opportunity to write why *you* think the FFA was started."
Show transparency 1: "Purpose of the FFA"	*Summary* Thus, the reason the FFA was started was to 1. "Give students something with which to identify and affiliate." 2. "Help students studying agriculture improve their feelings of self-worth."
Students check notes:	"Today we identified the purpose of the FFA. If you look back in your notes at the front page of the unit of instruction, you see that tomorrow we will be examining the timeline of events in the FFA history."

Analysis of Case. 6-1. In analyzing Case 6-1, first look at what was recorded in the teacher's plan. The key content for this interest approach is a story (written in the right column). The directions to self are the commands in bold before and after the story (written in the left column). These directions are sequenced so that they flow with the story.

The final phase of planning the lecture is to plan for leading through the class drawing conclusions at the end of the story about the purpose of the FFA. Teachers must remember that the lecture needs to be designed in a logical way to lead students through the study of a particular problem area. The teacher then needs to plan some bridging comments that link what was taught today with what was taught yesterday and with what will be taught tomorrow.

Presenting Good Lectures

Proper planning is the foundation for success in using the lecture. However, the plan has to be implemented effectively for the final positive and appropriate result to be archived.

A good way to grasp many of the salient points of presenting a good lecture is to consider the characteristics of good lecturers. Table 6-2 presents these characteristics. Teachers should seriously consider their own characteristics of speech and delivery and compare them to those listed in Table 6-2. The point is that it is not advisable to make heavy use of a teaching technique if it draws attention to one's weaknesses rather than allowing one to build on strengths.

Furthermore, as teachers consider giving lectures, they need to organize the notes and supporting materials such as visuals or props, so that everything is in proper sequence. The majority of agriculture teachers' lectures will greatly improve if they give illustrated lectures. An illustrated lecture is more than teacher talk. It is teacher talk supplemented with such things as key notes on the writing surface, transparencies, slides, charts, diagrams, specimens, PowerPoint, and other aids that help the teacher effectively teach the concepts under consideration.

Another key part of improving the presentation of a lecture is to make use of interim summary techniques. At critical points in the lecture, pause to point out the "big ideas" that have been presented thus far in the lesson. For example, during a lecture on "Managing Your Time," stop after presenting the topic "using calendars effectively" and indicate that, so far you have learned

Table 6-2 Characteristics of Good Lecturers

- Know their audience
- Are enthusiastic
- Speak clearly and fluently, and are able to vary the rate and intensity of speech
- Avoid using distracting mannerisms
- Illustrate their major points well
- Do a good job of summarizing—both at the end and in interim fashion

about "prioritizing projects" and "using calendars effectively." Remind them of the key points from each topic and then indicate that you are now moving into "establishing organization techniques." At the end of the lecture, be sure to develop very clear and specific conclusions. These should be recorded on the writing surface and in student notes.

DISCUSSION

Types of Discussion

Class Discussion. A major type of discussion is what is generally referred to as class discussion. This is a situation in which the teacher poses well-thought-out (prepared beforehand) leading questions to the class and the students offer answers. This dialogue continues in a purposeful vein until the major points to be developed have been brought out and clearly explained. The teacher serves as the facilitator of the discussion.

Brainstorming. Another way of promoting group discussion is through brainstorming. Brainstorming is a procedure whereby the whole group or several small groups within the class are given a topic, question, or problem and asked to come up with as many ideas, answers, and solutions as possible. When using brainstorming one is not interested in quality as much as quantity. It is important that the teacher not squelch participation or creativity by passing judgment on student responses. Instead, the teacher includes all responses that are morally worthy of consideration. The idea is to generate as many possibilities as the group can. Once the possibilities have been developed then they are evaluated and pared down as is appropriate. Brainstorming might well be used to

1. Develop a list of traits of good speakers
2. Develop a list of advantages of a given tillage system
3. Identify possible plants to include in a greenhouse cropping schedule
4. Suggest causes of a lawn problem
5. List types of habitats needed for the wildlife in a given region
6. Propose solutions to a given engine problem

Buzz Groups. Class discussion can also be facilitated using buzz groups. Buzz groups are small groups of three to seven students within the class. The buzz groups are asked to discuss a given problem or topic and come up with the best possible solutions or answers. Unlike brainstorming, in which quantity was the goal, buzz groups are more focused, purposeful, and businesslike. Buzz groups may rely on their combined current level of knowledge and experience in shaping their suggested answers, or they may use additional sources of information and data, such as references, magazines, and the like. The whole group might brainstorm to come up with proposed solutions to a

Table 6-3 Reasons Students Will Not Discuss a Topic

- The topic is beyond the range of their interest, knowledge, or experience.
- The topic is not relevant to them.
- The leader's introduction of the problem does not make it easy to approach.
- The discussion rambles too much.
- The discussion of one point becomes worn out.
- Argument or debate takes the place of group thinking.
- The student does not wish to reveal personal thoughts.
- The physical surroundings are not favorable to discussion.
- The students fear ridicule or disapproval.
- One person "knows it all."
- The leader is "teachy" and monopolizes the discussion into a lecture.
- The leader is dogmatic, unfair, or intolerant.
- The teacher answers too quickly without giving students the opportunity to think.
- The students are pressed or hurried.
- The leading questions are ambiguously stated.
- The question is one on which all are fully agreed.
- The question is the "yes and no" variety.

Source: Adapted from a handout prepared by the Department of Agricultural Education at South Dakota State University.

stream contamination problem. The class could then be divided into buzz groups to think of the best possible solution, which they would then present and defend.

Once the buzz groups have finished their discussions, which are supervised and guided by the teacher, then each small group reports its conclusions back to the total group, where the teacher directs the drawing of group conclusions. Each buzz group may have worked on the same assignment, or each buzz group may have received a unique assignment.

Teachers who wish to use discussion must remember that there are times when students will not discuss. Table 6-3 lists possible reasons why students will not discuss a topic.

Pair-Share. Class discussion can be conducted with large or small class sizes using Pair-Share. The individual student is asked to discuss (on paper) a given problem or topic and come up with his or her own idea or opinion. Then each student pairs up with a second student so the two can hear each other's ideas or opinions. They discuss their responses and try to determine best solutions. Additional sources of information and data such as references, magazines, and the like may be used in Pair-Share.

Pair-Share provides a quick and easy way to add variability to a class situation of any size. Because a writing component is used, quiet, reflective time is built into the technique, so it can be used as a transition or a bridge between an active discussion and a reflective conclusion to the class.

The Importance of Questioning to Good Discussion

A fundamental element for the success of a good discussion is the effective use of questioning. Of course, questioning is a fundamental skill of teaching and is also used to varying degrees with most teaching techniques. Questioning is complicated but can be mastered. Without good questioning, learning is apt to be superficial and transitory. With good questioning, learning becomes deep, insightful, and permanent. One measure of the success of the superior teacher is the manner in which questions are used in the conduct of instruction. Teachers can improve questioning. A prerequisite is an understanding of the function of questions. Dr. Richard H. Wilson, in a handout titled "On Questions and Questioning," offers the following suggestions:

1. Stimulate learning by questions to reveal the need to know, explore the benefit or advantage of knowing, provoke desire to know, promote the acceptance of knowledge offered, or evoke willingness to work to seek knowledge.

2. Direct learning by questions to guide the search for knowledge, identify what needs to be known, consider most likely sources of knowledge, weigh relative worth of available knowledge, relate new knowledge to that previously known, secure information possessed by specific individuals, seek recall of specific facts, cause participation in discussion, check accuracy of study in relation to knowledge needed, provoke deduction by logical analysis, focus attention, or rebuff inattention.

3. Evaluate learning by questions to gauge capacity or readiness to assimilate new knowledge, test recognition of knowledge needed, evaluate understanding of new conceptions, reveal grasp of new knowledge in relation to previous study, measure new knowledge gained, test comprehension of relationships, or appraise the growth in perception of generalization, broad conclusions, or general principles.

Questioning and questions are to be used in several ways. The teacher can ask the students questions either as a group or as specifically identified students. The students are then forced to think, reason, judge, and respond. When the questions asked them adhere to the guidelines on questioning presented in Table 6-4, students are able to make meaningful contributions to the solution of good problems. In addition, they develop lasting abilities in the whole area of thinking and responding, and of using previous information and experience to help solve new and different problems.

Another way for students to be involved in class is for them to ask the teacher questions. Thus, if the teacher is presenting some new information, explaining a new concept, or illustrating a key point, students should be given sincere opportunity to stop the teacher and raise a question (students will not respond if the teacher simply says, "Are there any questions" and quickly moves on). A student might seek clarification, want to know why something

Chap. 6 Group Teaching Techniques

Table 6–4 Do's and Don'ts of Questioning

- *Do* evaluate the teaching continuously as the lesson unfolds. Ask questions that test comprehension, understanding, grasp of an idea, or a relationship.
- *Do* provoke and direct thinking by a series of questions asked in a logical sequence, each building on the preceding premises.
- *Do* phrase questions precisely and carefully so students understand what you want answered.
- *Do* ask challenging questions. Avoid the trite or ridiculously simple.
- *Do* get more "mileage" from questions. Ask several students before acknowledging the correct answer.
- *Don't* ask questions students could not be expected to answer. The teacher must inevitably supply the answer. Students build the lazy habit of quietly awaiting the teacher's answer to questions they could answer. Students may come to question a teacher's good sense if he or she persists in asking questions the students should not yet know.
- *Don't* name students to respond before asking a question. You telegraph the idea that all others can relax. *Exception:* When you have overparticipation of students, identifying who is to answer the question before it is asked aids in classroom management.
- *Don't* always reject the first wrong answer. Continue testing it on others who identify it as wrong rather than you doing so.
- *Don't* supply answers to questions students should be able to answer. Use "wait time" appropriately so they can have ample time to think and respond.
- *Do* ask questions that will cause students to have to work. Require every student to write a response before continuing the oral dialogue.
- *Don't* identify correct answers by facial expression if you wish to keep students in doubt.
- *Don't* ask questions leading to simple "yes" or "no." They provoke limited thought and little discussion. If such questions are asked, follow them with "why?"
- *Don't* overquestion on one point. Cease when sufficient answers have stimulated thought, directed thought, or tested thought. To continue the questioning exhausts students' patience and interest.
- *Do* raise questions when lecturing that premise the teacher's answer. Phrase questions as though they were student raised: "Now, you may ask . . . " followed by the teacher's answer.

works the way it does, or see if an example has the same meaning as the teacher means. Obviously, this question asking by the student can become a good dialogue or conversation among those who want to know. The alert teacher sees students ask questions without words when students frown or appear perplexed and confused. These nonverbal questions also need to be pursued, brought out verbally, and answered.

However, the masterful agriculture teacher does not always give specific answers to the questions raised by students. Rather, the master teacher will often refer the question back to the class members and purposely guide them to the correct answer. Such teachers may also answer a student's question with a series of small leading questions that cause the students to discover for themselves the answer to the original question. Such procedures are not used because the teacher does not know the answer but because the teacher wants to encourage and even demand total class participation and improved reasoning ability. Teachers who operate in this fashion feel that students must practice reasoning if they are to be able to reason well.

Teachers need to remember that students asking questions of other students can also gain student involvement in a discussion. In fact, this communication among fellow learners is a highly desirable outcome of class discussions. Of course, such discussions must be focused and directed by the teacher.

Planning a Discussion

Before the specifics of planning a discussion are presented, one needs to ask the question, "What are some factors I should consider in deciding whether to use a discussion?" Familiarization with such factors helps the teacher automatically carry into the planning session pertinent information that helps make the planning more meaningful.

The first question teachers of agriculture need to ask is, "Why am I using discussion rather than some other technique?" Teachers should use one technique over another for some particular reason even if the reason is nothing more than to provide variety. Oftentimes, choice of technique is based on instructional intent or learning outcome, that is, "What do I want to accomplish in terms of learning by using this particular technique?" If one wishes to present completely new information, discussion is not a good technique to use because it is difficult for students to contribute to meaningful discussion if they have little or no knowledge in an area. However, if the problem is, "How can we stop erosion in the ditch located south of the school?" and the students are familiar with the ditch in question and have some previous experiences with controlling erosion, then it is quite reasonable to engage the class in some discussion in order to resolve the problem.

Another factor to consider is the amount of time available. Discussion takes more time than many other techniques. If one is extremely pressed for time, a decision to use some other teaching technique may be appropriate.

At least one other factor is very important in deciding whether to plan to use discussion for a given problem; that is, teacher background and experience. If the teacher is not very knowledgeable in an area, it is difficult to direct a meaningful discussion. If the problem is, "Which irrigation system should we use?" and the teacher knows very little about irrigation systems, the discussion that would result would probably be superficial and little more than a pooling of ignorance.

Once you have decided to use discussion, some planning needs to be recorded on paper. The following categories of information should be reflected in the plan.

1. *Sequence.* The general order in which the discussion will proceed should be reflected on paper. By giving thought to sequence, the discussion is more apt to be logical and, thereby, clearer and easier for the students to follow. There is nothing quite so frustrating as a ram-

bling discussion that appears to lead nowhere. With proper planning, this can be avoided.

2. *Important subject matter or key points to be brought out.* The purpose of most discussions is to help students discover definite ideas, options, pieces of information, concepts, or, in the final analysis, solutions to the problems. Unless the teacher records on paper the key points that must be made clear through the discussion, important information is likely to be omitted.

3. *Leading questions.* Teachers need to write down some leading questions. These are questions that cause students to think of the sought-after response. Questioning is a very difficult teaching skill, and the more inexperienced the teacher is, the more necessary it becomes to write down leading questions.

4. *Directions to the teacher.* As with any kind of good planning, teachers must include directions to themselves. Leading questions are one type of direction. Other directions might include writing down beside a particular key point the following direction: "Tell about Hawthorne example." This direction reminds the teacher to share an example that helps clarify a point, thereby enriching the instruction. Other types of directions to the teacher include: (a) Show transparency A-12, (b) Put cycle on board, (c) Explain meiosis, or (d) Show cross-section slide.

Thus, by writing down directions to self, leading questions, and key points (content) that must be brought out, all in a sequential order, the teacher is able to conduct a better discussion. Study Case 6-2 to see if you can identify the parts previously presented.

Case 6-2. *Sample Plan for a Discussion.*

Directions to Self	Key Content
Show slides A and B	A. Well-mulched landscape planting; very neat, crisp, weed-free.
	B. Landscape planting without mulch; weeds, dried soil and clods, generally unappealing to the eye.
Ask class:	"Which planting do you prefer?"
Take vote	
List student reasons on board (have students do this)	"What are your reasons for preferring landscape A?"
Brainstorm to come up with functions mulch performs and why	

(continued)

Write main ideas on board	Functions Mulch Performs	Reasons Why
	1. Enhances appearance	Neat, clean, crisp edges, no debris
	2. Controls weeds	Lack of light
	3. Conserves moisture	Protective layer, returns evaporation
	4. Improves habitat	Azaleas, increased humus, and so on
Remind students	"Make sure you get this in your notebooks."	
Write objective on board	"You have come up with some good reasons... Today our objective is to **identify functions of mulch and its benefits**."	
Ask class	"Now how does mulch do each of these things you have listed on the board?"	
Write responses on board	(Record key points beside reasons for preferring slide A.)	

Analysis of Case 6-2. Notice that the plan has sequence and purpose. The teacher started by gaining interest with the two slides. Then the students provided reasons for their choices. The teacher's plan includes all desired reasons (anticipates responses from students plus the list needed for full grasp of the content) so that any reasons omitted by the class could be added by the teacher. The same was true for explanations of how mulch controls weeds, for example. Also notice that the teacher provided directions to self:

- Show slides
- Write on board
- Remind students
- Ask class

Procedures to Follow in Conducting a Discussion

There are some general procedures teachers should follow when conducting a discussion that are so integral to conducting the discussion that the teacher should not have to write them down. One might call these requisite skills for conducting a good discussion.

If a discussion is to be meaningful, the teacher needs to be sure the students understand the task at hand. The teacher has to write the problem or topic on the board or otherwise make it clear to the students what it is they are discussing and why. Without creating such a mental set, a discussion is apt to lack good focus and a sense of purpose.

Another element that is fundamental to the success of the discussion is the effective use of questioning, previously discussed.

Teachers who use discussion effectively stop at the end of the discussion and take time to develop clear and concise concluding points. If this procedure is omitted, the quality of the discussion and the problem solution itself will be substantially diminished. After students have presented their divergent viewpoints in arriving at answers to the teacher's leading questions, a time of pulling together and sorting out the most important points is mandatory for students to achieve success in understanding. Students can help with this sorting out and conclusion-drawing process. The major concluding points should be summarized on the writing surface or in some other manner and the students should write them in their agriculture notebooks. This development of concluding points also serves as an excellent summary of what has been learned.

DEMONSTRATIONS

In agricultural instruction the psychomotor domain of learning is developed through hands-on application of key concepts. In every instructional area of agriculture there are psychomotor skills that need to be practiced if students are to successfully solve their problems. Consider the abilities for the various instructional areas listed in Table 6-5. For each example with which you are familiar, think about how you would teach students the given ability (see Case 6-3 for teaching the ability to use a torque wrench).

In many of these examples in Table 6-5, talking about it will not produce the ability. Students need to see and practice a step-by-step procedure that they must follow to accomplish a given task.

Demonstrations are best suited for teaching students how to do certain tasks. One must realize that demonstrations must usually be accompanied by

- Explanations (and perhaps some discussion)
- Illustrations (media)
- Practice on the part of the students (hands-on)

Demonstrations are used to show students how to accomplish a given process or task.

Teachers must remember, however, that demonstrations do not automatically work in all situations or for all students. They can be confusing, poorly organized, and frustrating if they are presented improperly or when students are not prepared for them. However, demonstrations, when properly used, can be a very effective teaching technique.

Requisites of a Successful Demonstration

Before a demonstration can be a successful group teaching technique, the teacher must meet several requisites. First, teachers must determine what they wish to accomplish. What one wishes to accomplish must correspond with what is possible. In other words, the teaching objective must be able to be demonstrated

Table 6-5 Abilities to Be Taught in the Various Instructional Programs in Agriculture

Program Area	Ability
Production agriculture	• Preparing a seedbed • Welding a piece of machinery • Calibrating a sprayer
Horticulture	• Disbudding a chrysanthemum • Making a corsage • Pruning a tree • Building a patio
Agricultural mechanics laboratory	• Grinding a valve • Timing an engine • Replacing a radiator • Adjusting a combine
Natural resources	• Laying out a spillway • Constructing a campsite • Tagging wildlife • Testing a water sample
Forestry	• Cruising timber • Felling a tree • Controlling a fire • Sharpening a chain saw
Agricultural business	• Ringing up a sale • Constructing a display • Serving a customer's needs
Food processing	• Cutting steaks • Sealing containers • Cleaning equipment • Sharpening a knife
Small animal care	• Grooming a dog • Setting up an aquarium • Handling a snake

(shown step by step); it has to be of the appropriate scope (i.e., small enough to fit into the classroom or simple enough to be accomplished in the allotted time) and at an appropriate achievement level for the students. If the scope is too large, the students will become confused and frustrated and will give up. Likewise, if the skill to be demonstrated is one for which students do not possess the prerequisite skills necessary to understand the demonstration, it is doomed to failure.

Second, given that the demonstration is clearly outlined, of the appropriate scope, and written at the appropriate level of readiness for the class, then the next important requisite the teacher must meet is to properly plan the demonstration. All demonstrations that are to be formally presented to the class need to be planned in advance. Granted, there are teachable moments in the laboratory where spontaneous individualized instruction is given in the form of a private demonstration. Otherwise planning is essential.

A third requisite of a good demonstration is to carefully select and assemble the necessary items. One could argue that this is a part of planning and preparing for class. Without this a teacher is asking for trouble. Nothing sabotages a demonstration quite so well as for teachers to reach a crucial point in the demonstration only to discover they do not have a certain tool that is needed. This disrupts the instructional process.

Finally the teacher must rehearse each important demonstration. By following this practice teachers can be sure that everything is operating properly and that they have the skill and prowess needed to masterfully show their students the process they must learn. The teacher who makes sure these requisites are always met need only plan and execute demonstrations when this is the best technique for the problem being taught.

Planning a Demonstration

The first thing a teacher must do in planning a demonstration is to be sure to select a demonstration of the appropriate scope. That is, select a phase of a process or an operation that can be comprehended and remembered by the students with reasonable effort. One cannot very well demonstrate how to overhaul an engine but one can break that block of instruction down into a number of manageable tasks. An agricultural mechanics teacher can demonstrate how to gap a spark plug, grind a valve, bore a cylinder, or time an engine. Likewise, a horticulture teacher cannot demonstrate how to grow a crop of poinsettias, but he or she can demonstrate how to prepare the soil, pot the rooted cuttings, or apply the proper drench.

When the teacher has selected a task of appropriate scope to demonstrate to the class, it is important to write down the basic steps to follow in order to accomplish the task. These steps should take the class through the process in a logical order and at a pace that is comfortable for the students. Teachers must be sure not to omit steps, take shortcuts, or go off on tangents. The primary reason for writing down the steps is for the teacher to be sure he or she has been complete and logical.

Once the steps have been listed, the teacher needs to decide what key information is needed for each step. There are reasons why a step is important, safety precautions one must take, and basic theory to remember to explain. Agriculture teachers refer to this supporting information as key points. Teachers need to not only present the steps but also to elaborate on the steps, why the steps work, principles of science to remember, points of safety to follow, and special techniques (tricks of the trade) to develop.

As the steps and key points are developed, the teacher must decide how best to present each point. By this we do not mean how to show them, but how to be sure the students understand the process and supporting information. Sometimes teachers list the steps on the writing surface, on a handout, or on an overhead transparency. Another dimension of this part of the planning involves deciding if a story, an example, or an analogy can be used to clarify certain key points. The teacher's options for assisting students in learning the key information in a demonstration range from simply telling them, to providing the information written out in detail, or in any other logical manner.

Once the steps and key points are recorded, the teacher should review the process and prepare a list of materials, tools, and aids that will be needed to give the demonstration. By developing this list of materials, oversights can be avoided the first time the demonstration is presented and in succeeding times.

It is also necessary that teachers develop an effective interest approach for demonstrations. The chief aim of the interest approach is to create a "felt need to know" (see Chapter 4) and to create mental set. Many teachers will want this to be the first activity they attend to in planning demonstrations, but it is presented last in this sequence to make a point. The point is that sometimes teachers have difficulty thinking of an interest approach they believe will be effective. When this occurs, a teacher should go ahead and plan the content and delivery then return to the interest approach. Oftentimes, a teacher has many more good ideas on gaining the students' interest after working through the subject matter. The interest approach for a demonstration may be asking one or two students to come forward and show the class "how to do it." If they cannot, they then have a felt need to know the necessary information for them to be successful. A teacher could also use a problem the students have encountered (or will encounter) in lab as a way of gaining interest. Yet another possibility is to demonstrate the required skill with great finesse, thereby intriguing the students who wish to be able to perform just as skillfully. When using demonstrations, as with all instruction, students must be motivated to learn. The preceding points regarding planning a good demonstration can be summarized through a sample plan for giving a demonstration shown in Case 6–3.

Case 6–3. *Question to Be Answered: How Do I Use a Torque Wrench?*

Tell students what they are about to see	Interest approach
Steps	**Key Points**
1. **Select** proper hand tool (torque wrench)	"We have to begin with defenitions so that we are all communicating clearly as we move through the process."
Define torque	A twisting or rotating force. Expressed in lb/ft or lb/in.
Identify parts of a torque wrench **Handout A**: Parts of a torque wrench	a. Head b. Indicating beam c. Scale d. Pivoted handle e. Pointer f. Square drive g. Beam h. Lever length

Chap. 6 Group Teaching Techniques

2. **Determine** torque requirements
Handout B: "Torque requirements"
	1. Use repair manual or chart; if not available, use both size and grade of bolt to determine torque.
	2. Determine run-down resistance and add to desired torque.
	3. Make necessary conversions.
	   a. lb/in = lb/ft × 12
	   b. lb/ft = lb/in ÷ 12
	4. Determine tightening sequence if applicable.

3. **Torque** to specifications	1. Tighten until you obtain 75% desired torque, then use single sweeping motion to obtain 100% desired torque.
	2. Reduce set or seizure.
	   a. Remove foreign material
	   b. Remove burrs
	   c. Lubricate threads and underhead
	   d. Use constant sweeping motion to complete torquing

Repeat the demonstration while **calling on** students to give you the steps in order	
Ask students what questions they have	
Students demonstrate use of torque wrench	Opportunity to show what they have learned.
Summarize	Today we had the chance to identify tools and define their use. We also had the chance to use a torque wrench. Tomorrow we will apply this knowledge and skill to small engine repair.

Source: Courtesey of Dr. Joe Gleim, The Ohio State University. Department of Human and Community Resource Development.

Procedures to Follow in Giving a Demonstration

After a teacher has the demonstration planned, the next step is to successfully give the demonstration. There are a number of basic procedures that a teacher must follow in order to successfully demonstrate a manipulative skill.

    The first procedure is to assemble the materials, tools, and equipment that will be needed. For example, if one is going to demonstrate how to weld a stringer bead, it is necessary to have the demonstration welder ready, as well as the proper electrodes, gloves, apron, helmet, pliers, chipping hammer, prepared metal to be welded, and safety glasses for all students who will watch the demonstration. Unless all this paraphernalia is ready to go and available at the point in the demonstration when it is needed, the demonstration will become fragmented, unclear, and frustrating for students as the teacher has to interrupt the flow of the demonstration to go back and get supplies and tools that were not assembled in advance for the demonstration.

Prior to demonstrating the skill to be learned the teacher needs to be sure to create a felt need on the part of the students so that they are motivated and have the proper mental set for learning the skill. Teachers need only refer to their plan to remind themselves of how they have decided to go about generating this interest.

The second procedure is to give the students an overview of the demonstration they are about to view. It is here that the teacher can point out important steps, techniques, or operations to which the students should pay particular attention. Likewise, a verbal overview of the process aids in clarifying the procedure and provides for built-in repetition.

Third, teachers present the demonstration in a step-by-step fashion. Students must be able to clearly discern the components of the process and special manipulative techniques that are needed to carry out the procedure. As each step is presented, it needs to be clearly labeled with words, with board work, on an overhead transparency, or in some other acceptable way. This delineation of crucial steps is vital to effective learning. Following such a practice substantially reduces misunderstandings and omitted steps. As these steps are presented the teacher also needs to explain:

- How each step is performed
- Why it is needed
- What it accomplishes
- What important safety points need to be understood and remembered

The reader will recall that these are referred to as key points in the plan the teacher is following as the demonstration is given.

Sometimes a given step may need to be shown more than once to help students see intricacies or hard-to-conceive procedures. Then, once the process or skill has been demonstrated and explained, it is often helpful to repeat the demonstration. This repetition of the process can add familiarity and clarity, and help to tie the entire procedure together clearly because less explaining and redemonstrating of the technique will be needed. The students are now ready to try it for themselves. Whenever conditions permit, the teacher should have all students practice under the teacher's supervision. This provides important repetition and helps students gain at least a small measure of confidence by being able to see for themselves that they can perform the skill, even if it is at a less-than-masterful level, on their first attempt. If this is not possible, and students will actually use the skill that has been demonstrated at varying times in the next several days or weeks, then students will usually

Chap. 6 Group Teaching Techniques

need refresher demonstrations as they are ready to try the new skill or process the first time. Following is a summary of procedures to follow in giving a demonstration:

1. Assemble materials
2. Develop a felt need on the part of students to learn the skill
3. Give an overview of the process to be demonstrated
4. Present each step
5. Repeat difficult-to-understand steps
6. Repeat the entire process
7. Have students practice

Pitfalls to Avoid

Too many agriculture teachers believe that the only way to teach a manipulative skill is to give a demonstration. As mentioned previously, this teaching technique is not always the best one to choose. There are a number of common pitfalls agriculture teachers need to be aware of and make every effort to avoid when giving a demonstration.

One pitfall is not having mastery of the skill to be demonstrated. Agriculture teachers have to teach a wide array of skills and procedures, even in highly specialized programs. It is difficult, if not impossible, to have complete mastery, including smooth techniques, for every skill that must be taught. However, it is imperative that teachers practice skills in which they are "rusty" to be able to give a credible demonstration. There is nothing to be gained by teachers who give a demonstration knowing that they have not practiced the demonstration with particular emphasis on technique.

Another problem to avoid is situations where students cannot see or hear the demonstration. The larger the group, the more difficult but also more important this task becomes. If students cannot clearly see and hear the details of the demonstration they then become confused, bored, and frustrated. Teachers usually need to be in front of the group with students arranged in a semicircular fashion. With large classes, twenty to thirty students, and demonstrations involving minute detail, such as planting impatiens seeds in a horticulture class, the teacher may find it best to give the demonstration several times to groups of a more manageable size. However, if this option is used, the teacher must be sure the groups who are waiting to have the demonstration, or those who have already seen it, must have other meaningful work to do. Otherwise pandemonium results. A real help in ensuring that everyone can see well is the use of a videotape recorder with a zoom lens. More about this possibility will be presented later in this chapter.

A fundamental problem with demonstrations is that students see a mirror image of the process when the teacher faces the group and demonstrates a skill. Teachers must be aware of the inversion of right and left directions. In order to solve this problem, teachers can use a demonstration table with a large mirror mounted above it. Students then observe the demonstration through the mirror to obtain the appropriate perspective of the visual image. Another solution involves teachers standing with their back to the class and holding their hands above their head as they demonstrate. This allows the students to view the demonstration from the same perspective as they will operate when they practice the skill. Of course, there are instances where these ideas are not workable, such as demonstrating a skill on a tractor engine that is mounted in the tractor.

Another common problem that teachers have with demonstrations is that they are sometimes so familiar with the process that they omit or fail to mention certain steps and fail to adequately explain some steps. When this happens students are at a real loss. The students do not know that the teacher failed to mention a certain step or that the teacher did not explain a key point. All the students know is that the demonstration does not make sense and that they do not understand how to perform the skill in question. The best way to avoid this problem is for the teacher to plan the demonstration in advance and then follow the steps and key points to give a complete demonstration.

One final potential pitfall in giving demonstrations is the failure of the teacher to get feedback from the students. It does little good to whiz through a demonstration assuming all is going well, only to discover at the conclusion of the demonstration that the students are confused. To prevent this problem, teachers need to pause to ask specific questions of students to be sure they understand. Another recommended practice is for the teacher to have students explain a given step as it is redemonstrated. Best of all, have at least one student try the skill once the demonstration has been given. In this way teachers can secure specific student feedback that helps them determine whether their instruction has been successful.

Providing for Student Practice

If an agriculture teacher gave a demonstration to the class and then moved on to the next item on the instructional agenda, then the instruction would be poor indeed. A vital part of the demonstration as a group teaching technique is that it merely sets the stage for real learning. Crucial to the success of demonstrations is providing time for student practice. The reason for the demonstration is to help students learn enough so they can practice the skill until it has been mastered.

A management decision for the teacher is deciding when students can practice what has been demonstrated. As was mentioned previously, the ideal situation would be one in which all the students could practice the skill under the teacher's supervision as soon as the demonstration was concluded. However, because of facilities, equipment, tools, and supplies limitations, this is often not possible. Once it has been determined that all students cannot immediately practice a skill, a rotation procedure needs to be established to ensure that over a given period of time all students get to practice the skill. The specifics of setting up such rotation schedules are presented in Chapter 9.

When students practice skills that have been demonstrated, they need structure. Teachers need to remember that directed learning is more effective than undirected learning (see Chapter 2). Furthermore, students learn better when there is organization and structure (see Chapter 2). These are important principles of learning that apply to the development of manipulative skills just as much as to stereotyped classroom learning. This structure can be provided in the form of computer-aided instruction, assignment sheets, task sheets, project plans, or through teaching assistants, volunteers, and resource people. The key is that the learner has to know specifically what is to be done, when and where it is to be done, what procedures are to be followed, and what the final product should look like.

Given that the student has been provided with this structure, it is essential that the teacher guide this practice through good supervision. The teacher needs to observe the students as they practice the skill. Encouragement must be offered, errors must be corrected, and further explanations must be given. Without this guidance the process of learning loses efficiency and appeal. Students fail to realize why their work is acceptable or unacceptable. In general, the outcomes are far below what can be expected when there is prudent teacher guidance of student practice following a class demonstration.

FIELD TRIPS

Field trips as a group teaching technique are fundamental to successful teaching in agriculture. Because of the uniqueness of agricultural subject matter, it is imperative for teachers of agriculture to take their students into the field to observe and participate in real-world situations. Every school cannot operate a commercial farm, greenhouse, or park. However, some career and technical centers have labs that come nearer to simulating actual industry conditions than others. In the case of production agriculture, horticulture, natural resources, and forestry to name a few, it is essential that good use be made of real-world settings through well-planned field trips.

Another value of field trips is that they provide variety and a change of pace; they get both the teacher and the students out of the routine of the class and laboratory and into a new environment. Field trips also offer tremendous enrichment through sensual stimulation; that is, students get to use eyes, ears, noses, and hands as they go about the task of learning. An agricultural instruction program without field trips is sterile indeed.

When teachers take their classes on field trips they are able to provide concrete frames of reference. For example, if the class is studying different types of wetlands, students who go on a field trip and study various types of actual wetlands no longer have to wonder exactly what the teacher means when referring to a certain term or concept. The students and teacher now have the same frame of reference and are able to develop a deeper understanding of the concepts being studied.

Situations for Which Field Trips Can Be Used

There are literally hundreds of specific teaching–learning situations for which field trips as a group teaching technique are useful. However, these specific ideas that agriculture teachers have for using field trips can be characterized into three general categories.

One good use for field trips is to create interest at the beginning of a unit of instruction. Oftentimes, when students can see a group of problems firsthand, they are more interested in learning how to solve such problems. For example, a horticulture teacher could take a class to a new home that is to be completely landscaped as a class project. This initial field trip to the site would generate interest. Additionally, the students would discover the specific problems to be solved. All the students would then have a common frame of reference and a high degree of readiness for learning.

If a teacher was beginning a unit on establishing a wildlife habitat, a field trip to a site where such a project is to be initiated provides a perfect way of creating interest and a felt need to know. For someone beginning a unit of instruction dealing with laying out hiking trails, a field trip to a nature study area where such a trail needed to be planned would provide a fine introduction to and backdrop for the study that was to follow.

Field trips need not be limited to introducing a unit. They can be used very effectively for gathering factual information to use in solving problems. Horticulture students learning how to plan and construct a patio might well use a field trip to a ready-mix concrete plant to study types of concrete, aggregate, prepared mixes, costs, and delivery limitations in order to more accurately learn how to order and use concrete. Many useful field trips are made to agribusiness firms where students learn specific marketing facts firsthand. For example, students in agribusiness classes can visit grain elevators to learn how to test grain moisture content; study marketing procedures; learn how to use the futures market; and learn how grain is received, processed, and shipped.

Another category of uses for field trips is in concluding a unit of instruction. Once students have answered a group of related questions in class, developed a list of approved practices for success, and prepared individual plans of practice when appropriate, they can take a field trip to a site where these approved practices are routinely used. Take the case of a class that has completed the unit of instruction on caring for your reptiles. An excellent way to conclude such a unit would be to take a field trip to a zoo where the class could study in a firsthand way how a reptile management program is operated. This is an effective way to gain closure and to help reinforce the relevance of the instruction being offered in the agricultural education classroom.

Planning a Field Trip

Too often teachers decide to take field trips at the last minute, and too often they also do far too little specific planning for a field trip. Whenever this happens the resulting quality of the educational experience is predictable. With the fiscal restraints faced by local boards of education, it will be difficult to have haphazardly planned field trips approved.

The first step in planning a field trip is to establish the goals for the trip. The teacher must decide why a field trip is necessary as the teaching technique rather than some other technique. Specific objectives (learning outcomes) that need to be achieved via the field trip should be developed. The field trip should contribute to the quality of learning outcomes in ways other techniques could not. For example the educational value of a field trip to the elementary school to present an agricultural literacy program cannot be replaced by any of the other group teaching techniques.

Once necessity for the field trip is established, the second step in planning the trip is to select an appropriate site. Factors to consider in making this decision include

- Quality of the experience to be seen or participated in
- Willingness of the host to have the class visit
- Distance
- Cost

Once those two steps, establishing goals and locating the preferred site, have been determined the teacher must secure approval from the appropriate administrator to schedule the trip. Be sure to work ahead. In some schools, a field trip may have to be approved at a school board meeting one month or more ahead of the trip, and in some cases, the item needs to be on the board meeting agenda one week or more before the meeting.

Once the trip is approved, the teacher has to make specific arrangements with the host at the field trip site. These specific arrangements include selecting a date and time; outlining specifically what the teacher and students need to see or do in accordance with the objectives of the trip; inquiring about parking the bus or other vehicle; and deciding who will explain, direct, and demonstrate. The two parties also need to agree on the order in which the events of the field trip will proceed, what the students may actually do (participation), and at what time the field trip will be summarized and by whom. As a part of making these arrangements with the cooperators, the teacher is also able to make a "dry run" to determine the best transportation routing and to develop a reasonable time schedule.

Finally, the agriculture teacher should thoroughly prepare students for the field trip. The teacher should work with students to develop a list of questions to be answered by the field trip experience—students will be responsible for the answers to those questions when they return from the field trip. The students need to understand specifically what objectives are to be accomplished (what is to be learned) from the field trip, rules of conduct, exact dates, times and places of departure and return, eating arrangements, amount of money to bring, and type of clothing to wear. Parental permission will be needed. This process varies, and each teacher must conform to the practices required by the local board of education. In cases where field trips require absence from other classes, students will need to make arrangements with other teachers to make up missed work. A follow-up debriefing, summary paper, discussion, or other assignment is necessary to reinforce the learning that took place on the field trip.

Field trips that are planned as outlined here have a high probability of success. When they are not planned in this manner, they may become outings that do little more than provide a change of pace.

Conducting a Field Trip

Once a field trip has been well planned, it needs to be conducted with expertise. In addition to the logistical arrangements made as a part of planning the field trip, the agriculture teacher needs to be sure the bus is scheduled, the driver is ready (with a map to the destination), and everyone involved is prepared and ready to leave the school at the agreed-on time of departure. The teacher may make a last-minute phone call to be sure the people at the location to be visited are expecting the class.

Another essential dimension of conducting an educational field trip is to be sure students' behavior is well managed. Students quite naturally get excited about such an activity but their excitement should not be allowed to exceed reasonable limits. The best way to manage student behavior on field trips is to make it very clear from the start that taking a field trip as a class or going on a field trip as an individual student in the class is not a right; rather it is an opportunity or a privilege. Those who abuse this privilege will have it revoked. If a given student cannot behave appropriately on the bus or at the site, that student should be denied the opportunity to take additional field trips. It is to the teacher's benefit to have precisely this understanding with the appropriate school administrator and with students and their parents in advance. Do not make threats, make promises. If a student cannot behave appropriately and has been given a reasonable chance to correct inappropriate behavior, that student should be denied field trip privileges in the future. The same logic holds for the class. If a sizeable cadre (15 to 20 percent) of the class cannot behave sensibly, field trips should cease. A teacher should not hesitate to have a bus stopped en route and returned to the school if students cannot conduct themselves appropriately. Once students have observed such action, the task of managing behavior on field trips becomes simple for the agriculture teacher.

It is essential that teachers know and reinforce all policies and procedures of the school in conducting a field trip. For example, schools have strict policies about students driving to school functions. In most school systems students are forbidden to drive themselves on field trips or other class-related functions. The teacher must reinforce this policy.

Sometimes agriculture teachers take sizeable numbers of students on all-day or even overnight field trips, for example, career development events, extension-sponsored programs, or FFA conventions. Whenever this is the case, the teacher should solicit additional help from other adults, such as parents, FFA alumni, school counselors, or young farmers. This ensures adequate supervision for students who are on a trip for quite some time.

When a teacher maintains appropriate student behavior while on the field trip, learning can take place. If there is appropriate behavior, the teacher's efforts should be directed to student learning on the scene. It is the teacher's job to be sure students notice crucial operations, skills, practices, facilities, equipment, duties of workers, and the like. The teacher needs to raise questions with the students—questions that stimulate probing and keen analysis. The teacher may also give a demonstration, for example, on how to dehorn a calf, grade roses, or collect a water sample from a stream. Many times the teacher will arrange for the host to present such demonstrations. Even then it is the agriculture teacher's responsibility to make sure key points are explained or clarified.

Another responsibility of the teacher is to be sure the students express appreciation to the host(s) for taking time to help them become better prepared for their future careers. It would also be appropriate to share such appreciation with the proper school authorities. In fact, the teacher as well as the students should show such courtesy. The host should be thanked verbally on site and also be sent a letter of appreciation following the field trip.

Summarizing What Was Learned from a Field Trip

Too many field trips end when the class boards the bus to return to the school building. It is essential that the teacher develop with the class a list of conclusions to be drawn from a field trip. Major key points need to be summarized. This should happen as soon after the field trip as possible. Sometimes it can be done at the site of the field trip. Other times such summarizing can be accomplished en route back to the school. However, quite often the best place for helping students clearly pull together what they have learned is back in the classroom.

Another vital activity is to have students develop plans for applying what they learned from the field trip whenever possible. Students should make plans to prune their own shade trees, vaccinate their livestock, maintain their machinery, establish a wildlife habitat, or whatever else is an appropriate application as a result of a given field trip. This application of learning is just as important following a field trip as with any other teaching technique.

ROLE-PLAY

In teaching the course of study for most agricultural instruction programs, teachers find that role-playing is a basic group teaching technique that has a number of uses. Role-playing, as the name indicates, involves having class participants (students or the teacher), and sometimes guests as well, play or portray a given role. The playing of the role and the subsequent analysis of the episode provide the information or concept development that is sought.

Types of Problems for Which Role-Playing Is Suited

There are three general classifications of problems for which role-playing is best suited. When one seeks to teach concepts and skills related to human relations, role-playing is often quite useful. As basic ways of getting along with others are introduced, it is helpful to demonstrate what is meant. One might show the class how to greet others; offer suggestions to people; or start a conversation the correct way, the incorrect way, or somewhere in between.

The development of leadership skills is another area that lends itself to the effective use of role-playing. Role-plays can be used to show how to make a speech, present a motion, give an award, escort someone to the platform, chair a meeting, or any of a whole host of other leadership abilities.

When teaching agriculture students basic sales abilities, role-playing is effective. Students need an opportunity to practice abilities, such as approaching a customer, making suggestions to a customer, registering a sale, and making change. Practicing such abilities in the security of a classroom or laboratory setting under the expert guidance of an agriculture teacher before attempting such tasks in the real world provides students with that extra edge they need to compete effectively in society. This assumes, of course, that the agriculture teacher has worked with class members to the extent that they understand role-playing as well as the need to be supportive of their classmates even if they are not good actors. Role-playing cannot flourish as a method of teaching in a classroom where students tease one another, laugh at one another's mistakes, and ridicule some students for whatever reasons. Thus, it is important for the teacher to work with students by providing clear expectations, written agreements, operational guidelines, praise, or other means to help them learn to respect the dignity of each individual. This is important for role-playing and other forms of student participation to be plausible techniques for learning. It is also important to teach students in their relationships to others throughout life.

Role-playing is also a good way to approximate real-life situations in a secure environment. Students are often self-conscious about speaking in front of others, developing mastery of social graces, properly using a telephone, introducing strangers, learning proper hygiene and professional grooming, and other important skills as well. By being able to practice such abilities in a caring climate, much structured learning can take place.

Planning a Role-Play

Before a role-play can be planned, the teacher must select subject matter that is suited to this technique. One cannot very well role-play the events of a four-stroke cycle, the nitrogen cycle, classifications of soil, or how to start geraniums from cuttings. Even if it were attempted, there are other teaching techniques that are much more appropriate to teach such content. As indicated previously, most role-plays deal with human relations, leadership, and sales skills.

Once the area of the role-play has been identified, the teacher then needs to decide on the specific objectives to be accomplished through the use of role-playing. If this is not done, the role-playing can turn into a chaotic "dog and pony show" or a dysfunctional dialogue of little educational value.

The next step is to design the role to be played that will accomplish the preestablished objectives or solve the problem being taught. For example, if a teacher wants to teach the class how to properly introduce strangers, the teacher will need to decide on the specifics of this role. Perhaps the class has learned that when one introduces an older person to a younger person the older person is introduced first. A good role-play could have a student play the role of someone who is making introductions while two other students play the roles of the older and younger strangers to be introduced. Thus, the role-play has been designed; there is an objective, there is a role to be played, and characters are identified.

The next step in the planning process is to decide on which students should actively participate (all students in the class participate in some way, but active participants are those who are before the class). There is no preferred way for deciding who plays roles. It depends on the students' personalities, time to get ready, their individual needs for such experiences, and many other considerations as well.

Once the persons who will conduct the role-play have been identified, they usually need to be prepared to play their respective roles. Granted, there will be times when the teacher decides impromptu performance is best, but this is generally not the case. Role-players need to at least be briefed on what the role-play is, what educational objectives the teacher wishes to accomplish, and how the role-play will be handled. Sometimes it is advisable to have scripts, but most often that requires more time and effort than either teachers or students are able to give. In some cases, it is possible to find prewritten role-plays that can be adapted to fit the objectives you are seeking to accomplish. For example the role-play, "They are not like us!" is an interactive exercise that helps students understand stereotypes and biases.[1] Other ideas can be found in books such as, "*Role-Plays,*" from the School Mediation Association in Watertown, Massachusetts.

A final aspect of planning for using a role-play as a teaching technique is to plan an analysis and summary of the activity and to identify important key points. This planning should consist of a few lead questions. These could be given to the class in advance or raised verbally at the conclusion of the role-play. The teacher also needs to develop a list of the key points that must be made. Of course, the teacher will need to decide how to get the key points across and will probably use lecture-discussion plus the appropriate visuals.

---

[1] Benoit, B. Center for Adolescent Studies. Indiana University.

Conducting a Role-Play

If a role-play is to be used effectively, the teacher must satisfactorily prepare the class for it. The students need to know why the role-play is being used, what they need to look for, and what their role is once the role-play is concluded.

Next the teacher or a student introduces the role-play, and the players act out their roles. The teacher may provide coaching from time to time in order to keep it going. A role-play can be created to be as short as five minutes or designed to be as long as thirty minutes (this allows ample time to introduce the role-play at the beginning of the class period and summarize the role-play at the end of the class period). An example of such a role-play might be, "Not in My Backyard,"[2] where students play the role of the mayor, township trustees, concerned citizens advocates, farmers, and real estate agents during the citing of a large-scale agriculture production facility. Once the role-play is completed the teacher terminates the action and secures the class's attention so that a productive analysis of the role-play can be made.

Analyzing the Learning from a Role-Play

Perhaps the most important phase of a role-play as a way of learning is the thorough analysis that follows. The opportunity for reaching the highest cognitive levels of thinking exists when role-play is properly conducted. The first step in a good analysis is to lead the class in focusing on what happened. Be sure everyone has seen the same thing. A videotape of important role-plays provides opportunities for playback that could be very helpful in this regard. Once the major events that transpired have been identified, a detailed discussion of these events is needed. Through this discussion the teacher ensures that key points are discussed and recorded in the students' notebooks. The teacher may also want to further illustrate the concepts being developed by using analogies, reminding students of real events the class may have witnessed previously, or reenacting a given ability themselves.

Finally, the teacher guides the class in a summary. Here the teacher can use students to help summarize the important learnings that were gleaned from the role-play.

RESOURCE PEOPLE

When to Use a Resource Person

When an agriculture teacher seeks to provide factual information to a group of students to aid them in solving a problem, he or she may choose to provide that information through an expert individual who can be invited to the class to serve as a resource person. Instead of the teacher directly providing the

---

[2] Adapted from Conservation Leadership School Exercise. The Pennsylvania State University.

needed information, the teacher uses a resource person as the source of information needed by the students.

Even the best teachers cannot be expert in *all* phases of the course of study they teach. However, they can know their limits and they can know the full array of resources, including people, available in the community. The job of the teacher, then, is to wisely draw on the needed resources, including individuals, at the appropriate time.

Resource people also ought to be used to introduce students to key individuals in their communities whom they will need to know and rely on as they make decisions in the future. Without full knowledge of competent people whom students can and should contact for help after they have graduated from high school, students will not be able to be as independent as they need to be.

Teachers may also decide to use certain resource people in order to add relevance to what is being advocated or taught. The agriculture teacher may have complete command of the subject matter, practices, or process being studied but may discover that students need the boost an outside resource person can add to convince students that what they are learning is important, useful, and current.

The Role of a Resource Person

The basic role of a resource person is to serve as a consultant to the agriculture teacher. The resource person must be more than a guest speaker in order to be optimally effective. The teacher needs to work with the resource person in advance to be sure the resource person is specifically briefed regarding "where the class is" and what the students need to learn. For example, the resource person who is helping with a unit of instruction on filing income taxes should know current information needed for IRS purposes and what information the teacher has already provided. In essence, the teacher and the resource person team up to provide the best possible learning outcomes in the time available.

The resource person may address a specific problem and help the class find the best solution. For example, the resource person may teach the class how to use several methods of calculating depreciation. In such a role the resource person would be handling the entire problem solution (even though the resource person is in charge of the class at that time, the teacher is always present in the room). However, the resource person might also be used in other ways to accomplish different objectives.

For example, an agriculture teacher might use the resource person to supplement what the class is learning. This additional information could provide additional factual information, such as explaining to students how to make use of highly specialized tax forms or how to average income. It could also be adding new insight, presenting a different perspective, or adding richness to the study of the problem. This richness could come from citing actual examples of tax audits on clients with poor records versus good records and the outcomes of such audits.

The resource person could also be used to clarify. When studying a complex problem, such as animal diseases, students are often able to gain a clearer understanding when a well-selected resource person contributes to the class's deliberations.

A resource person might also be used to endorse a process or practice. When a naturalist who has used a new practice attests to its value, the students are more apt to seriously consider using the practice than if only the agriculture teacher has encouraged adoption of the practice. When a potential employer of students indicates he or she wants workers who do things the way they have been taught, the teaching of the agriculture teacher has been very effectively reinforced.

The Teacher's Role When Using a Resource Person

The most gross error committed by agriculture teachers in making use of resource people is that they want to invite them in, introduce them, and then sit back with their brain in neutral. Such practice of using a resource person is unsound and indefensible. In using a resource person, teachers must follow the same phases of preparation as when they use any other teaching technique.

Just as if the teacher were going to teach the problem himself or herself, the teacher must decide what objectives must be met by the resource person and these objectives must be clearly communicated to the resource person. Teachers must also be sure the prospective resource person understands the context in which the unit is taught, the instruction that preceded it, and the specific points that need to be discussed when the resource person is in class. One way to accomplish this is to guide the students, in class, prior to the resource person's visit, in developing a list of questions for students to ask during the visit. Unless this is done, there is no way the effectiveness of the resource person can be maximized.

The teacher must contact the resource person to set a specific time for the meeting with the class. At this time the agriculture teacher also communicates the types of information discussed in the previous sessions.

Once the resource person is at the school and introduced to the class, it is the responsibility of the agriculture teacher to be sure the resource person focuses on the previously agreed-on objectives. Furthermore, the teacher needs to promote the raising of questions. The teacher may also politely interrupt the resource person to ensure that information the resource person is presenting is clear to the students. Another role of the teacher involves helping the resource person focus specifically on the situations that exist in the supervised experience programs of the students in the class.

The teacher should devise ways to cause students to apply what the resource person has taught whenever this is appropriate. This does not necessarily have to be done while the resource person is present. For example, the following day students could be asked to write a one-paragraph summary of the major discussions. The point is that the agriculture teacher specifically causes the students to look at what they have learned from a resource person and to devise ways of applying it whenever possible. Such instruction provides an excellent way of summarizing the presentation. Otherwise, the resource person is in and out of the class, and little permanent change occurs in the students as a result of his or her presence.

Of course, teachers of agriculture need to be sure their classes are courteous to resource people. The teacher and students should be on time and ready to learn; students should engage in behavior appropriate for learning; and everyone should respond to the resource person's requests for discussion or examples. Also the students, or a class representative, should send a note of thanks to the resource person the next day (not a week or a month later).

COOPERATIVE LEARNING

Cooperative learning[3] provides an opportunity to transition classrooms and learners into active learning modes. Some goals of cooperative learning include empowering students to assume greater responsibility for their own learning, and developing important group processing and social skills among the community of learners in the classroom.

Cooperative learning is a learner-centered instructional process in which intentionally selected groups of three to five students work together on a well-defined learning task for the primary purpose of mastering content. Cognitively, cooperative learning is designed to assist with time-on-task, to reinforce information processing, and to provide critical thinking and reflective thought.

There are numerous cooperative learning tools. One example is the traveling file. The teacher prepares traveling files by developing one question related to the current unit of instruction for each traveling file (a total of five questions if there are five cooperative learning teams). The traveling file is a manilla folder. The tabs of the folders are labeled "Learning Team 1," "Learning Team 2," and so on until all teams have a traveling folder labeled for their use. On the front of the folder is a list of all the teams, so each team can be checked off as the traveling file "travels" (is circulated) between teams.

The teacher prepares one traveling file for each cooperative learning team (enough for about three to five students per team). During class the teacher distributes one traveling file to each cooperative learning team. Each team chooses a recorder. They discuss the question, and as they discuss, the recorder, with the team's assistance, prepares a short written response and includes the response in the traveling file. After ten to twelve minutes (depending on the nature of the question in the traveling file) of discussion and writing time, students check off their team name from the front of the traveling file, and pass the traveling file to the next cooperative learning team, so that each

---

[3] Adapted from Karre, I. *Busy, Noisy, and Powerfully Effective*. Department of Speech Communication, University of Northern Colorado, 1991.

team gets a new file and a new question. The new team now responds to the question in the file that just "traveled" to them. Additional resources may be used for the students to research questions before recording their responses in the traveling folder. After teams have responded to the question in two or three traveling files, the teacher reads and discusses the questions and responses in the traveling files as a class. The teacher will lead the students to compare and contrast responses, and to ultimately draw conclusions about the questions.

This and other cooperative learning techniques are appropriately used for reviewing for an exam, gathering data for a project, concluding a unit of instruction, establishing a felt need to know more information, teambuilding, icebreaking, or researching a topic.

FACTORS TO CONSIDER IN SELECTING GROUP TEACHING TECHNIQUES

Because there are a number of possible teaching techniques to use when the agriculture teacher uses a group forum, the teacher must decide which specific technique will benefit any given period of instruction. Following are some factors that need to be considered in making such a decision.

Nature of the Subject Matter

One way of considering the nature of the subject matter in determining which group teaching technique to use involves assessing the domains of learning (cognitive, psychomotor, affective) to which the subject matter is directed. For example, if we are teaching subject matter in the cognitive domain, such as the parts of a plant, certain teaching techniques will be plausible whereas others will not. In the case of teaching the parts of a plant, a field trip or role-play is probably not the best choice of technique. However, lecture, discussion, or using a resource person could readily be used to teach the factual information (subject matter) that is needed by the students to solve a problem.

If the teacher is teaching students how to mix a soilless medium or how to prepare Petri dishes for experiments, both of which are psychomotor abilities, a demonstration would probably be in order. Certainly, the demonstration could be given by a resource person or on a field trip, but the technique being used would be the demonstration. The demonstration would probably be supplemented with some lecture and discussion (and of course, practice!).

In the case of teaching in the affective domain, demonstrations would probably be of little value but a role-play might be a very appropriate choice. Using a good resource person or taking an appropriate field trip could further certain affective abilities.

At any rate, as teachers reflect on what is being taught, certain techniques are logical choices. Likewise, other techniques clearly are not plausible.

Objectives to Be Achieved

By the time teachers develop specific behavioral objectives that will be somewhat analogous to the questions to be answered, they already know the domain of learning in which they are operating. Furthermore, they are clear as to the specifics that must be covered.

These specifics, or key points, will further guide teachers in the selection of an appropriate teaching technique. If we want students to identify specific ornamental shrubs, the students should be shown these shrubs (field trip or illustrated lecture) and provided with key information to help them remember the shrubs (lecture, discussion, or resource person).

However, if we want students to be able to design a nature trail for their school laboratory, then group techniques may not be appropriate beyond a given point. Rather a decision would need to be made to use one or more of the individualized teaching techniques presented in Chapter 7.

Nature of the Learners

If teachers only considered the first two factors that were presented (nature of the subject matter and objectives) and failed to consider the nature of the students in the class, many poor decisions on technique would be made. The background experiences of the students greatly influence what technique is best. If a class has a rich background of experience in forestry and the teacher is teaching control of forest insects, it is reasonable to plan to use considerable discussion. However, if the class is studying meiosis and no one has studied meiosis previously, little could be gained from class discussion.

Another student factor involves how vocal the group is. If students are shy and do not wish to verbalize much, this could affect the teacher's choice of techniques. Attention span of the students would also influence the choice of technique. If the class has an attention span of only five minutes, using more of a variety of techniques would be in order. Another example of the nature of the learners influencing the choice of technique would be whether they are trustworthy. Teachers who have students they cannot trust will probably take fewer field trips. If the students are "hams" (born actors), more role-playing is in order. And so it goes. Teachers' decisions of which techniques to use must be made with the students in their class in mind.

Resource Material Available

If transportation is unavailable, field trips are out. But the teacher, with the help of the advisory committee, FFA alumni members, and others, must work to gain the right to use this teaching technique. If there are no appropriate resource people available, this technique will not be used much. However, appropriate people can be identified with the help of community leaders. In the case of lectures, discussions, role-plays, and demonstrations, they cannot be enriched with supportive media if such media are not available. But here again this must only be a short-term limitation, for successful agriculture teachers must also be successful program builders.

Amount of Time Available

Time is a practical constraint. If there is not enough time to take what would be a very valuable field trip, then that technique ceases to be a viable possibility. If a teacher waits too long to engage a resource person, that technique is no longer a real possibility. If it takes more time to conduct a good role-play and analyze it than the teacher has, another technique has to be selected. Thus, teachers must plan ahead, anticipate, and use time efficiently; otherwise, they will be dependent on so few teaching techniques that the results will be unacceptable.

The Teacher's Preference

Perhaps of all the factors affecting the choice of the teaching technique to use, the most persuasive one is teacher preference. If a teacher feels uncomfortable using role-plays, he or she will probably elect not to consider a role-play as a suitable technique. Hopefully, teachers will learn to master the ability of planning and use of the various techniques so that their students are not hampered by teachers' reluctance to master at least the basic techniques. Likewise, teachers who wish not to be hassled with setting up field trips use very few field trips, yet such teachers must resist the temptation to abandon a good technique just because it requires extra effort on the part of the teacher. So, in a real sense, the wishes of individual teachers greatly influence their decisions as to what group teaching techniques they will use to help their class solve a problem as a group. Likewise, teachers' determination to be master teachers is a desire that is so potent it can become the controlling desire for excellence in teaching.

SUMMARY

There are a number of group techniques from which agriculture teachers can select to provide meaningful instruction for their students. Teachers need to use a variety of techniques, each of which must be appropriate for the situation in which it is used, so that instruction given in the group setting is dynamic and interesting to students.

Teachers need to be sure that technique does not supplant purpose. The real goal is to provide students with the knowledge, skills, and attitudes that they must have in order to enter the agriculture industry or postsecondary education and compete successfully. To reach that end, students must be able to use problem solving in future life situations. Thus, the agriculture teacher

Chap. 6 Group Teaching Techniques **149**

must use the teaching techniques presented in this chapter in a problem-solving fashion.

It must be emphasized that none of the teaching techniques discussed herein will be very effective unless the teacher observes the principles of learning presented in Chapter 2 and unless the teacher plans for the instruction that the technique is capable of providing.

FOR FURTHER STUDY

1. Visit an agriculture department. Make note of each teaching technique used. What purposes were addressed? What outcomes were experienced? What changes in the use of the technique are desirable?
2. You are to teach students how to calibrate a sprayer. What method(s) will you use, and what are is your reasons for selecting the method or methods you selected?
3. Outline the procedure a teacher should follow in order to give an effective demonstration.
4. List five techniques you could use to generate a high-quality discussion.
5. Your superintendent has issued a memorandum indicating that field trips can no longer be approved unless they are essential to the promotion of the learning outcomes specified in the teacher's course of study and are shown to be educationally sound. How would you go about justifying your need to have selected field trips approved?

7

Individualized Teaching Techniques

Consider the following teaching episode.

> The agriculture class had been studying the unit "Selecting and Using Hydraulic Machinery." The specific question being answered focused on the case of a particular student in the class. His name was David, and the question was, "What plow should David buy for his moderate-size tractor?"
>
> During the previous class the teacher had used group teaching techniques consisting of lecture and discussion with PowerPoint and handouts. On this day the teacher reviewed, through discussion, yesterday's findings and placed on the chalkboard a summary of factors to consider in selecting the proper plow for the job. The teacher then had David list on the board his tractor size, number of acres his family plows, tire width, type of soil, and types of fields they plow.
>
> The teacher then helped the students review the importance of these factors to David's specific choice, based on their previous study. He then suggested that "we" ought to refer to the agricultural engineering bulletin that "we" used earlier in this unit of instruction and specifically the charts on pages 19 and 20 in order to help David choose the best machinery for his situation.
>
> At that point he divided the class of twenty-one students into study groups of three and had a student give everyone a copy of the agricultural engineering bulletin. The teacher then placed the following statement on the board and asked each student to study the tables in the bulletin and complete the statement: "David should buy a _____

plow because _____ ." Another student distributed blue half sheets of paper for students to use to record their recommendations.

Students were again reminded of David's specific factors and allowed to begin the study on which their recommendation to David would be based. As the seven groups studied, the teacher assisted them. The teacher helped some figure out how to read the reference tables and charts. Others were reminded of factors they needed to consider. The teacher reviewed the conclusions previously developed by the class for others. At one point the teacher asked for everyone's attention while David explained whether he planned to trade tractors or change the tires on his tractor. In one case the teacher praised a student for her performance in the basketball game the night before. Of course, the teacher also answered many questions raised by the students.

When the teacher sensed that most students were ready to offer their advice to David, the class was called back together. Then the teacher had the students hand in their sheets of paper containing their recommendations. As the teacher went through the recommendations a student kept a tally. The most popular recommendation was for David to buy a hydraulic lift, five-bottom, 16-inch plow.

The teacher then used group discussion to list on the board the reasons for this recommended decision and completed the blanks in the statement given to the students before they embarked on their study.

The teacher revealed that David had checked with an agricultural engineer and two local equipment dealers. In addition, David had already developed a tentative decision. David was then asked to present these results. All of the advice David had received agreed with the class's majority recommendation.

In concluding the problem, the teacher then had each student use the agricultural engineering bulletin to make the same decision for his or her own situation. Students without tractors or land helped classmates with their individual problems. Once again the teacher supervised each student's progress.

Notice that in this case the teacher builds on some of the group teaching techniques examined in Chapter 6. However, the teacher also adds new dimensions to the students' learning experiences. In this instance, students pursue learning as individuals and in small groups.

It has been pointed out previously that a variety of learning experiences are important if optimum learning is to be fostered. Not only should teachers use a variety of group teaching techniques but they should also use individualized teaching techniques. This provides variety as well as diversified learning.

There are other important reasons for using individualized teaching techniques. In the case presented, each learner had a different situation and, as a way of drawing the problem solution to a conclusion, each learner worked on his or her specific problem.

There are other differences among learners that can also be addressed by using techniques that are designed for use by each learner as an individual. Differences in interests, reading level, speed of learning, background experiences, interest span, learning styles, and learning goals can be better addressed when students pursue learning individually rather than when all the students in a class get the same lesson regardless of whether they need it.

OBJECTIVES

After studying this chapter, you will be able to

1. Explain the reasons for selecting individualized teaching techniques.
2. List and explain the use of the basic individualized teaching techniques.
3. Successfully plan for instruction using individualized teaching techniques.
4. Successfully use individualized teaching techniques.
5. Explain the factors to consider in selecting an appropriate individualized teaching technique for a given situation

WHY USE INDIVIDUALIZED TEACHING TECHNIQUES?

Before selecting a teaching technique that is more individually centered than group centered, a teacher of agriculture should consider why individualized teaching techniques are best suited in some cases. Based on such a consideration, the teacher can match the needs of the students with techniques that best meet those needs.

Individualize the Learning

If there is one truism about learning, it is that each individual is different and should be taken from where he or she is to where he or she is capable of being. Unfortunately, teachers frequently have a difficult time of individualizing instruction such that it actually meets the specific needs of each learner. The ideal situation would be one in which specific goals are established for the course and each learner can achieve these goals in a manner consistent with his or her needs and abilities. For each goal, each learner could have the best materials, approaches, and levels of assistance available and an expert diagnostician would individually prescribe learning experiences. The student would then be individually and personally guided through the prescribed learning experiences. However, such a utopia does not exist. Nevertheless, most teachers of agriculture can use a number of basic individualized teaching techniques with at least some variety of learning resources and personal assistance to help students learn as individuals.

Foremost in the agriculture teacher's mind should be the realization that learners are different. They have different learning styles and varying levels of aptitude, interest, energy, previous experience, opportunity to use given information, and long-term goals. Although teachers may not have precise measures of the extent to which these differences exist among students, they do have a general sense of the nature of these differences. Of course, there are some areas of difference for which there is specific information available, for example, reading level, general intelligence, verbal ability, and quantitative ability. Wise use of such information should be made.

Students also have different needs, calling for individual or at least varying treatment in the development of the learning process they are to follow. Students' goals differ. Some may want to become purebred animal breeders, whereas others only want and need to know the basics of animal breeding. Some have extensive projects in an area and need to go beyond the basics the teacher typically teaches. Then there are students who have a unique interest for which the teacher will want to provide special opportunities for them to study, but the whole class need not study that specific area. For example, there may be one student in a horticulture class that is tremendously interested in orchids. This student may already be growing a few orchids at home and desires to study in depth the cultural practices that are currently recommended. Such a desire should be accommodated if possible. In a production agriculture class located in a community that has almost no poultry, there may be a student who has an insatiable interest in poultry and, in fact, is quite involved with poultry at home. Here again this student's need for studying poultry is quite different from the rest of the class. This need can be addressed using individual teaching techniques. The prudent agriculture teacher recognizes these individual differences and meets as many as possible.

Help Students Learn to Inquire into Subject Matter

Another laudable goal of individualized teaching techniques is that they help students learn to inquire into subject matter. The principle of learning to which this refers is important because it increases learning. Teaching must help students learn more and better.

Whenever most individualized teaching techniques are used, students are individually involved with seeking information. They are not cast into roles of waiting for the teacher to tell them what they need to know. They are not limited by what someone else decides to tell them. Rather they control their own learning. They discover as much as their current level of capability allows.

Through teaching techniques aimed at the individual rather than the group, students gain valuable practice at inquiring into subject matter. They thereby begin to instruct themselves and become independent problem solvers.

Promote Independence

A primary aim of all teachers should be to help students become increasingly independent of the teacher. Otherwise students leave school with the mind-set that teachers are the source of knowledge; teachers solve problems; teachers make decisions. Obviously teachers do not want to foster such a mentality.

The best way to avoid this dilemma is to ensure that students gradually learn how to learn in ways other than directly from the teacher. Using individual teaching techniques, this goal can be realized. As students begin to work individually and have precalculated successful experiences, they become increasingly less dependent on the teacher as a source of knowledge.

Learn to Make Use of More than One Opinion

When students pursue learning using teaching techniques designed to accommodate individual learners, they quickly discover that there are many sources of knowledge and many opinions to consider in solving their problems and arriving at individual decisions. This is crucial to the future success of students. If such strategies are used in solving problems in school, it is likely that students will follow similar models later in life. Thus, instead of functioning with one opinion or source of knowledge, students will learn to consult a variety of sources of information in arriving at important decisions and in solving the problems they encounter in life.

Learn How to Evaluate and Apply Information

Teachers of agriculture should teach students how to evaluate and apply information using individualized teaching techniques. No teacher wants students to leave a program routinely accepting whatever information is presented. Therefore, as a part of teaching students how to learn independently by using individualized teaching techniques, the teacher must teach students how to properly evaluate factual information.

Teachers should teach students to distinguish between objective information and subjective information. Textbooks, experiment station bulletins, extension documents, and curriculum material service references provide objective information. Such information has been developed based on objective data and analyses.

Subjective information is highly opinionated and has little basis in fact. Examples of subjective information include hearsay stories, personal claims of others, some Internet sites, and any other information that cannot be supported by specific documentation. Teachers also need to teach students to be wary of accepting information in advertisements and promotions without thoughtful evaluation.

Clearly, the Internet is a valuable source of information. However, caution must be exercised in selecting the information that we use from this source. Check the validity and reliability of all sites before judging the information to be objective and factual.

Teach Students How to Learn

A final role of individualized teaching techniques involves teaching students to learn independently; the teacher is not always available to personally assist them. There are many times in the classroom and laboratory when the teacher cannot immediately assist a given student. Granted, the very nature of providing for individual needs requires teachers to provide individual assistance to students. However, while the teacher is helping one student or a small group of students, the remaining students must be able to continue to learn and to behave in an acceptable fashion.

Students develop this ability when they practice using individualized teaching techniques and realize that they cannot simply wait for the teacher. Rather they must move on to other tasks until the teacher is available. Otherwise, the class would always be in chaos because while the teacher helps one student the remainder of the students cease to learn until the teacher can attend to them.

BASIC INDIVIDUALIZED TEACHING TECHNIQUES

In this chapter only the fundamental individualized teaching techniques that are frequently used are presented and explained. Readers are given insight into the purposes for which they are suited, explained how to plan for using them, and provided practices to follow in the use of each. The techniques presented include supervised study; experiments; independent study; the use of student notebooks; and using information sheets, assignment sheets, and skill sheets.

SUPERVISED STUDY

Case 7–1A. *If You Had to Teach One of the Following Problem Areas, How Would You Do It?*

Questions to Be Answered
1. How much fertilizer (and what analysis) should be applied to my corn?
2. Which Christmas tree varieties should we start this year to maximize our market potential?
3. What should be the timing setting on the tractor I am working on in lab?

Whatever your teaching specialty, or whichever question you selected, there are obviously any number of ways you could proceed to answer the question. However, in each case, there is an element that suggests that learning where to find the current information makes the most sense, either because each learner has a different situation or because the specific answers change from one year to another.

In order to analyze the charge that was given, question 3 from Case 7–1A will be used, and a daily plan that can be followed to teach its solution will be developed. The reader is encouraged to sketch out a personal daily plan before studying the authors' proposed daily plan.

Case 7–1B. *Proposed Plan for Case 7–1.*

If You Had to Teach One of the Following Problem Areas, How Would You Do It?	
Directions to Self	**Key Content**
List on the board the makes and models of the tractors currently in the lab	1. John Deere 2. International 3. Ford
Distribute handout 3–1	Makes and models of student-owned tractors (attached)
Ask students	"What is the correct timing for each of these tractors?" Possible student responses "Don't know" "It differs"
Ask students	"Why do timing specifications differ?" Possible student responses "Models and types of tractors" "What tractor is used for"
Ask students	"Can we use a rule of thumb?" Student response, "no."
Ask students	"Why not?" Possible student response "Won't work for every tractor" "Not specific enough"
Ask students	"Then how will we be able to have the answer to this question whenever it arises?" Possible student response "Use operator's manual"
Show Transparency 3a	Content is written on transparency.
Distribute manuals	OK. Let's each look up the specifications for your tractor in lab or at home
Assist each student	(*Note to the reader:* Early in the year the teacher must teach students how to efficiently use all the types of references to be used in the course.)
List several timings on board	Pull from page 5 in owner's manual (see attached notes 3–N–1)
Brainstorm on board a list of places students can locate this information whenever they need it	Possible student responses "Tractor dealership" "Extension office" "Owner's manual" (see attached notes 3–N–2)
Develop on board a list of problems created if they don't take time to look up the specifications	Possible student responses "Tractor won't run" "It will damage other parts" (see attached notes 3–N–3)

Analysis of Solution to Case 7–1B. Some group teaching techniques were used (see Chapter 6), and there was also time when each individual studied or learned on his or her own. The individual study used in the preceding solution is called *supervised study*. It allows each student to learn to use basic reference materials, to find answers for themselves rather than depend on the teacher as their source of knowledge, and to obtain the specific information they need without being burdened by the information needed by everyone else in the class. It also allows the teacher to have some time to work with each student individually.

But what does it take to make the use of supervised study a success? This question can best be answered by considering this sequence: planning the supervised study, conducting the supervised study, terminating the supervised study, and developing conclusions based on the supervised study.

Planning the Supervised Study

If a supervised study is to be effective, the teacher must plan it just as any other technique requires teacher planning. Teachers must identify what they wish to accomplish through such study, decide on the resources to be used, be sure of the subject matter outcomes that are desired, and plan how they will conduct the supervised study. The objectives selected must be achievable via supervised study. Most of the objectives to be achieved through supervised study will be cognitive or affective. However, in the case of the development of psychomotor skills, a demonstration (a group technique unless it is shown to one student at a time) followed by actual practice by the student (the individual practice could be considered as supervised study) could be used. In the case of the psychomotor skills there could be prior study of a supervised study type that would prepare students for a demonstration followed by actual practice by the students.

Teachers also need to plan ways to gain the interest of students and create, on the part of students, a "felt need to know" (see Chapter 4) so the students will be willing to invest time and energy in finding answers such that they remove the felt need and actually solve the problem. Following the procedures described in Chapter 5 regarding the interest approach may help. For example, a teacher wanting students to solve the problem, "Which timber stand improvement practices should we use for our school woodlot?" might generate interest as follows.

Interest Approach. (1) "The FFA has been given the authority to manage the school laboratory woodlot however we see fit. Any earnings from improvements will be shared equally between the school board and the FFA. Our class has the responsibility of deciding what practices to use and then following through during class and lab time." (2) Lead students to realize they do not know what the best practices are. (3) Suggest gathering information and developing a plan of action.

The teacher should plan the lesson with structure and direction. This is usually accomplished by developing study questions to guide students as they seek information needed to solve the problem. Oftentimes, a problem is too broad to be the only thing students focus on as they gather factual information. By breaking the study down into a series of phases, through the use of study questions, students experience success and enjoy supervised study. In operationalizing supervised study, a teacher may decide to use group discussion to develop the questions for the supervised study. Nevertheless, teachers need to know in advance the study questions they wish to draw out through group discussion; therefore, they still need to have them recorded in their unit of instruction. In the example of timberstand improvement practices, study questions might include.

- What are the recommended timberstand improvement practices?
- What factors should be considered in deciding which practice or practices to use?

Another planning decision is what type of supervised study should be used. There are at least six basic types of supervised study that could be used:

1. All students study the same problem using the same reference.
2. All students study the same problem using different references.
3. Small groups of students (often called study groups) study the same problem using the same references.
4. Small groups study the same problem using different references.
5. Small groups study different problems using the same references.
6. Small groups study different problems using different references.

The decision to have students study individually or in groups will be based on several criteria. If references are limited, the teacher may have to use group work. Perhaps the teacher believes that there are many times when students need to practice working in small groups and arriving at small-group decisions because such activities are so much a part of the real world. Sometimes teachers have students work individually so they can foster independent inquiry. Times like these allow the teacher to give specific attention and to help students as individuals.

The choice of uniform or varying references may have to be made based on what is available. However, many times different references are purposely used to allow the teacher to better match the reading level of the material to the students' reading needs, to challenge the more capable students, or to provide students with the opportunity to learn that there are a number of sources of information, or to learn that everyone does not agree on how some things should be done. In the case of the timberstand study questions, the teacher could go either way, depending on the secondary learning outcomes he or she wished to promote.

Teachers also need to be familiar with the resources available for supervised study in order to make good planning decisions. The major materials the teacher needs to effectively conduct supervised study serve as references for

Chap. 7 Individualized Teaching Techniques **159**

students to use in their study. Printed materials are usually associated with supervised study. However, one needs to realize that supervised study could be set up and conducted using a field trip, an experiment, a resource person, a film, a slide presentation, an Internet activity, a telephone interview to gather information, or any number of other ways. The key notions are

- The problem is clearly defined.
- Students understand what information they need to discover.
- The teacher directs their efforts toward the desired end.

The authors of this book do not subscribe to a narrow view of supervised study as consisting of only reading in class, but rather they argue that supervised study can involve essentially any resources from which students can learn. The resources listed in the accompanying display can be used effectively with supervised study.

Printed Materials

- Books
- Curriculum materials known as student references
- Extension bulletins
- Commercial flyers
- Magazines and journals
- Activity guides
- Job sheets
- Internet printouts

Visual Materials

- Videos
- Slides
- Pictures
- Transparencies
- Internet sites
- CD-ROMs
- DVDs

Audio Materials

- Tape recordings
- Telephone inquiries

People

- Resource people
- Talking to experts in the field

Experiential Resources

- Field trips
- Experiments
- Observations (such as field studies in natural resources)

Additionally, teachers need to develop and write into the unit of instruction the conclusions students must arrive at through supervised study. This means the teachers need the correct answers to the study questions as well as final conclusions for the problem. Otherwise the discussion that must follow a period of supervised study will be too vague and fragmented. In the timberstand example the conclusions to the problem might be listed in the unit of instruction as follows:

Case 7-2. *Listing Possible Student Responses.*

Directions to Self	Key Content
Develop conclusions with class:	
Write on blackboard	Types of practices recommended for our area: (Possible student responses) 1. clear cut 2. selective cut 3. leave natural for wildlife habitat
Show Transparency 3	Factors to consider: 1. topography of the land 2. traditions in the area 3. environmental considerations
Brainstorm options on board and choose one	Our decision is, "we should _____ in the woodlot." (To be developed by students based on the facts relative to their woodlot.) (Possible student responses) 1. cut down the trees and sell them 2. leave the trees natural 3. Sell only the black walnut trees
Tell students	Write this in your notes with your rationale.

Conducting the Supervised Study

Although thorough planning is essential to having a good supervised study, teachers must also use sound practices in the actual conduct of supervised study. The first important activity in conducting quality supervised study is to get students started in a smooth and purposeful fashion. This is an impossible task unless students have been taught how to use a supervised study session. The following suggestions can be used in teaching students how to use supervised study.

Suggestions to Follow in Teaching Students How to Use Supervised Study

1. Teach students what resources are available, where they are located, how to find them, and when and how you expect them to be returned.

2. Teach students how to use basic printed reference materials. This requires teaching students how to effectively use a table of contents and an index. In addition to a verbal explanation of how to use these two tools, try planning several exercises whereby students need to find a reference and locate certain material. To add a little excitement, let the class compete to see who can complete the exercise in the least amount of time. Following the first exercise or so, show the students how you would have located the desired information, thereby giving them a frame of reference against which to compare the procedure to follow.

3. Teach students to access search engines on the Internet and to locate information on the World Wide Web.

4. Teach students how to record notes. If students are not taught how to take notes during supervised study, they tend to write down everything verbatim. You must help students learn how to glance at headings, skim the passage, and identify the major concepts. Then the students need to learn to record only the most pertinent key points that apply to the question on which they are working. A good way to help students master the ability to select only the key ideas to record in writing is to provide a practice exercise. Have them look at the appropriate reading and, as a group, discuss what they believed was important. You can distribute notes to the class that indicate what information you would have recorded. Following the next few assignments you could again distribute model notes against which the students can compare their efforts. Early in the year the students' notes should be collected and evaluated (not necessarily graded in these initial trials) so students develop good note taking techniques. Unless this degree of emphasis is given to note taking, supervised study will never reach its potential. Teachers also need to give students supervised practice at taking notes when viewing a video or taking a field trip.

5. Furthermore, students need to learn what to record in their notebooks and how to record it. Once the class has developed conclusions for a problem, these class conclusions also go in the notebook and are labeled as such.

6. Finally, teachers need to teach students how to report the findings of their supervised study in class. This is also the beginning of teaching students how to speak in public. It is rather routine when students simply report their answers to a study question either on the board or orally. However, teachers often have students complete a more detailed study and in varying subareas of the problem or problem area. In these cases students need to give minioral reports. They need to be coached on how to do this effectively.

Once students begin their supervised study, the teacher needs to focus on the task of supervising the study. Actually, the teacher is overseeing and directing the learning of each individual student.

Suggestions to Follow in Supervising the Study of Individual Students

- Stay in the room. This is not a time to check the mail, place phone calls, or visit with other teachers.
- Direct the study of students. Do not try to plan lessons, grade papers, or attend to correspondence. During supervised study, the teacher's responsibility is to direct the learning of each student, and this is a full-time responsibility.
- Observe students as they work. Spot problems and offer help. For example, if a student has great difficulty locating pertinent information, offer assistance. If a student has trouble reading, provide help (you or another student can read with the student).
- Keep students productively involved. Nip misbehavior in the bud. Keep yourself positioned so you can see most of the class whenever assisting a given student.
- Question findings; challenge more capable students.
- Use this as a time to offer recognition to students who crave attention. Pass on an encouraging word to students who are feeling down.
- Spot situations in which students are on the wrong track and correct the problem.
- Ask why. Probe.
- Point out poor note taking and help the student do better.

If teachers attend to tasks such as these, supervising the students' study will be effective and will take all the time available to the teacher. If these guidelines are ignored, supervised study will be chaotic and ineffective.

Providing for Individual Differences. During periods of supervised study, teachers have an excellent opportunity to provide for individual differences. If there are students with differing reading abilities, the teacher can assign a resource that is appropriate for each student's ability. Students who are poor readers might use a simple leaflet, whereas more accomplished readers might use advanced reference works or journal articles.

Another way of providing for differences is in the scope and depth of the assignment. For students with limited previous experiences in the area being studied, basic questions may be assigned. In the case of students with considerable background, questions that require synthesis may be assigned.

In the case of both reading ability and extent of background, the teacher may elect to mix students who read well with those who need help reading, so as to let students help one another.

Supervised study is a good time to provide individual instruction for the student who learns at a slower rate than the others. It is also a time to counsel, to encourage, and to otherwise address personal and emotional differences of students.

Terminating the Supervised Study

Teachers who actively direct learning during a period of supervised study will have a very good sense of when to terminate the study period and move toward obtaining closure on the problem. Seldom can one wait until everyone finishes, and yet one dares not conclude the study period as soon as the first few students have completed their assignment. The middle ground must be worked. That is when most students have answered the study questions or otherwise completed the assignment, terminate the study. Also, consider having additional work ready for students who finish early. In this way, all students continue to be actively engaged during the entire period.

Developing Conclusions Based on the Supervised Study

Once the students have completed their supervised study, either individually or in small groups, it is important for the teacher to work with the class to develop final conclusions. This is particularly true in situations in which the whole class is working on a common problem or problem area. If this is not done, there is not enough opportunity to clear up misconceptions, elaborate on crucial concepts, or put the class's findings together to determine the most accurate and complete conclusions.

This time of developing conclusions with the class allows for students to express themselves as well as to listen to the viewpoints of others. It also gives the teacher a good opportunity for quality instruction based on the students' informed study. Of course, the development of these conclusions provides an excellent summary for the problem as well. However, good lead questions must have been developed by the teacher previously and written into the unit of instruction.

As the teacher works with the class to develop conclusions, it is important that time be taken to point out specific practices learned in the supervised study that students need to use in their experience programs. Such practices will later be summarized in the unit of instruction as a list of approved practices for the problem area.

But the teacher must help the students develop another important dimension to their conclusions: general principles. It is not enough to only conclude the specific facts learned or how certain problems are solved. It is also crucial to be sure the students understand *why* certain concepts are true, *why* certain practices are best, and *why* certain information learned in one problem area is applicable to other situations.

Recall the principles of teaching and learning related to transfer of learning (see Chapter 2). It is the responsibility of the teacher to guide students in understanding why certain answers or solutions are correct because understanding why means that underlying concepts and principles are made explicit and their meaning understood. When these procedures are followed in using supervised study to solve problems, students will learn and teachers will be successful in their teaching.

EXPERIMENTS

Consider the examples in Cases 7–3, 7–4, and 7–5.

Case 7-3. *Baby chick experiment.*

> A teacher in a small animal care program was beginning a unit of instruction on balancing rations. The students did not believe they needed to worry about learning how to balance rations. Their view was that it did not matter as long as you give the animals enough food.
>
> Rather than arguing about it, the teacher said, "OK, let's see." Louis, one of the students, was then instructed to purchase two baby chicks at the local cooperative and was given a purchase order. The next day Louis came to school with two baby chicks, and the teacher had him put them in separate cages located side by side.
>
> The teacher prescribed (and furnished) a balanced ration for chick A. The students were allowed to develop whatever feeding plan they chose for chick B. The students decided to feed chick B the food they enjoyed. So they gave it potato chips, cola, and milk chocolate candy.
>
> Then the class developed a record sheet for each chick to be used each day in recording the date, the food the chick received, how much it consumed (grams and milliliters), and a description of each chick's general appearance. Within a matter of days the students asked the teacher to call off the experiment.
>
> Chick B was not looking healthy. Its wings were drooped, its legs were weak, and it could not open its eyes or hold up its head. The chick on the balanced ration had gained weight and in general looked healthy. The class concluded on its own that diet makes a big difference. The students readily agreed they needed to learn more. The use of a simple experiment had created a felt need to know more about animal nutrition.

Case 7-4. *Feed testing analysis.*

> Later in this problem area the same teacher had the students each take a measured quantity of gerbil food and separate out the various ingredients. Each component was identified with the help of the teacher and selected reference materials, then weighed. The students calculated the percentage of the food that each ingredient provided.
>
> The teacher then led the class in deducing the formulation of the feed through testing with chemical kits. By using this quantitative experiment, students discovered specific factual information through their own inquiry.

Case 7-5. *Poinsettia experiment.*

> When a horticulture teacher was challenged by three students as to the necessity and value of shading poinsettias, the teacher did not argue. Rather, she had each of the three students select one poinsettia from the crop and separate it from the rest, which were shaded. When their plants were the only three not showing color for Christmas, it was clear to them that the practice of shading was necessary and the reasons for it as taught in class were valid. Thus, they were able to decide for themselves what practice was best. As a result they changed their behavior.

Now, consider these examples of using experiments to promote learning. What is the value of having students use experiments as a way of learning?

The Value of Using Experiments

When students conduct experiments, either individually or as a group, there is a high degree of real involvement and hence interest. Students who are actually involved in their learning learn more and better. In the case of conducting an experiment, students must use their minds. In addition, they are also physically involved in setting up the experiment, making observations and collecting data, and developing their conclusions. They are, in essence, creating their own information. When this is done well, it is a very success-oriented and student-centered form of learning.

Another value of experiments as a teaching technique is that it is a way of graphically illustrating important concepts, theories, or approved practices. When students try different ways of growing a plant, shrub, animal, or bird, and then are able to see the difference one way makes over others, a very lasting impression is formed. Students no longer have vague ideas about the problem being studied. They have a very specific understanding of the results of solving the problem in various ways.

The use of experiments in teaching agriculture also helps students learn to think systematically. In teaching students how to conduct an experiment, it is very important that the teacher emphasize the importance of being systematic, clear, and thorough. Students learn to think things through, to be specific, to keep careful records, and to be precise in carrying out their plans. They also learn to proceed in an orderly, step-by-step fashion. These are important traits for students to develop as a result of their schooling.

A final value of using experiments in teaching agricultural subject matter is that it causes students to have to study the facts they gather and then draw conclusions. They learn that conclusions cannot be properly developed without considering all the facts. They also learn that a conclusion is different from the facts. In other words, one must study the facts and then develop a conclusion that is based on the facts. For example, if a student in horticulture decided to see if using bottom heat when propagating chrysanthemum cuttings was superior to propagating without bottom heat, the student would have to collect data on which to formulate a conclusion. The data may reveal that the mums with bottom heat had a 90 percent rate of rooting, whereas those without bottom heat had a 65 percent rate of rooting. Based on this data the student could conclude that the use of bottom heat gives one a higher percentage of rooting than if bottom heat is not used. The facts were not reiterated; rather, a basic conclusion was advanced. Notice that the conclusion was specific and did not go beyond the evidence on which it was based. The student did not conclude that the use of bottom heat was best and rightly so because, for example, there were no data available on the added cost of using bottom heat. Such a conclusion would have gone beyond the scope of the data. When students develop sound conclusions based on the analysis of the pertinent information that is available, an important ability has been developed.

Planning for Using Experiments as a Teaching Technique

When teachers decide that they have an objective or a problem with which an experiment can best be used to meet the objective or solve the problem, then the nature of their planning for the students to meet the objective or solve the problem is quite a bit different from their planning to conduct a supervised study or give a demonstration. In the case of planning a supervised study or demonstration, for example, a teacher records key points to be developed and how to go about developing this information. In the case of planning a problem solution when an experiment is the teaching technique, the teacher needs to plan how to direct the process to be used by an individual student, if only one student will do an experiment, or a process to be used by all of the students, if the experiment is being done by the whole class or specific groups within the class.

Thus, the basic planning teachers need to do is to identify an appropriate experiment or experiments that will provide the information needed to solve the problem. They then need to outline the procedure to follow in conducting the experiment (see Case 7-6) and develop a list of supplies and materials needed. Teachers should also plan a system for gathering the data. It would also be wise to jot down the basic findings and conclusions the experiment is intended to reveal to the students. Perhaps the sample plan in Case 7-6 will provide a model to consider when planning the use of an experiment as the technique for teaching the problem solution.

Case 7-6. *Sample of the Type of Plan Needed to Use an Experiment.*

Directions to Self	Key Content
Materials needed: three engines, timing lights, wrench set	**Question to be answered**: Why do we need to time an engine to a specific degree of accuracy?
Conduct trial discussion to see how much the class knows (Create divisions of opinions as to reasons why careful timing is needed.)	"John, what do you think?" "Jodi, do you agree?"
Suggest that students try several ways and learn firsthand why accurate timing is important	
	Anticipated Findings
	Tractor / Timing / Timing / HP After Timing HP Before Timing 1 50 specs 50 2 50 10° up 45 3 50 10° down 46
Ask students	"What is the horsepower output of engines?" • Then time one engine specifically as outlined in the manual. • Then time one engine 10 degrees before manufacturer's specifications. • Then time one engine 10 degrees after manufacturer's specifications.
Ask students	"Now, what is the horsepower output of all three engines?"
Compare results (see anticipated conclusions)	"Let's look at what happened."
Draw conclusions (see anticipated conclusions)	**Anticipated Conclusions.** If you time an engine 10 degrees or more different from the manufacturer's specifications, you can expect to lose horsepower.

Procedures to Follow in Having Students Conduct Experiments to Solve Problems

The first procedure for teachers to follow in using experiments to solve problems is to create interest. Suggestions on how to do this are discussed in Chapter 5.

The teacher then needs to help the individual student or the group, depending on the setting in which experimentation is used, plan how to conduct the experiment. The amount of assistance the students need will vary, as is the case with any technique. The teacher's role is to help the students think through the procedures to be followed. Of course, there may be times when teachers choose to distribute a handout that lists the procedures. Whichever procedure is used, it is important that the students are clear as to what the goal is. They should also be encouraged to use reasoning to predict the outcomes of the experiment. This is important because students need to learn to use prior knowledge in anticipating answers to new problems. They need to be willing to predict and then check the accuracy of their predictions.

Once the students know how to proceed and what materials are needed, the teacher supervises the experimentation. This is true for experiments students conduct at home as well as at school. Teachers need to be sure procedures are correctly followed, questions are answered, and data are correctly recorded.

Once the data are collected, the teacher needs to help students analyze it and develop sound conclusions. This may be done on an individual basis or during group work. Whatever the procedure followed, if the problem is of interest to more than one or two individuals, the results need to be presented and discussed with the group.

Reporting the Results of the Experiment

Once the experiment has been completed, and if the information derived from the experiment is needed by the entire class, the student or students involved should make a report to the class. There are several reasons for this reporting. First, it is a way of sharing knowledge; all the students benefit from what was learned in the experiment. Second, the student(s) reporting will learn to sort out the important key points that need to be shared with others. Third, by reporting to the class, those giving the report have one more experience of speaking to and communicating with a group.

If this reporting to the class is to be successful, teachers need to teach students how to develop and give a report. Case 7-7 presents an outline students can use in organizing their report.

Case 7–7. *Sample Format to Follow in Reporting on Experiments.*

FORMAT TO FOLLOW IN REPORTING ON EXPERIMENTS
Title: (What was the experiment about? Keep it simple and clear.) Example—"How does pH affect the growth of my _____?"
Purpose: (What did you try to accomplish?) Example—"The purpose of this experiment was to see if too low a pH would reduce the yield of my _____."
Procedures: (A step-by-step synopsis of how the experiment was conducted.)
Findings: (The important data presented in the form of tables, graphs, or charts.)
Conclusions: (Based on this experiment, what can we now conclude?) Example—"As pH drops below the recommended level, _____."

Once reports are given, the teacher must be sure that what has been learned is clearly summarized. One needs to refer to the problem on which the students were working and make sure the key conclusions are written in each student's notebook.

Using Data from the Research of Others in Teaching the Solution to the Problem

Agriculture teachers must rely on the findings of the research in agriculture if they are to teach current information. From the time research is conducted until it is reflected in basic texts, as many as five years may have passed. In an industry that is as technologically dependent as agriculture, teachers must keep their instruction up-to-date. Once a teacher has students using experiments as a way of learning, it is very natural to also bring in the results of experiment station research and other research as well.

It is best if the teacher relates the basic notion of the experiment in story fashion rather than being overly formal. In other words, put the research in nontechnical terms and at the level of the students.

Once the background, setting, and procedures have been explained, the good teacher will ask the students to guess what happened. This generating of possible answers (hypothesizing) is important because students must develop the ability to analyze a situation and deduce possible outcomes.

Findings should be presented in clear and simple terms. High school students need not be burdened with statistical jargon or research design lingo. Rather, the teacher can merely indicate, for example, that the researchers found that if people do things a certain way they make more money than if they do them other ways.

In presenting numerical findings in the form of charts, the teacher may want to leave some cells blank and have the student estimate the numbers that go therein. This forces students to spot trends and identify relationships. The same notion can be used with a graph. Draw it to a certain stage, then ask the students what they think the remainder of the graph will look like.

Some Suggested Experiments to Use in Teaching Agriculture

Following is a list of ideas for experiments to be conducted in agricultural instruction. It has been developed to help you begin to think of other good ideas for using experiments in your teaching. These experiments may be completed at school or in another part of the students' supervised experience programs.

Teaching Specialty	Ideas for Experiments
Production agriculture	The effect of amperage on strength of welds.
	How does the plant population per acre affect yield?
	What herbicide controls weeds best?
	Which ration is best for a given species of livestock?
	Which practice controls a given disease best?
	How does irrigation affect profit?
	Which fasteners are best for wood?
	How does date of planting affect yield?
Horticulture	What is the most effective way to propagate a given plant?
	The effect of rate of fertilization on plant quality.
	The effect of temperature on flower color.
	Which variety is best for a given situation?
	What is the best way to control a given insect or disease?
	The effect of sterilizing versus not sterilizing soil on plant growth.

Agricultural/industrial mechanics	The effect of temperature on oil viscosity.
	The effect of combine cylinder speed on corn harvest losses.
	The effect of tractor weighting on wheel slippage.
	How does the size of the gap of a spark plug affect engine starting performance?
Small animal care	What is the best temperature for the water in an aquarium?
	The effect of a nonbalanced diet on guinea pigs.
	The effect of a given chemical on internal parasite infestation.
	The effect of temperature on the spread of disease.
Natural resources	How to attract wildlife to a given area.
	Comparison of erosion control practices.
	What is the most effective way to control pond algae?
	The effect of given chemicals on soil life populations.

INDEPENDENT STUDY

Independent study is any form of study that is conducted by an individual student. Supervised study where each student works on his or her own or an experiment that a student conducts alone are both considered independent study. The same is true to a certain extent for the supervised experience program of students outside of school laboratory time. However, there can be good independent study apart from these examples.

The Role of Independent Study

The primary role of independent study is to meet the needs of individual students. Students have different interests, abilities, and rates of learning, and independent study can speak to all of these differences. For the student who learns quickly, independent study offers the chance to go beyond minimums, to expand and enrich one's basic ability. In the case of a student who from time to time cannot participate in a given learning experience because of a physical disability, independent study offers a potentially meaningful alternative. This independent study may take the form of students reading on their own. It could also involve the use of programmed materials, self-paced instructional units, computer-assisted instruction, or any of a number of other alternatives. For example, independent study could consist of job placement, interviewing selected agribusiness people, or doing a laboratory project on one's own.

Just as in the case of supervised study, the use of independent study fosters independence on the part of the student. During the use of independent study the teacher takes a facilitating role. The student is, in essence, in charge of his or her own learning. Learning to learn on one's own is a highly desirable outcome from using independent study as a teaching technique.

Another role of independent study as a teaching technique is that it offers variety. Students get tired of discussion, supervised study, and any other technique if it is overused. If independent study is made exciting and within the range of challenge for each student, it helps greatly in bringing a fresh perspective to the learning environment. However, one must help students learn how to study independently and to experience success from doing it.

How to Promote the Use of Independent Study

The first requisite to promoting independent study is that the teacher be sensitive to unique interests and abilities of individuals in a class. Teachers have to work at knowing individual differences. Individual interests can be determined in a number of ways. We can survey the students to get their list of special interests in agriculture. We also use home visits, FFA activities, and other times when we are with a student or small group of students to learn their particular areas of interest. Of course, every teacher needs to constantly be on the lookout for special interests of students. During the year, a student may encounter a problem in a supervised experience program that warrants time for independent study. This independent study may be carried out at school or at home, in class or lab, or outside of class or lab.

Teachers who decide to promote independent study will need to teach students how to study independently. Many of the study skills associated with supervised study, for example, how to use a table of contents and an index, and how to take notes, will be used in conducting independent study. In helping students engage in an independent study, teachers will need to provide the students with some degree of structure. This structure offers direction and meaning. A good way of providing this structure is to use a format for students to follow in organizing their study. The format in Case 7-8 is suggested.

Case 7-8. *Sample Format to Use for Independent Study.*

What is the problem to be solved?
What are the questions that must be answered?
What references will I use?
What steps will I follow in conducting my independent study?
What will I learn from my study (a summary)?

The teacher needs to suggest independent study for students at times and in ways that make it clear that independent study is a reward and not a penalty. Independent study involves freedom, a chance to do something different from the other students, and an opportunity for students to study areas in which they are most interested. Teachers ought to be willing to allow well-conducted independent study to substitute for what would otherwise be some "normal" work. Otherwise, independent study becomes a burden for many students.

Planning for Students to Study Independently

The teacher planning required for using independent study will be different from the planning needed when using other teaching techniques. As students indicate a desire to do independent study, teachers will need to draw on their personal knowledge of available resources to help each student decide where to look for information related to the independent study. Teachers also need to help develop questions to guide the study. The least teachers should do is to check the list of study questions to be sure the student is following the proper course of action and is being thorough in the study.

Teachers of agriculture should have a place where students can study independently. Such a place can also be used for students to study for career development events during study halls or other free time. A very satisfactory arrangement is to have a study carrel at the back of the room. It should be equipped with an electrical outlet and be capable of housing a slide projector, an audio tape recorder, a CD player with headphones, and possibly a computer.

During the course of the independent study there needs to be interaction between the student and the teacher. Questions need to be aired, opinions offered, and progress checked. Once the student completes the study, the teacher needs to take time to thoroughly evaluate the student's efforts and offer specific comments that indicate what the student did well, as well as areas for future growth.

Making Use of What Students Learn from Independent Study

Independent studies need to involve more than students studying on their own and writing reports. As with all career and technology instruction, it is important that the learning be applied. The teacher should help the student find opportunities in which to use the key points from the independent study. For example, if a student uses independent study to learn how to build a brooder for quail or pheasant chicks, the teacher could have the student follow up this study by constructing a brooder in lab and then raising a batch of chicks to be released as a supervised agricultural experience.

There will also be times when the class will want to hear about the specifics of a student's study. When this is the case, the teacher should provide an opportunity for the student to report to the class. The teacher also needs to advise and coach the student on how to deliver the report effectively.

THE USE OF STUDENT NOTEBOOKS

An essential aid to learning is for students to have high-quality student notebooks. Getting students to have high-quality notebooks requires a high degree of commitment on the part of the teacher. It also requires that the teacher use student notebooks in such ways that students see the value of keeping good notes. When this happens, teachers of agriculture have developed another important teaching technique they can use.

Why Notebooks?

Student notebooks provide a way for students to organize and accumulate their learning. Through the very process of organizing learning, students are able to begin to see how the questions to be answered in each unit of instruction, and how each unit of instruction in the course, fit together to make the whole. As the notes in the course begin to accumulate, students begin to realize how much they have learned. This helps them recognize that the agriculture course is a place where serious learning occurs.

Taking and keeping good notes promotes learning. Consider the educational implications of the following chain of events. At the beginning of the class the teacher focuses everyone's attention on the problem being studied. Students have to locate the logical place in their notebook for today's notes. They see that what is being studied today is connected with what they learned yesterday. Then as the day's lesson proceeds, students hear important ideas; they see key points on the writing surface, overhead projector, computer projection, or other media; and they write these important key points in their notebooks. As they write their notes, they probably repeat them silently and certainly see the key points again as the notes are recorded on paper. This repetition and use of multiple senses increases the amount that is learned and retained.

A final justification for student notebooks is that they become an excellent reference for future use. As students encounter difficulties in lab, in supervised study, conducting experiments, or engaging in independent studies, they are able to refer to their notes for help. This should also be true when they are placed on a job or completing a project. Likewise, notebooks are an essential reference whenever students are studying for quizzes, tests, or examinations.

Suggestions to Follow in Getting Students to Keep Quality Notebooks

The first requisite for quality student notebooks involves teaching students how to set up their notebooks for their studies. The teacher can suggest a common system of organizing the notebooks. It is a good idea if all students have the same type of notebook. In fact, teachers will experience much more success with notebooks if every student has the same type notebook. Suitable three-ring binders can be bought in bulk for the class, and the cost of these can be included in the fees for the course. The teacher also needs to look at the course of study and decide how the students' notebooks should be organized. Tab dividers can be used for the major divisions of the course.

In addition to being sure all students have notebooks with the same subdivisions, the teacher needs to teach the students a format for their notes. Many teachers insist that notes be kept in ink. Such decisions reflect the preference of the teacher. Each day's entry should be dated. The title of the problem area should appear at the head of each new section of notes. Each problem needs to be labeled. Then, under each problem, students should record factual information, and, before moving to a new problem, the conclusion or a decision needs to be labeled and recorded. In order to be sure students use the format desired, the teacher should distribute an outline of the format at the beginning of the year. Teachers also need to distribute sample sets of notes as models of how they want notes to be kept.

Teachers need to devote some time to explaining to students why they want good notebooks. Former students can come to class to attest to the usefulness of a good notebook. Teachers should include notebook grades in determining grades for each marking period. With this kind of importance attached to notebooks, students are apt to be willing to do a good job of keeping an agricultural education notebook.

It is also important for teachers to help students learn to take good notes. This is best accomplished by teaching so as to facilitate good note taking. In essence, anything that needs to appear in the students' notebooks needs to be visually displayed before the class. This could be written on the writing surface, projected onto a screen using various media, displayed on a flip chart, or hung on the wall or bulletin board. The information should be organized on the writing surface precisely the way the teacher wants it to appear in the students' notes because invariably the notes of students reflect that from which they have recorded. If the teacher underscores captions, so do the students. If the teacher uses poor grammar or misspells, so do the students. If the teacher uses a header to organize sections of notes, so will the students. Thus, teachers need to provide the model from which their students' notes will be derived.

Suggestions to Follow in Making Effective Use of Student Notebooks

Students will not have good notebooks unless they are taught to record and use the notes in meaningful ways. It is crucial that teachers refer students to their notes whenever they raise questions that have been answered in their previous studies. This is true whether the question is raised during class or lab. Get students into the habit of referring to their notes as a source of useful information.

Another good way for students to use their notes meaningfully is to allow them to be used as a reference for solving problems and answering questions on quizzes and tests. When notes are used, they are more likely to be well kept. Nevertheless, unless the teacher takes time to do a good job of periodically evaluating students' notes, they will probably do a less than superior job of note taking and note keeping.

Evaluating Student Notebooks

Student notebooks should be graded at least once each marking period, but doing so more often will bring positive results. Rather than wait until the last minute, teachers should systematically grade a few notebooks each day over a period of several days. This way the teacher does a much more thorough job, and students can receive more detailed feedback than when grading is done in a crash effort.

Teachers should decide how much of the grade for the marking period will go toward notebooks. The percentage must be high enough to make a difference in the grade but should not be ridiculously high such that the student can earn a respectable grade from his or her notebook alone. Somewhere between 15 percent and 25 percent of the course grade is suggested. Students should know this at the outset of the course.

Furthermore, a score sheet should be developed and made available to the students and followed by the teacher as notebooks are graded. Scoresheets allow teachers to bring objectivity to the grading process and to provide specific feedback to students as to their strengths and weaknesses. Case 7-9 shows an example of the kind of score sheet a teacher could use.

Case 7-9. *Score Sheet for Grading Agricultural Notebooks.*

SCORE SHEET FOR GRADING AGRICULTURAL NOTEBOOKS		
Categories	Possible Points	Points Received
Neatness	10	_____
Completeness of notes (all days)	20	_____
Thoroughness (pages dated, problems listed, key points recorded, conclusions included, handouts included)	30	_____
Accuracy of information	30	_____
Plans of practice recorded	10	_____
Total:	100	_____

During the first few weeks of school the teacher needs to intermittently check each student's notes to be sure the student is developing appropriate note taking habits and skills. Otherwise, students may be penalized for errors they did not realize they were making. In addition by waiting until the end of a grading period, students will have had time to form bad habits that are difficult to change.

USING INFORMATION SHEETS, ASSIGNMENT SHEETS, AND SKILL SHEETS

Another way of individualizing learning is to have students use specific study sheets that are designed to guide each student's learning experience. These types of sheets are most often used with laboratory learning. Three different types of these sheets are suggested: information sheets, assignment sheets, and skill sheets.

Information Sheets

Information sheets are short handouts that provide basic information (the what, why, and how) a learner needs in order to perform at a particular skill level or to complete a given job. Such sheets could introduce the learner to information for the first time, or more likely they could be used as a synopsis or summary of previous classroom study. Figure 7-1 shows an example of an information sheet.

Assignment Sheets

An assignment sheet is used to tell the learners the assignment(s) they are to complete. This sheet also includes information about how to complete the assignment and how to check one's work on completion of the assignment to ensure nothing was overlooked. Figure 7-2 illustrates an assignment sheet.

Skill Sheets

Skill sheets are much like assignment sheets, except they are more limited in scope. Skill sheets focus on a specific skill and provide guidance to the learner in developing and mastering the skill. In addition to skill sheets, teachers use pictorial guides, illustrations from textbooks, Internet sites, and CD-ROMs to accomplish the guided step-by-step process for which skill sheets are used. Figure 7-3 shows a sample skill sheet.

The Values of These Resources

The primary value of using these resources is that they allow students to learn at their own rate and to progress from one assignment or skill to another when they are ready to progress. These aids are narrow in scope and written at a reading level suitable for the majority of students, and they allow the students to be very actively involved in their learning.

> **STARTING MOTORS AND SWITCHES**
>
> The starting motor is designed to turn the engine crankshaft with enough speed to get the engine started. To get this job done, a large amount of current is needed for a short period of time. The wires, switches, and starting motor must be heavy duty in order to handle this large amount of current. The lightweight switches and wires used in the rest of the electrical system would be destroyed immediately if they were used to carry the large amount of current used by the starter.
>
> Switches used in the starter circuit are always low-resistance, high-capacity units but they may be either manually or electrically operated. Through the use of a heavy-duty switch mounted on or near the starter, extra lengths of heavy-duty wire from the battery to the control panel and back to the starter can be eliminated. The heavy-duty wire can then go directly from the battery to the starting motor switch and the starter. The switch may then be controlled by a mechanical control or by current carried by lightweight ignition wire and a remote switch.
>
> The electrically operated switches use the principles of the electromagnet for their operation. They are designed like an electromagnet except that the core is movable. When the current is not flowing through the coil, the core is held part way out of the coil by a spring. As current flows through the coil, the core is drawn into the center of the coil. When the current is turned "off," the spring forces the core back out of the coil. This electromagnet with a movable core is called a solenoid. The movable core of the solenoid is then connected to the movable part of the heavy-duty starter switch. In this way, the heavy-duty starter switch can be controlled from almost any location merely by using lightweight ignition wire.
>
> Basically, the starting motor, or starter, is a direct current motor designed to provide high power for a short amount of time. It is designed to use a low-voltage source of electrical power, such as a storage battery. Because the voltage of the electrical energy used is low, the amount of current (amperes) used must be very high to produce the amount of power that is needed.
>
> Although the starter is highly complex in design, it works on the same basic principles as the electric motors discussed Unit IV, "How Magnetic Forces Make an Electric Motor Work." The same electrical forces that made the motor you built run make the starting motor do its job. Naturally, the starter is much more complicated but it also produces much more power. If the starting motor is properly maintained, it is almost trouble free in operation.
>
> The various components of a starter are identified in Illustration XI-A.
>
> Annually or after every 500 hours of operation some maintenance work should be performed on the starter. The following operations should be carried out.

Figure 7–1 Information sheet.
Source: Printed with permission of The Ohio Curriculum Materials Service. From the Ohio Agricultural Education Curriculum Materials Service (1970). *Individual Study Guide on Electrical Systems for Spark–Ignition Engines.* Columbus: Author.

Chap. 7 Individualized Teaching Techniques 179

> 1. Remove and check the brushes. If they are worn to less than half of their original length, they should be replaced.
> 2. Remove glaze from the commutator. This can be done easily by using a strip of fine sandpaper held against the commutator while the armature is turned by hand.
> 3. Check for signs of overheating. If such signs exist, for example, solder thrown out of the commutator or charred insulation on the winding wire, take the starter to a service shop for expert repair.
> 4. Ensure that all connections on the heavy-duty wire from the battery to the starter are clean and securely connected. If the insulation on the heavy-duty wire is cracked or shows signs of deterioration, the wire should be replaced.
>
> Starting motors and switches are designed to do very difficult jobs, especially in cold weather. They should be kept in top condition if they are to work properly. When they are used, they must carry a large amount of current. Dirt and worn parts will reduce the performance of the starter and increase the drain they place on the battery.
> Cutaways of two starters are shown in Illustration XI-A.

Figure 7-1 Continued

Another benefit offered by these aids is that they provide structure for learning. The learning task is clearly stated. The outcomes to be accomplished by completing the sheet are clear. The materials that are needed to complete the sheet are spelled out. The steps to follow are logically sequenced, and often they are accompanied by illustrations designed to improve the clarity of instruction. Finally, these aids often provide the students with a self-rating form. Each of these features contributes to bringing structure to the learning activity.

Use for These Resources

One of the best uses of these resources is to teach specific skills. Once students have had specific instruction for a skill and have been shown the basic demonstrations, appropriate sheets (information, assignment, or skill sheets) can be used as they practice the skill. These aids provide repetition of previous instruction and guidance to help the students pace themselves through the activity.

The guidance provided by these sheets is very important in laboratory learning because the teacher can seldom be with each student for as much time as one would prefer. Thus, students use the given sheet for supplemental instruction and are able to progress at their own rates until specific help from the teacher is absolutely essential.

1. Assume that you are working in the floral shop at your school. Given a list of delivery and pick-up addresses, several special deliveries, and a map of your hometown, plan a delivery route.

Resources	Procedure
Map	• Locate general delivery area
Telephone directory	• Locate and mark each delivery address
List of delivery addresses	
Several special deliveries (specific times)	• Locate and mark each pick-up address
List of pick-up addresses	• Plot the most efficient delivery and pick-up route and trace the order of delivery
Pen, pencil, paper	

Performance Checklist
- You are using a detailed, up-to-date city map.
- Delivery addresses and pick-up addresses are distinguished by different symbols and colors.
- No routing is made on opposite directions on a one-way street.
- Most pick-up addresses are planned for the return trip to the shop.
- No backtracking is planned.
- Special deliveries are given top priority in the delivery schedule.
- You have planned the most efficient use of delivery time with the least amount of mileage.

2. Plot several delivery or pick-up locations around your school grounds (building). Practice loading the delivery vehicle with floral designs and unloading these designs at the assigned locations. (Plan the most efficient and time-saving route.)

Resources	Procedure
List of destinations	• Load floral products and secure with holding devices
Clip board	
Floral design (tagged)	• Double check loaded products with list of "addresses"
Holding devices	• Unload floral products
Delivery vehicle	• Pick up used products
Floral kit	• Unload used products (at shop)

Figure 7-2 Practical application of assignment sheet.
Source: Printed with permission of The Ohio Curriculum Materials Service. From Stratman, T.S. (1976). *Retail Floriculture.* Columbus: The Ohio Agricultural Education Curriculum Materials Service.

Performance Checklist
- All floral products are secured by holding devices of some kind.
- Floral products are arranged in order of delivery (first to be delivered in front; last to be delivered in back).
- All the delivery bags face the front of the truck.
- Floral products are carried upright to and from the truck.
- Pick-ups are loaded last.
- Seat belt is fastened whenever delivery vehicle is running.

Glossary

C.O.D.—Cash on delivery; the customer will pay for the product when it is delivered.

Holding devices—Racks, braces, sandbags, etc., used to hold or support floral products upright during deliveries.

Order routing—The process of mapping out a path of travel for delivery according to destination, delivery date, and delivery time.

Supporting References

Pfahl, Peter B., and P. Blair Pfahl, Jr. *The Retail Florist Business,* Fourth Edition, Danville, Ill.: The Interstate Printers & Publishers, Inc.

State Highway Department

Mapquest.com

Figure 7-2 Continued

FACTORS TO CONSIDER IN SELECTING INDIVIDUALIZED TEACHING TECHNIQUES

Certainly, the factors considered in Chapter 6 for use in selecting group teaching techniques also apply to individual teaching techniques. The same notions about the nature of the subject matter, objectives to be achieved, characteristics of the learner, resource materials available, time available, and teacher preference also apply to selecting individual teaching techniques.

However, there are some additional factors one needs to reflect on when deciding whether to use an individual-centered teaching technique, as well as when determining which particular individual-centered technique is most appropriate.

One important factor is the student's readiness (See Chapter 2) for using an individualized technique. In order for a student to pursue independent study, for example, he or she must be able to organize thoughts, read, manage personal behavior, and synthesize information. For some students these conditions will never be met, whereas others may be able to function well under such conditions early in the first year of the agricultural instruction program. The background of the student also influences readiness to use individized teaching techniques. For example, a student with considerable background and skill can probably learn by using a skill sheet, whereas someone with no prerequisite skills could not. Students who are unable to read efficiently would

HORTICULTURE LEARNING CENTER

Instructional Program Area — *Floriculture Production*

Duty Area — D — Managing Greenhouse Crops, Poinsettia Production

POTTING POINSETTIAS

Task 03

PURPOSE:

The roots of poinsettias are exceptionally brittle, therefore care must be used when handling while potting. Carelessness and rough handling will result in a large number of dead plants. Therefore by using as much care as possible, you should be able to successfully pot as many poinsettias with a 100% live factor as specified by the instructor.

REFERENCES:

- Commercial Flower Forcing, Laurie, Kiplinger, and Nelson, (5)
- The Greenhouse Worker, Wotowiec, (1)
- Producing Poinsettias Commercially, (3)

MATERIALS:

- rooted poinsettia cutting or poinsettia cutting in 2¼" pots
- watering devices
- sterilized potting soil
- pots
- empty flats
- waterproof markers
- plastic labels

PROCEDURE:

Your instructor will tell you how many pots you are expected to do. He will also specify the pot size and the number of plants per pot.

1. Begin by locating the assigned area in which you are to work.
2. Move sufficient quantities of pots, labels, and flats to the potting area. Be sure there is a sufficient supply of sterilized potting soil available to complete the job.
3. Determine the cultivar name and the color of the poinsettias you will be potting.
4. Gently remove the rooted cutting from the propagation flat or the shipping carton. If you are transferring from 2¼" pots, tap the cutting from the pot. *Do Not* remove more cuttings than you can grade and pot in 15 minutes. If root systems dry out the plants are dead.
5. Before potting, it is essential that you grade the poinsettias for uniformity in height and development. If one of the 3 poinsettias placed in the pot is 2" shorter than the other, the finished product will be the same. Good growing starts at the potting table!
6. Using an empty flat, transfer your graded cuttings to the potting table. Place the flat of rooted cuttings to the right of the soil pile. The pots should be arranged on top of the soil pile with several empty flats to your left.
7. Pick up a container with your left hand and scoop in soil to fill the container approximately ⅔ full.
8. Pick up a cutting in your right hand and insert it into the pot. With your left hand, place enough soil around the base of the cutting to hold it in place. Do not fill the pot completely full with soil. *Caution!* Poinsettia roots are very brittle — handle gently.

Figure 7-3 Skill sheet.

Source: Printed with permission of The Ohio Curriculum Materials Service. From the Ohio Agricultural Education Curriculum Materials Service (1978). *Poinsettia Production.* Columbus: Author.

9. Insert 2 more cuttings into the pot, using the same procedure described above.
10. Your 3 cuttings should be evenly spaced in the pot, leaning very slightly towards the outside of the pot, and at least 1" from the rim of the pot.

11. When all 3 cuttings have been placed, use the thumb and forefinger of both hands and firm the soil in the container around the base of each cutting.

12. Using a waterproof marker and plastic pot labels, label each pot with the following information: cultivar name, color, and potting date.
13. Transfer the potted cutting to the empty flat at your left. When the flat is full transfer it to the greenhouse and water immediately.
14. Your instructor will indicate your final bench placement for these plants.

EVALUATIVE CRITERIA:

1. The cuttings were handled with sufficient care.
2. Correctly graded the poinsettias for uniformity of height and development.
3. Cuttings were evenly spaced, correctly sloped, and not too close to the rim of the pot.
4. All cuttings were tight in the pot.
5. Correctly labeled all pots.
6. All pots were watered and placed at correct bench location.
7. Completed the assigned number of pots.

	Student	Teacher
1.		
2.		
3.		
4.		
5.		
6.		
7.		
Final Grade		

Figure 7-3 Continued

be less able to productively use much supervised study where textual material was the source of learning than would more efficient readers.

Another factor to consider when deciding whether to use individualized teaching is the spread of interest in a class. When there are homogeneous interests and abilities, there is less pressure for individual-centered techniques than when there is a wide variety of interests and abilities represented in the class. For example, in production agriculture, if the class consists of 95 percent farm students who have beef cattle projects, there will be less need for individual study on beef cattle than in a class of 50 percent suburban-, 25 percent farm-, and 25 percent city-dwelling students. The same is true for a class in which 35 percent own beef cattle, 10 percent have pocket pets projects, 35 percent are experiencing agribusiness placement, 5 percent are engaged in bedding plant production, and 15 percent own horses.

Teachers who have learning packets, skill sheets, and other resources for supervised study and independent study will use techniques predicated on the availability of such resources more so than teachers without them. It may take teachers a number of years to develop a sufficient variety of study guides, lab sheets, and experiments such that they are able to truly infuse individualized techniques into their teaching.

Finally, teachers must consider the need that both they and their students have for a variety of ways of learning. Group techniques alone are not enough. Students want to function as individuals, and they also enjoy the rewards of being able to pursue their study of agricultural education in a variety of ways.

SUMMARY

Agriculture teachers should use supervised study; experiments; independent study; student notebooks; and information sheets, assignment sheets, and skill sheets. By using these teaching techniques, you will be able to individualize learning, help students learn to inquire into subject matter (See Chapter 2), promote student independence, and help students in general learn how to learn in a discriminating way

There are no shortcuts to success if teachers want to masterfully use individualized techniques of teaching. You must do a thorough job of planning, conducting, and terminating the particular type of study being used. You must also lead students to develop clear conclusions to the problems being studied.

FOR FURTHER STUDY

1. List the basic educational goals furthered by the use of individualized teaching techniques.
2. You are to teach a unit on fish diseases. What individualized teaching techniques would you use, and why?

Chap. 7 Individualized Teaching Techniques

3. Outline the procedure you would follow in planning and conducting a period of supervised study.
4. Explain under what circumstances you would use individualized teaching techniques as opposed to group teaching techniques.
5. Visit agriculture teachers and observe their use of individualized teaching techniques. Peruse the teaching resources they are drawing on for the use of individualized teaching techniques.

8

Managing Student Behavior

Recently, a day was spent in a local department of agriculture videotaping the techniques used by the teacher in that department. It was a wonderful experience. Anyone who was interested in teaching agriculture could not help but enter the profession if she or he could see the admirable job of directing learning that was taking place in this desirable environment for learning.

Of prime importance, the teacher acted like a teacher. She was in the room, ready for class, and showing her personal interest in each individual student well before each class began and well after each class ended. She was warm and caring. She came across as being genuine, being herself and enjoying it. Her techniques of teaching and dealing with students fit her personality rather than running contrary to her real self.

Another striking element of what contributed to this atmosphere was that the teacher knew her subject. She was technically competent and caused students to clearly understand the points of her lesson.

The teacher's rapport with her students was as near perfect as possible. The teacher and students knew one another and respected one another. They listened to one another. Each party contributed to the development of the lesson. The students and teacher could and did laugh together while balancing the allotted class time with on-task productivity.

The lesson was organized and clear. The students had been taught how to study and learn according to the approach used by the teacher. Students clearly knew what behavior they were expected to display.

The teacher did not shout, nor did she publicly confront her students. A word, a look, a direction was all that was needed.

It was obvious that the teacher personally knew the students and their parents and home situations. She built on this knowledge and skill-

Chap. 8 Managing Student Behavior

fully provided for individual differences. The students were the center of their own learning and were responsible for their own actions.

Truly, this was a desirable learning environment, and one that most teachers can achieve. But many teachers do not have such an environment, and very often it is the teacher's practices as much as the actions of the students that contribute to an undesirable learning environment. Consider the following case:

This was a class where the students were in the room walking on top of the tables when the agriculture teacher arrived. The teacher screamed at the students until they began to get to their seats. After a five-minute delay while the teacher tried to get some notes together, class began.

There was a no interest approach. The teacher talked at the students but the flow of the talk was confusing. It was obvious that the teacher did not care for the students and they shared this feeling. As the class continued there was a continuous verbal war between the teacher and the students. The teacher was unhappy and so were the students.

Now this situation did not always exist. The situation evolved to this point.

How could the teacher with the desirable learning environment have it so good, and the teacher with the undesirable learning environment have it so bad? Certainly, there is no clear-cut answer. Yet, those who study these matters know that both situations are primarily the results of the teachers' actions, attitudes, and abilities.

OBJECTIVES

After studying this chapter, you will be able to

1. Identify the techniques of successful teaching that are positively related to successfully managing student behavior.
2. Identify and use an array of specific strategies to effectively manage student behavior.
3. Develop ways of using parents in promoting acceptable student behavior.
4. Help students learn to accept responsibility for controlling their own behavior.

THE TEACHER AS THE KEY INGREDIENT
FOR AN ACCEPTABLE CLASSROOM ATMOSPHERE

Without a learning environment that includes students whose behavior is acceptable, it is virtually impossible for students to learn. Students cannot give their attention to the task of learning and simultaneously view the performance of the class clown. Likewise, it is impossible for teachers to involve students in a clear and organized presentation if they must constantly stop to restore order. Without good class control, teachers cannot be effective teachers. Thus, a major challenge for teachers is to manage their classrooms so as to create a desirable learning environment.

School administrators and the public as well are critical if teachers cannot control their students' behaviors. The annual Phi Delta Kappan Gallup Poll consistently reveals that one of the public's chief concerns with education is discipline in the schools.

Not only is the effective management of the classroom important because of its impact on learning and its necessity in the eyes of the administration and the public but effective classroom management is also closely associated with one's satisfaction with teaching or lack of such satisfaction. No matter how competent teachers are at teaching, they must manage student behavior successfully or they will dread entering the classroom because of impending discipline problems. In such cases, they will not be able to perform as adequately as they are capable of performing.

Remember, just as the first situation referred to in this chapter did not occur overnight, neither did the second situation. Rather, classroom atmospheres are generated; they evolve. In the case of discipline problems, they often evolve so slowly that teachers are not fully aware of what is happening until conditions have deteriorated to a state of crisis.

Perhaps the following analogy will help to illustrate the point of not waiting for a crisis to develop before taking positive action. Frogs, unlike most animals, are not able to perceive gradual changes in the temperature surrounding them. In fact, if a frog is placed into a pan of water, and the water is gradually heated, the frog will complacently sit in the water as the heat rises to the boiling point. The frog does not feel the heat and thus does not make a change in the environment. The same seems to be true of many teachers. They step into a typical classroom and are so naive that they fail to detect the gradual "heating up" of the environment until at last they are "sacrificed." An early detection system is essential in maintaining classroom discipline. This is part of the art of teaching for which teachers must develop a "sense." There are no ready-made recipes. However, there are some specific skills we can master that will help in successfully managing student behavior.

THE RELATIONSHIP OF TEACHING PERFORMANCE TO PROMOTING APPROPRIATE STUDENT BEHAVIOR

The Effect of Technical Competence

Before teachers can teach well they must be very knowledgeable about their subject areas. If the teacher does not have a firm understanding of the subject matter, it is difficult to perform well as a teacher. Teachers must be technically competent. No university can teach an agriculture teacher all that is ever needed. Agriculture teachers must accept the responsibility to continue learning what they do not know as such knowledge and skill is needed.

Teachers who clearly know their stuff are able to teach with confidence and authority. This sense of being taught by a pro does much to dissuade students from being tempted to behave unproductively during class because the students are able to sense that such teachers know their subject matter and probably know how to handle students who behave inappropriately. However, when students sense that a teacher is not well informed or up-to-date, they find it hard to resist exploiting such a weakness.

The Effect of Rapport on Maintaining Acceptable Student Behavior

Another crucial ingredient in teaching performance that enhances appropriate classroom and laboratory control is good rapport. Teachers who are able to develop good rapport with their students are apt to have fewer problems than teachers who lack this quality.

For example, there was a young agriculture teacher who enrolled in a first-year teacher program where a member of the agricultural education department at the university visited the teacher regularly. As the year progressed the teacher educator sensed that the teacher's classroom control was deteriorating. After some questioning, it was learned that this teacher, who was single, spent time most evenings at the local cafe. This was also the gathering place for the local students, and progressively the students and the teacher became overly familiar with one another to the point that this familiarity transferred into the classroom. Once the teacher realized what was happening and ceased to visit the cafe, the classroom situation improved. Professional distance was restored. This professional distance is necessary.

The relationship that teachers seek with their students is one of reasonable give and take. Teachers must listen, evaluate, and then act rather than jumping on students before they have had a chance to adequately explain a situation. Teachers must also have a sense of humor and be able to laugh at themselves and with their students. For example, if a teacher makes a mistake at the board, he or she ought to be willing to let students have a laugh without becoming defensive. Likewise, when students say something that is in error, the teacher and the other students ought to be able to laugh with the student without the student feeling that he or she is being laughed at. It is imperative that teachers remember that students have feelings and that sarcastic remarks can hurt. Students also have egos, and many of them do not hesitate to defend their egos when they are confronted or put on the defensive in public, especially in front of their peers.

The Effect of Interest on Student Behavior

Given that one has the proper climate for good class discipline, there are a number of points regarding teaching techniques that affect discipline. Students are not apt to cause difficulties if (1) what they are taught is interesting and (2) it is taught in an interesting way. It is unrealistic to expect good behavior from students when they are being taught material that is not relevant to them presently or in the foreseeable future. Likewise, class behavior is generally not good when students are asked to learn more than they need to know in order to satisfy their own goals.

Even if what is taught is important and interesting to students, it must be taught in a manner that is interesting. Students must be mentally set in order for learning to occur. Interest must be gained at the beginning of every teaching session. The few minutes it takes to think of an interesting way to begin a lesson is much more productive than the time it will take to keep order during a boring class. One must have control of the class's behavior and must invest the time and energy required to gain and maintain the interest and appropriate behavior of the students. However, interest is not permanent. Therefore, students have to be "reinterested" during the class session.

There are many ways to generate and maintain interest, but the total design will certainly include and demand several changes of pace each hour. Consider Case 8-1.

Case 8-1. *Generating and Maintaining Interest.*

QUESTION TO BE ANSWERED : "HOW DO I DETERMINE THE SIZE OF A FRESHWATER FISH POND?"	
DIRECTION TO SELF	**KEY CONTENT**
Distribute the following problem to the class	"We have a problem. Our class is to 1. Fertilize the fish pond in our nature preserve and 2. Apply herbicides to control undesirable vegetation. Before we can figure how much fertilizer and herbicide to buy we must know the (a) acreage and (b) volume. "John would you explain how to figure the acreage of a circular pond as well as its volume?" **Assuming he doesn't know how**—"Do we need to figure out how to calculate this before we can proceed?"

Divide class into two groups A. Have one-half of the class read pp.13–14 of "Managing Freshwater Ponds" and determine how to figure surface area. B. Have the other half of the class read pp. 14–16 of "Managing Freshwater Ponds" and determine how to calculate volume.	pp. 13–14 of "Managing Freshwater Ponds" (attached) pp. 14–16 of "Managing Freshwater Ponds" (attached)
Show transparency 1	Formula: Surface area = $\dfrac{(\text{Circumference in feet})^2}{547{,}390}$
Show transparency 2	Formula: Volume = Surface area $\times$ Average depth
Students write: After study, have one person from each group put an answer on the writing surface.	Problem 1: Circumference = 780' Average depth = 11' Solution Surface area = $\dfrac{(780)^2}{547{,}390}$ $\dfrac{608{,}400}{547{,}390}$
Overhead: Work through a sample problem on overhead	Answer 1.1 acres
Handout second problem for practice	Problem 2: Circumference = 500' Average depth = 7' Answer Surface area = 0.46 acre *Note to reader:* You would proceed with volume in the same fashion. Answers will be stated in acre feet.
Students calculate surface area and volume of school pond during laboratory	(Opportunity to show they learned the content)
Summary	"We have learned to calculate pond average and volume. Now we are ready to proceed with solving our problem. What else do we need to know?"

Analysis of Case 8–1. Notice that distributing a real problem generated interest. Because the students will actually do the work referred to, they will likely want to know how to proceed and will generate quality questions to be answered because they can see themselves in the situation.

In teaching the solution, notice that the teacher involves students in supervised study (reading and solving problems), putting answers on the board, and taking measurements during lab. Each of these activities, as well as the teacher using the overhead projector and helping all the students as they do sample problems, is an effective change of pace. Thus students are not required to maintain their interest for long periods of time. The teacher has built into the teaching relief from naturally occurring boredom.

Students have high energy levels and they must have ample opportunity to expend this energy. If the teacher does not sufficiently involve students in class to help them productively expend their energy, the students will create their own ways to expend their energy, and their creative ways of being energetic generally are troublesome for the teacher. Thus, the teacher must realize this wholesome aspect of the adolescent and teach with sufficient variety and involvement such that discipline problems are avoided rather than created.

The Effect of Organization and Clarity

Not only must the instruction that is provided be taught in an interesting manner, it must also be presented in a clear and organized manner. Teachers can be filled with enthusiasm and change the pace often, but if they confuse their students, the students will soon give up in frustration. It is not enough to entertain. A teacher must also help bring about a change in the behavior of students. Learning must take place, and organization that brings clarity (note the flow of events in Case 8-1) promotes effective learning.

Students in the class where Case 8-1 occurs have a clear frame of reference (their school pond problem) and adequate structure for learning (the formulas and practice using them). Without such a framework, the class becomes confusing. With confusion comes an increase in frustration and anxiety. As the level of frustration and anxiety rises within a student, the student becomes increasingly impatient and resentful. Once a student is in this state of mind he or she is easily provoked and may, in fact, create trouble in the classroom simply to relieve the frustration.

Teachers can make additional use of providing structure by reminding students that they are systematically developing increasing levels of competence. This can be accomplished, for example, by reminding students who have learned to sterilize soil, propagate poinsettia cuttings, and stick rooted cuttings that, after studying two more problems, namely growing poinsettias and marketing poinsettias, they will be able to produce a crop from start to finish. Another way to help students realize that they are systematically developing competence is by using skill charts and record books.

STRATEGIES FOR PREVENTING STUDENT MISBEHAVIOR

No one can prevent inappropriate student behavior for the teacher. Teachers must do it themselves. Only you can manage student behavior. Suggestions that consistently work will be offered, but the individual teacher has to imple-

Chap. 8 Managing Student Behavior

ment those suggestions. In using the various suggestions that are offered, teachers must also keep in mind the principles of learning presented in Chapter 2. Thousands of teachers learn what to do, when to do it, and how to do it, but they attempt to succeed by using only a small number of the numerous resources available, and they usually fail.

Maintaining desirable student behavior is heavily predicated on prevention. If student misbehavior is not allowed to begin, it does not have to be stopped. Thus, the more you manage the classroom, the less you have to discipline the students.

General Guidelines to Follow in Promoting Acceptable Behavior

Before suggesting specific strategies to use in securing appropriate student behavior, consider the following general guidelines that teachers find helpful. These are guidelines on which there is rather common agreement by practitioners and theorists alike.

In addition to these guidelines, teachers must remember that they are a part of the school system and as such must operate within school policies. Teachers cannot expect the administration to allow them to violate general school policy, even if the teacher disagrees with it. Such actions would create an unmanageable situation.

1. *Start out firm.* One can always ease up later, but the opposite is not true. Remember that people are greatly influenced by first impressions. Teachers must realize that once a pattern of operation and behavior has been established, it becomes the norm and is very difficult to change. So, the questions, "What will I do?" and "What will I say?" on opening day are critical. Start planning for these questions now.

2. *Be prepared to teach well.* Keep students busy in a meaningful way. In other words, do not simply assign busywork; rather, have students actively involved in doing things that make sense to them. Many behavior problems are created because teachers are late in arriving to class, do not have definite instructions planned for the day, and allow students to become bored.

3. *Have a definite routine by which each class is started.* A routine may simply involve taking the roll, making announcements, and explaining what will be accomplished that day. A "bell assignment" (a question for students to answer, a reflection for students to write from the previous day's notes, a seat activity, or other short, quiet writing assignment) written on the chalkboard, whiteboard, or projected on the overhead works well for this time period. A part of this routine should include the teacher always being in the classroom or laboratory *before* students arrive to greet students as they enter the room. The existence of such a procedure creates an environment of certainty and security. Students quickly come to know what to expect, and they get in to a habit of acting accordingly.

4. *Make generous use of praise.* Encourage good behavior and let students know that they are appreciated. All people enjoy the feeling of being admired, appreciated, and acknowledged in public. Students who experience praise for good behavior and good schoolwork will tend to repeat such behavior. Know your students so you know whom you can praise publicly and whom you must praise privately.

5. *Do not have favorites.* Each student is a person of worth. Students dislike getting the sense that the teacher has predetermined their destinies as angels or demons.

6. *Be consistent, yet not predictable.* Although this may seem ironic, it is not. Teachers need to be consistent. When inappropriate behaviors occur, the students involved must know that such misbehaviors will (a) be dealt with and (b) be dealt with fairly. However, it is not in a teacher's best interest for students to be able to predict specifically how a given inappropriate action will be handled; they then know how much risk is involved whenever they misbehave. Knowing the precise degree of risk prompts some students to decide, "Well, that is not so bad—it will be worth it just so I can . . .". For example, if a student broke a glass in the greenhouse on purpose he or she should know that the teacher will punish him or her. However, the student should not know whether the punishment will consist of paying for the damage, working extra after school, or having the parents in for a conference. A teacher must consistently punish such misbehavior, but not in predictable ways.

7. *Take action whenever a problem arises.* Granted, a teacher should not discipline students when the teacher is angry. However, it is unwise to postpone handling problems because students interpret this as weakness and indecisiveness. Although teachers may want to wait until a planning period to specifically deal with a problem, they should let the student know at the time of the misbehavior that it will be dealt with at a certain time.

8. *Learn to separate the action of the student from the person of the student.* What is meant here is that the teacher needs to communicate to students that the teacher dislikes a certain behavior and will not tolerate it but does not dislike the student. The student is a person of worth and is appreciated, but the behavior will not be tolerated.

9. *Never make threats, only make promises.* Students soon learn the idleness of threats. When teachers say, "If you do that again, I'll break your neck," students know full well it is not going to happen. However, when students learn that if a teacher says something will happen then they can count on it happening, students are much less likely to press the issue. For example, if a teacher tells students that there will be no more field trips if they throw objects out of the bus again and they disobey, then end that field trip and future field trips as well.

10. *Set a good example.* Students need role models. They also appreciate people who practice what they preach. If teachers want students to be serious, productive, on time, and well mannered, teachers must act accordingly.

11. *Be sure the penalty fits the offense.* Teachers must learn to gauge their penalties to the degree of seriousness of their students' misbehaviors. No matter who the student is who misbehaves in a given way, that student should be dealt with at the same level of severity as other students who may have misbehaved similarly.

12. *Be attentive to all behavior in the classroom or laboratory.* Students are quick to take advantage of teachers who are able to see only one thing at a time. Teachers must learn to be very observant; otherwise, irritating little problems increase until they are out of control.

13. *Learn to forgive and forget.* Teachers who hold grudges will not be very successful. All people make mistakes. Once a student's mistake has been dealt with, the student should not be constantly reminded of the past failure. If a student curses you, then once you have satisfactorily handled the misbehavior, be willing to say to the student, "OK, now let's forget it," and do so.

Beyond these guidelines there are some specific strategies that have been found to be helpful in promoting appropriate discipline. A real cornerstone of a discipline program is the establishing of clear expectations regarding student behavior.

Setting Expectations

The most important time to discuss discipline is the first day of class. This is true whether one is a student teacher taking over a class, a first-year teacher on a new job, or an experienced teacher meeting a class for the first time.

A firm but fair beginning is crucial. The tone that is set at the beginning will temper all that follows. A procedure, delineating points to be discussed during the opening day expectations, should be presented. However, keep in mind that what each individual teacher does must be consistent with established policy in the school. Likewise, each teacher must use an approach that is congruent with his or her own personality and mode of operation.

In opening a discussion of discipline with a class, it makes sense to begin by letting the students know that you are just as normal and human as they are and that you have a personal life outside the school. If you are married, tell them. Chat with them briefly about your spouse and children. Share a few of your background experiences and current hobbies and interests. You might be surprised how interested students are to learn that their teacher played a saxophone in high school, enjoys scuba diving, is an accomplished photographer, and so on. An additional benefit of this kind of a beginning with a class is that it helps break the ice. Naturally, you should not discuss private, personal details or reveal so much that you lose what is called professional distance.

After the period of breaking the ice, open a discussion that encourages students to provide a list of what they expect from the course. This is easily accomplished by asking students questions such as, "What do you expect to get out of this course?" "Why are you enrolled in this course?" "What are some things you expect to learn here this year?"

Students generally give some rather sensible responses to such questions, but they may have to be encouraged or prompted to speak up. They will probably say things such as, "I want to learn how to be a better . . . (whatever career for which your program is geared)," "I want to become a state FFA degree winner," "I want to learn how to . . . (perform some specific task)." This list should be recorded on the writing surface (and in their notebooks) so that students realize (1) you are serious, (2) you are businesslike, and (3) they are expected to take notes and pay attention from the very first day of class to the last day.

Students should realize that unless they have some rather valid reasons for being in a class and unless they are expecting some general kinds of things to be delivered, they may be disappointed. Notice the psychological importance of this strategy. Students will develop the feeling that they are able to affect the direction of the class, and such a feeling provides important psychological comfort. Be sure to maintain firm control of the class during this dialogue.

Once students have revealed their course expectations, it is important to move to a more personal level. Find out from the class what they expect of you. This will require even more probing, but it is essential that students express realistic expectations of the teacher. One strategy is to ask each student to write at least one realistic expectation on notebook paper before opening a dialogue. Once again, use leading questions that will prompt the types of responses that are needed. It may be necessary to point out that this is the chance of their lifetime; but do not become too informal at this point because students may misinterpret such informality as an indication that this is a big game and may thereby be encouraged to be too rambunctious too soon. Also do not accept any smart answers at this point. Students will probably suggest things such as "be fair," "don't be too hard," and "keep us interested." These responses should also be listed on the board for the same reasons you had them record their expectations for the class. Again, the psychology of this step is very important. Students have had a chance to say this is what they want their teacher to be, which builds a beautiful foundation for the next step.

The next step is to present the class with this question: "What should be the mode of operation for our class?" Here, have the students collectively formulate some general operating procedures. Once again this can probably best be done by asking some leading questions such as, "What should be the general way in which we will operate our class?" "Would you suggest some general guidelines that we should follow as we conduct our class during this year?"

However, avoid of a list of rules. Remember, for every rule there must be penalties, exceptions, enforcement, and many other associated problems. Being overly specific only highlights areas of possible violation and, in fact, can provide an irresistible inducement for some students. According to Smith and

Smith, "This new rule creates a new environment which must be tested. A child must break the rule, usually several times, before he [sic] is sure it is truly in effect."[1] Thus, it seems appropriate to create an atmosphere of general understandings as to what is acceptable and unacceptable behavior rather than to prescribe in legislative fashion every activity that is considered inappropriate.

When generating this idea of "How should our class operate?" a student or students may offer specific suggestions, such as "We shouldn't throw spit wads." This is a good time for the teacher to help the class see that we as mature students (this lets students know you realize they are more than little school children) know right from wrong and that we must be mature. Thank the student for the suggestion, acknowledging that it is a correct notion, but point out that we need only list more general understandings such as abiding by school rules, treating one another courteously, or putting forth our best effort.

Finally, and most importantly, teachers need to clearly and positively explain their own expectations of the students. This is the place to discuss absolutes, whatever they might be, for the teacher. Some examples are arriving on time (and explain that you believe on time means in your seat with your notebook and a pencil <u>before</u> the bell rings), being prepared (i.e., assignment is read, pencil is sharpened), doing honest work (i.e., do your own homework), using only appropriate language (i.e., only use words you would write in a graded English essay), not giving passes to the restroom, or whatever absolutes you feel strongly about. Teachers need to explain their idiosyncrasies and personal biases so that the students realize that the teacher knows this is his or her issue, but that if students prey on that issue, they can expect quick action. Teachers need to let students know that they realize this may be something peculiar to them, but that is the way it is. By alerting students of such specifics, they can stay clear of them rather than encounter them, thereby provoking the teacher unnecessarily. Discretion must be used at this point. Only the most important items should be discussed, and teachers should be sure they do not establish absolutes that they have no personal right to establish.

Once teachers have established this basic set of expectations and understandings, it is important to do some teaching. Even if they have to deal with a rather short lesson, they should do so. Remember the importance of first impressions. The teacher should teach as well as possible. An important psychological advantage of providing some masterful instruction at that first meeting is that it lets students know the teacher is very serious about the course and that this teacher is there to teach.

In summary, a teacher's brief outline of the key points to discuss with the class in establishing a basis for managing student behavior might look like this:

1. Personally introduce yourself, your background, and the nature of the course.

---

[1] Smith, Judith M., and Smith, Donald E. *Child management.* Champaign, IL: Research Press Company, 1976, p. 16.

2. Have students tell you what they expect from the class.
3. Have students tell you what they expect of you.
4. Discuss how the class should operate.
5. Tell the class your expectations and the little "things" you will not tolerate.
6. Teach some agriculture as effectively as you know how.

This first session is not meant to be a sermon. Rather it should be a businesslike discussion that establishes the setting in which the students and the teacher should operate. Students should go home with the impression that this is a teacher who acts with authority but is also there to help students grow and develop into the kind of people they ought to be. If the items previously discussed are not implemented during the first class session, they become essentially useless. There is little value in waiting until Thanksgiving or Christmas to discuss the expectations of the teacher and student because by then it is too late and, therefore, seldom works.

Teachers who realize that good discipline is essential and use the procedure discussed for establishing expectations regarding student behavior are well on their way to developing the kind of basis that is needed to promote appropriate discipline. During this process we must also realize that another key ingredient for successful classroom management is humor. Teachers must be willing to laugh at themselves and with their students. Although teachers should and must conduct themselves in a businesslike manner, they must also establish rapport with the class. Effective disciplinarians do not have to be steel-hearted monsters with no personal concern for their students. Quite to the contrary, effective teachers must know their students, that they care about them, and earn their respect.

Knowing the Student as a Person

Real cooperation between students and their teacher becomes possible when they care for one another. Teachers must provide quality learning experiences that show they care for students. To do this, teachers have to get to know their students. There are many good ways to accomplish this objective. Perhaps the best way to really get to know your students so that you can show you care in personal ways is to acquaint yourself with their home situations. Case 8-2 illustrates the point.

Remember, too, that students have interests and talents that may not surface in the agriculture course. They may be artists, musicians, dramatists, or athletes. If they are, you can bet they want their agriculture teacher to see them perform and be proud of their accomplishments. Teachers of agriculture need to attend general school functions, thereby showing their students that they (the students) are persons of worth and are cared about.

Case 8-2. *The Case of the Long Walk.*

> An agriculture teacher found he was required to visit his students in their homes. From this experience he concluded (and you will too) that home visits are essential to providing the best instruction possible. One balmy spring day in the Blue Ridge Mountains a shy boy in the back of the room sheepishly raised his hand and said, "Would you come out to my home this afternoon and help me figure out what is wrong with my chickens?" What else could a young, inexperienced teacher say but "of course." Then the boy, Jackie, said, "You'll have to walk a long ways." Again, what could anyone who cares for people say except, "That's OK, no problem."
>
> They drove a pickup truck as far as possible and then walked an additional three miles to reach his house. His chickens lived under terrible conditions, and there was clearly the need for much help and individual instruction. They then went to the house to meet his mother. Entering the house was easy on that beautiful spring afternoon, for the doors and windows were open and there were no screens. They entered the kitchen and the teacher saw the student's two younger brothers sitting on the floor eating their supper—beans and biscuits. Green flies swarmed as they entered the room; his rooster was sitting on the refrigerator. His mother was standing by the sink. They walked over to meet her. She was a large lady, plainly but cleanly dressed. She was obviously apprehensive.
>
> Jackie introduced his agriculture teacher, and the teacher and Jackie's mother began to chat. Early in the conversation she said, "Oh, won't you stay and have a bite to eat with us?" She was a good, kind lady. She was a person of worth. She cared about her children. That and subsequent visits had quite an impact on the agriculture teacher. He taught Jackie differently after that because he could relate his instruction to him in a more personal way.
>
> Jackie and his agriculture teacher cared more about one another after that. The teacher was able to better understand why Jackie was the way he was. Thus, the teacher no longer expected Jackie to write a report from the farm magazines he read at home because there were none. Jackie was not hassled with having to do improvement projects that were unreasonable under his circumstances. Rather, his improvement projects were carefully selected based on his needs and resources. This personal knowledge of Jackie helped his teacher to be more effective.

Using the preceding discussion to gain a better understanding of students will result in an improved relationship between teachers and their students and, thereby, improve the climate for effective discipline. However, given all of the previously mentioned background and preventives, there will still be discipline problems that will arise and will require positive teacher action. Teachers can be sure that no principal wants teachers to send him or her their discipline problems. Sending anything but a real crisis case to the principal only tarnishes the image of the teacher in the eyes of the administration. When you have recurring problems with a student, you may want to counsel with the appropriate administrator regarding next steps. By showing the administration that you are aware of a problem, have taken specific action, and have sought advice, you build a foundation for strong support in the event that you must refer the case to the principal or other administrator later. There are no panaceas, no magical cures. Several strategies are suggested from which the teacher can make prudent selections. Teachers will need to develop their own fortes of discipline techniques that they find are successful for them with their own students.

STRATEGIES FOR DEALING WITH STUDENT MISBEHAVIOR

Withholding Reinforcement

Very often discipline problems arise when some students need to have extra attention. This type of student is often willing to take seemingly large risks in order to receive direct attention from the teacher, even if the attention is extremely negative. Likewise, such students are almost certain of receiving attention from their peers, for students often love the antics of a classmate.

Unfortunately, many teachers fall into the trap of reacting to a student's ploys and thereby satisfying (or reinforcing) the student's need for attention. When the teacher takes time to deal with such a student during class, the teacher has, in effect, done exactly what the student sought to have done, namely receive personal attention and not have to study agriculture. Thus, the behavior of the student, which was inappropriate, triggered the teacher to react, thereby doing precisely what the student set out to make happen. Naturally, the next time the student feels the need for teacher attention, all he or she has to do is misbehave and the vicious cycle is reenacted. Such a progression of action puts the student in the offensive court and forces the teacher into the position of reacting in a defensive posture.

Actually, the teacher would be better off ignoring some undesirable behaviors. By ignoring the behavior, at least publicly, the teacher does not provide reinforcement to the student in the form of the student gaining public attention, and the behavior will often become extinct.

The key for the teacher who desires to promote acceptable student behavior is to learn to perceive when misbehavior is aimed at gaining attention, then to refuse to fulfill such needs by reacting in public. Obviously, this does not apply to all behavior problems. There will be instances in which an individual or an entire class needs to immediately be corrected publicly when a disruptive behavior occurs. For example, if the class bully is physically abusing the class underdog, he or she must be stopped and dealt with then or after class. If the whole class is unruly, the teacher must stop the unruliness and regain the attention of the class. The essential ability or art that a teacher must possess is to distinguish between (1) behaviors that can be extinguished with firm action and (2) behaviors that are designed to fulfill one or two students' need for pampering or to merely direct the teacher's energies away from promoting learning and toward wasting class time.

Deal with students seeking recognition by this means on an individual basis. Keep such students after class, call them in during a planning period, or deal with them after school. One can effectively deal with the problem without wasting class time chastising a student before the whole class, when the student, at whom the comments or other strategies are aimed, cares little about the punishment and in reality enjoys the treatment.

The wise teacher will learn to provide personal attention and visibility to students when they are not misbehaving. This shifts reinforcement and teacher attention to appropriate behaviors while at the same time meeting a student's need to be accepted, recognized, and valued.

Using Suspense

In dealing with student behavior problems, often one of the best strategies to employ is the use of suspense. Keep students wondering what you will do and when. If they can predict a teacher's reactions, they know the amount of risk involved in deviating from acceptable patterns of behavior. This often encourages students to behave unsatisfactorily.

Likewise, in disciplining students, teachers must take their time. Allow students to wonder what is coming, to weigh the seriousness of their actions and what the proper consequences of such actions might be. Often the mental anguish is more effective than any specific disciplinary action the teacher might take. This course of action also gives the teacher time to think, and if the teacher is emotional, there is time to remove such emotion from the corrective actions that are taken.

Using the Individual Conference

One of the most effective strategies available to the teacher in managing student behavior is the use of the individual conference. There are several advantages offered by dealing with a student in private. First of all, the confrontation is on the grounds of the teacher. There is no audience for which the student can perform. Neither the student nor the teacher has an image to project or protect, and both parties will have the opportunity to be less emotional and more rational. For the most part, teachers come to regret dealing with class disruptions while they are angry.

Rather than deal with a nagging student in the classroom or laboratory, it is often to the teacher's advantage to tell the student, "See me after class." Then before the class is dismissed, be sure to remind the student to stay. Whatever the problem, if it will take more time than is available at the moment, the student should be told a specific time to see the teacher. The waiting that the student experiences from the time the conference has been arranged until it actually begins is often very helpful in changing the student's attitude and the teacher's emotional state. Students also have time to think about their actions and to vicariously act out their emotions before they and their teacher have the individual conference.

Once the conference begins, the way the teacher handles the conference is extremely important. For rather serious behavior problems the following procedure is suggested. Begin by questioning the students until they admit why they are there. This often takes as much as five or ten minutes. Do not rush it, and unless a student is unmalleable, do not tell the student why he or she is there. Unless the teacher has acted very capriciously, the student knows why he or she is there.

The beauty of this phase of the conference is that students have to admit that they made an error, that they knew better, and that there is justification for the conference. This in and of itself gives the teacher a psychological advantage. The students have in effect pleaded guilty. They have said, "I am wrong." Having such an advantage, the teacher should build on it. Students also need to understand why their behavior was wrong.

The next phase of the conference involves getting the students, through a series of leading questions, to realize that their behavior choices have been a disappointment to people who care about them, including parents, girlfriends or boyfriends, peers, and the student who is the focus of the session. Rarely should the teacher encourage or allow students to discuss whether they have disappointed the instructor conducting the conference. When students have very much respect for their teacher, they are not involved in a conference of this kind.

In order for this phase of the conference to be effective, it is imperative that the teacher be familiar with the home situation of the student. It will obviously do little good to attempt to get students to admit they have been a disappointment to their parents if there is no feeling of respect between the student and the parents.

The teacher is now at a point where consequences must be discussed. Rather than automatically doling out a penalty, have the student suggest what it should be. This is another area of the conference where there is no need to hurry. Time for the student to think of suitable consequences for the offense may be more important than any penalty per se. Having students make the suggestion for the penalty causes them to weigh the seriousness of their misbehavior. Additionally, students are forced to begin accepting responsibility for their behavior.

It is often desirable to reject many, if not all, of the consequences suggested by the student. There is a basic reason for this suggestion. Many of the penalties students suggest are not very realistic; they are either too severe, or not the type of penalties the teacher believes in (such as writing sentences or corporal punishment), or do not align with the misconduct. When such consequences are suggested, teachers should explain to the student why the teacher believes the penalty is inappropriate. Also, if the conference has caused the student to recognize his or her problem and admit the action was inappropriate and deserving of a penalty, the student's attitude has probably been changed, which is what the teacher seeks anyway.

After careful consideration, a consequence may be selected. However, if the conference has gone well, the teacher may decide there is no need for any penalty because the student's attitude has improved. In this case, teachers may tell students that they believe the situation has been taken care of, that it will not arise again, and that if it does arise again an appropriate consequence will be waiting. This cannot be a bluff. Teachers must become known for being fair and for acting decisively when a situation does not improve.

Regardless of whether a penalty for the misbehavior is assessed, it is important to have the student (you may have to help) review the major happenings of the conference. Thus, students, in reviewing the conference again, admit they were wrong, realize they have been a disappointment to some people they care about, and suggest a penalty. This helps crystallize for students what has happened and allows them to leave with a clear notion of what is expected and will be tolerated. An outline for an individual conference might look as follows.

Outline for an Individual Conference

1. Get the student to explain why he or she is in the conference—to admit he or she did something that should not have been done.
2. Get the student to realize his or her actions have been a disappointment to some people he or she cares for—parents, friends, class members, self.
3. Have the student suggest a consequence that will guarantee the action will not occur again.
4. Have the student review the major points of the conference.

Without a doubt, if the conference is handled well, it will alter the student's behavior. There have been no public scenes and no hideous actions. The student has essentially been held accountable and taken full responsibility for the behavior.

When students leave such a conference, their friends will ask what happened. What will their reply have to be? What can they tell their friends? Essentially they will say the teacher did nothing, especially if no penalty was given. The teacher did not give them detention or take away a privilege. "The teacher simply talked to me," they will say. However, if the student acts much differently in class the next day everyone will certainly wonder what happened. The only way other students can be sure of what happened is to receive a special invitation so they can experience it personally. For most students, the risk will be too big to take.

In conclusion, examine the psychological advantage this type of approach gives the teacher. Students admit they were wrong, that they did something that was unacceptable, that they are guilty. Then students admit some special people will not be very proud of them when they find out what happened. The students then agree they should be penalized for their actions and they personally try to identify an appropriate penalty. Then the teacher forgives all and assures the student that appropriate action will be taken if this happens again. Notice, the student has made all the tough decisions. This has a tremendous sobering effect on most students.

For times when something has to give, the individual conference works as well as any discipline strategy available. Obviously, the strategy is reserved for more serious situations. Examples of such situations are flagrant acts of destruction, such as marring furniture or equipment; fighting; verbal abuse of teacher or peers; outright belligerence, cheating, or stealing; or having tried you as a person one time too many. There are other techniques for less severe problems.

Use of Volume

The first thing that probably comes to the minds of most teachers when they think of using volume to manage student behavior is loudness. This is not necessarily so, for loudness only allows one to make use of half of the strategy of volume.

Certainly, a shout from the teacher can startle a class and cause the students to give the teacher their attention. Likewise, teachers may raise their voices and lower them to create interest, which is a positive way of preventing student misbehavior.

Teachers have to be careful, however, not to try to out shout a class. New teachers often discover that the sound level in their classroom has gradually grown to the point that the class discussion has become a shouting match. This type of environment only serves to create tremendous tension for the teacher and students alike. Loudness seems to work best where it is used as a shocking force. It should not be used constantly or it will lose its effectiveness. If the teacher and students are always loud, the teacher's plea for quiet will likely go unnoticed. It is much like hearing an air hammer working outside of an airport. It goes unnoticed. It is simply one more loud noise.

Certainly, one can use reduced volume just as effectively in controlling a class. When teachers develop enthusiastic discussions, they must realize that this method breeds noise. The noise level can often be controlled if teachers will merely drop their volume to just above a whisper. Students then find that they cannot hear unless they lower their voices, or better yet, stop talking.

Something that is very important for teachers to realize is that they have a propensity to use volume, both loud and not loud, in cyclical patterns. By "cyclical" we mean that if someone charted the noise level in a classroom, it would probably look somewhat like Figure 8–1.

That is, it would increase to a certain threshold level (A), the teacher would act to quiet the class (B), then the cycle would recur. Teachers need to be sure that the time between high points on the curve does not become too short; otherwise, teachers get to the point that their control of unnecessary loudness is ineffective and they, in fact, spend essentially all of their time pleading for quiet.

Perhaps a word about the loudness of classrooms is in order. Teachers have to remind themselves that any room in the United States with twenty or so people in it will indeed have a certain level of noise present. This is true for living rooms, waiting rooms, church rooms, and even funeral rooms. Over time a group of people chat, fidget, cough, shuffle, and clear their throats. All of this contributes to the general noise level of the room. High school classrooms will not be silent places. The key for the teacher is to be able to distinguish between

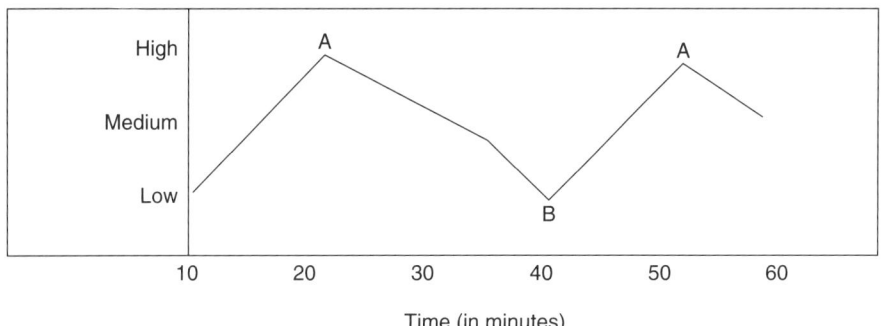

Figure 8-1 Cyclical pattern of classroom noise level.

productive noise and disruptive noise. Patience and tolerance are required because adolescents cannot be quiet very long. Also, when students are excited about learning, such excitement will be manifested audibly. The prudent teacher senses when to reign-in the group without hassling them unduly.

Of course, teachers do not always need to rely on words, either softly or loudly spoken, to control student behavior. They can use actions or nonverbal communication to correct inappropriate behavior.

Use of Nonverbal Communication Techniques

There is a whole array of discipline strategies that are used consciously as well as unconsciously by most teachers. These strategies employ the use of nonverbal cues. Teachers should be more aware of the potential of nonverbal expressions as deterrents to undesirable student behavior. Nonverbal means the use of actions, facial expressions, or body language to convey specific messages to others.

Facial Expressions. Many teachers effectively use a number of types of looks to obtain the kind of student behavior they desire. Foremost among these special looks is the silent stare. Penetrating eyes can quickly defuse potential behavior problems. Remember, if one decides to use a stare, it is important to know when to stop—otherwise, it can deteriorate into a silly game. Frowns and raised eyebrows are also effective for expressing disapproval of a behavior.

Teachers also use nonverbal cues, such as snapping fingers, pointing, or folded arms, coupled with silence. However, teachers should also attend to the nonverbal expressions of students, for this second source of nonverbal language provides potent feedback to the teacher. It is present and must be recognized and interpreted. Otherwise, teacher and student will not communicate as fully as they should.

Silence. The adage, "Your action speaks so loud I can't hear a word you say" holds true for teachers as much as for people in general. It may very well be that how we act as we seek appropriate discipline becomes more apparent than what we say. Many times desirable student behavior can be obtained by saying nothing. Indeed, a period of well-placed silence can be very effective. Students will generally continue the behavior for a short period of time but will soon realize something is wrong and indeed the silence will become a source of uncomfortableness. When becoming silent has the desired effect, nothing needs to be said. The teacher may proceed without comment. Here again, the length of the silence is critical for maximum effectiveness; too short and there will not be ample time for the students to process the cue and react to it. But too long a period of time encourages students to resume the inappropriate behavior. Practice is the best technique for realizing the most effective window of opportunity.

THE ROLE OF THE PARENT AND THE HOME IN PROMOTING ACCEPTABLE STUDENT BEHAVIOR

Without a doubt, the basis for what most educators would call appropriate discipline is the proper home environment. When students have been reared to behave properly and are punished by their parents or guardians if they do not, then the teacher has a real basis for securing appropriate behavior. When parents and teachers jointly seek to obtain the same goals and when both parties emphasize the same values, much more can be accomplished than when these two important parties in a child's life work at cross-purposes. Unfortunately, the solid base of home support that was once widely enjoyed by agriculture teachers and others has deteriorated. Perhaps today's teachers will have to work at cultivating that parental support that teachers were able to take for granted in past generations. The point is that this support is so valuable that it is worth the energy it takes to cultivate it. Teachers must realize that just because there are more broken homes and single parents today than ever in history, this does not mean there is less support for schooling. However, the existing support may be manifested differently. Educators must believe that the great majority of people care about their children's proper growth and development.

The real key to securing home support in this area of promoting acceptable student behavior is for teachers to develop the appropriate relationships with the parents or guardians of their students. Teachers should initiate the development of this desired relationship. Fortunately, this does not even require additional work or time. Because teachers of agriculture are already visiting incoming and already-involved students, the development of a relationship of openness, trust, and commonality of goals with parents can be accomplished during these same visits.

Teachers must be willing to explain not only their program and the FFA activities to the parent(s) in the presence of the child but also to explain their expectations regarding student productivity and behavior. This should be accomplished during the very first home visit. Then if the parent or the child does not desire what the program consists of and what the teacher expects, they can elect another course that is more in keeping with their preferences.

When all aspects of schooling in the agriculture program have been presented and agreed on in the presence of the teacher, the parent(s), and the student, then the real foundation for promoting acceptable student behavior and performance has been laid. Certainly, during such home visits the prudent teacher will allow for and encourage the input of the parent(s) and child. Of course, the teacher does not act in a high-handed manner during such meetings. Rather, the teacher is a good public relations person and operates in a kind and humanistic fashion.

Once this parent–teacher relationship has been developed, build on it. When making future home visits let the parent(s) know of the student's good efforts, positive accomplishments, and goals toward which to work. As long as the student is willing to respond effectively to the teacher's efforts to secure appropriate behavior, the teacher need not involve the parent(s). Teachers should not tattle on the adolescent each time they are in touch with the parent(s). However, if, after repeated attempts to change a student's behavior, the teacher has not been successful, then a visit with the parent(s) regarding the problem would be in order. In such cases the teacher and the parent(s) need to work jointly in deciding on the best course of action.

Teachers need to remember that parents are generally very proud and protective of their children. So do not approach parents suggesting that their Harry or Melanie is rotten to the core. Rather, approach them in a positive way, saying, "We all care about Melanie"; or "I need your help, ideas, and support." Another caution that should be observed is that teachers should beware of ever threatening students that they are going to see their parent(s). Just as with all other discipline strategies, teachers do not threaten. Rather, when teachers have carefully decided the parent(s) need to be informed, they simply inform the student that this is their selected course of action. Otherwise the idea that the teacher will personally visit with a student's parent(s) about a behavior problem will lose its potency.

Do not only use parents when there are problems. Teachers also should involve parents in the good times. Use them as chaperones at FFA activities and to assist with field trips, trips to career development events, and conventions. Parents do not lose their willingness to be involved with school activities just because their child has gone beyond the sixth grade. Secondary public school teachers simply have not done enough to capitalize on the vital link with the home that is initiated in the elementary years.

PROMOTING INDIVIDUAL RESPONSIBILITY FOR CONTROLLING BEHAVIOR

All of the previous discussions were aimed at helping students learn to behave in appropriate ways. One must not forget that students cannot always have teachers or others around to tell them what to do. A real challenge is to help students realize the need for and accept the challenge to become independent of the need for others to prompt them to behave appropriately. In a very real sense, teachers should have as a major goal helping their students become accustomed to being responsible for their own behavior. This necessitates discussing this goal and provides opportunities for students to accept more and more responsibility and then to be accountable for the results.

One technique, which can be used to help promote this individual responsibility for students' actions, is to constantly relate the concept of responsibility to the expectations students will find in the industry they enter after graduation. Teachers should not only talk about industry's insistence on each employee being responsible for his or her actions, but they should expose their students to their future type of work sites and let the students observe such expectations firsthand. Representatives from the industry periodically should be invited into the school classroom, where one of their themes will be to acquaint students with how the real world operates.

Students who plan to farm must realize early on that not putting forth energy because you are not in the mood will not lead to success. No one tolerates an employee or a partner who can only be depended on sporadically. Likewise, employers do not want workers who are undependable, unreliable, and tardy; cannot get along with their peers; and are unwilling to follow orders. Time after time employers stress that people do not lose their jobs because they cannot perform but because they cannot get along with others. Thus, in agriculture courses the issue is bigger than behaving in ways the teacher and school believe are appropriate. What teachers must really focus on is helping students learn how to behave successfully in life.

Teachers need to work to help students realize that in later life you either behave responsibly as an individual or you are made to behave by society through the laws society makes. Seize the opportunity to discuss current events in local communities as well as big time stories to help students discover the point that, as individuals, we are responsible for our actions and must accept that responsibility if society is to function. Rather than preaching to students, engage in meaningful dialogue. Another strategy for helping students learn the value of accepting responsibility for their actions is to involve civic leaders, law enforcement personnel, judicial personnel, and former juvenile delinquents in class discussions or FFA meetings. Teachers need reinforcement from others in society in their efforts to help students transfer learning from agricultural instruction to life in general.

In the case of promoting individual responsibility for controlling one's own behavior, agriculture teachers need to be sure to use the same model for learning as it applies to discipline, which makes the study of agricultural subject matter career focused, that is, providing for the use of application. The best way to promote individual responsibility for controlling behavior is to give students ever-increasing amounts of responsibility for their own behavior as they progress through the agriculture program. In class they can be given more time to study independently in a self-directed mode. In lab they can be assigned increasingly complex tasks for which they are held responsible. The same strategy can be followed with respect to FFA activities. As students are increasingly allowed to direct their own activity, the teacher needs to remind them of what is happening. Positive reinforcement is essential, but so is corrective feedback at a one-on-one level.

SUMMARY

Successfully managing class behavior is a prerequisite to success in teaching. The actions of the teacher during the first few class sessions greatly influence the type of behavior exhibited in a class. A firm, fair beginning is needed. One cannot let behavior become a crisis and then expect to restore order. Order was lost because it could not be restored in the increments by which it was lost. Generally, it is absurd to think one can control a riot when one cannot individually control students who were unusual.

Once the atmosphere has been established, there are a number of techniques that can be used. One must remember, however, that the teacher and students must understand the problem and why it is a problem before it can be resolved. This demands communication. Communication is usually effective in an individual conference.

Other strategies for obtaining desirable behavior focus on the actions of the teacher well before the class. These include withholding recognition of undesirable behaviors, using suspense, using voice, and using nonverbal gestures.

Finally, the teacher must realize and accept the responsibility for being his or her own disciplinarian. Parents are the teacher's best allies. The real goal is to have students accept the responsibility for directing their own behavior.

FOR FURTHER STUDY

1. Visit a local agriculture department and observe how the teacher manages student behavior. Note what works and what does not work; then figure out why. Discuss your observations with the teacher.

2. Discuss the basic requisites teachers must meet, in terms of pedagogy, if they hope to effectively manage their classrooms.

3. Explain how to go about setting expectations regarding appropriate classroom and laboratory behavior.

4. Give an example of the conditions under which the following discipline strategies should be used, and explain why: (a) individual conference, (b) withholding reinforcement.

5. How can you best involve parents in classroom management? Outline a program that will result in a successful relationship between you as a teacher and the parents of your students.

PART III Application of Learning

Learning has occurred when there is a change in beliefs, attitude, or behavior. When students have used in practice what has been learned, they are more likely to retain knowledge and skills. Classroom instruction can be applied in several ways. Teachers of agriculture use the school laboratory, supervised agricultural experience, and the FFA organization as vehicles for the application of learning. These vehicles assist in developing student motivation to learn, in providing practice of desirable competencies, in providing direction for learning, in providing real problems for study by students, and in enabling students to visualize the need for additional learning.

The laboratory for agricultural instruction is presented in Chapter 9 as a crucial component of the teaching-learning environment. It is a place where students learn and practice the skills needed for agricultural careers. Teachers will find this chapter helpful in planning and conducting laboratory instruction. Emphasis is also given to providing safety instruction and assessing student performance in laboratory work. Methods for teaching and learning within the laboratory environment are presented with the rationale for their use based on principles of teaching and learning.

Chapter 10 presents supervised agricultural experience as needed by students in order to learn the essential skills for careers in agriculture. The supervised agricultural experience of students is to be used for supervised practice of what is taught in the classroom and laboratory. Teaching and learning are more effective when there is supervised practice. Ownership programs, placement or cooperative programs, and improvement and skill development projects are presented as types of supervised practice. Methods are presented for developing individual program plans, conducting supervisory visits, and presenting records instruction.

Chapter 11 contains a discussion of how one might use the FFA organization as a laboratory for learning technical, leadership, and personal development skills. Suggestions are given for building FFA activities into the curriculum in a structural way. Career Development Events, proficiency awards, achievement awards, and the degree program are placed within the teaching-learning context. Methods are suggested for conducting chapter activities with a view of student learning as the primary goal.

9

Application of Learning: The Laboratory

> Recently, a teacher related the following situation. "I taught my students how to sharpen a lawn mower blade. I used a variety of techniques including supervised study and a demonstration illustrated with PowerPoint. I gave them a quiz, and they scored very well. Yet as I've made home visits, I've found that most of their lawn mower blades need sharpening. The students say they've tried but have not had much luck. Obviously, they have learned or they wouldn't have done so well on the test. What is the matter? Where did I go wrong?"

How should these questions be answered? What is missing?

Hopefully, you immediately recognize that passing a test does not mean one can perform in an actual situation. The learner may know how to do something but may not be able to do it. The teacher needs to carry the teaching one step further—to the application stage—and one place to apply what was taught about sharpening a lawn mower blade is in the agricultural mechanics laboratory.

OBJECTIVES

After studying this chapter, you will be able to

1. Explain (in terms of the basic principles of learning) why application of learning is an important phase of the teaching–learning process.
2. Demonstrate how to plan for laboratory instruction.

3. Develop a laboratory rotation schedule and its learning centers.
4. Explain how to use the laboratory to develop students' leadership abilities.
5. Outline a procedure to follow in successfully managing a laboratory session.
6. Develop a laboratory clean-up plan.
7. Outline the components of a laboratory evaluation scheme.
8. List the various types of laboratories used in agriculture programs.

RATIONALE FOR LABORATORY LEARNING

The teacher of agriculture must realize that laboratories for agricultural instruction do not exist based on tradition. Rather, laboratories are a crucial component of the teaching–learning program for education in agriculture.

The purpose of laboratories is to provide organized and systematic instruction. This instruction is of two types: (1) group instruction and (2) individual instruction.

This chapter is limited to examining school laboratories. Table 9–1 indicates some of the types of laboratories in several specialty areas of agricultural instruction that are addressed in this chapter. Laboratory instruction is well grounded in the fundamentals of learning. Without the laboratory, much of the effectiveness of agricultural instruction is lost.

The Need for Application of Learning

The whole notion of learning-by-doing is a very important psychological construct. Consider the following principles of learning (see Chapter 2) on which the agricultural laboratory is able to draw in improving the quantity and quality of learning

Principles of Learning Which Support the Value of Application

1. Students must be motivated to learn. Learning activities should be provided that consider the wants, needs, interests, and aspirations of students.
2. Students are motivated through their involvement in setting goals and planning learning activities.
3. Students are motivated when they attempt tasks that are challenging to a degree that success is perceived to be possible but not certain.
4. Students learn what they practice.
5. Supervised practice that is most effective occurs in a functional educational experience.

Chap. 9 Application of Learning: The Laboratory

Table 9-1 School Laboratories

Specialty Program	School Laboratories
Production agriculture	• School barns
	• School farm
	• School grounds
	• Research lab
Agribusiness	• Customer service area
	• Computer lab
Agricultural mechanics	• Ag mechanics learning laboratory
	• Farm shop
	• Greenhouse
	• Headhouse
	• Nursery arboretum
Horticulture	• Turf plots
	• School grounds
	• Customer service area
	• Hydroponics lab
	• Research lab
Natural resources	• Nature trail
	• Outdoor lab
	• School grounds
	• Wetland
Small animal care	• Clinic
	• Pet shop
	• Grooming area
	• Kennels
	• Wards
Aquaculture	• Aquaculture lab
	• Fish production area

6. Directed learning is more effective than undirected learning.
7. To maximize learning, students should inquire into rather than be instructed in the subject matter. Problem-oriented approaches to teaching improve learning.
8. When students have knowledge of their learning progress, performance is superior to what it would have been without such knowledge.

One must realize that agricultural instruction needs to go beyond learning about theory and practice. A career and technology education course should provide students with the material they are studying in realistic situations, solving actual problems. For example, knowing how to build a wildlife habitat is not enough. The student must be able to actually construct a wildlife habitat. In essence, unless a good deal of what is learned is used, the instruction falls short of being career oriented. Teachers of agriculture who do not provide mechanisms for their students to apply that which has been learned in a functional setting are, in reality, not effective agriculture teachers but are informational agriculture teachers.

The *informational* agriculture teacher teaches students the steps to follow in establishing a seedbed and then tests their mastery of the concepts. The effective agriculture teacher teaches this information also, but in a skill-development context. The teacher ensures the students have in mind a concrete frame of reference, such as their supervised experience or lab project. Once students have learned the essential concepts (or in some cases, as they learn them), the teacher makes sure that the students practice the use of these concepts in a real setting so that the students develop not only "head knowledge" but also the specific ability to *do* the task as well.

When students are able to practice what they have learned, they have completed the teaching–learning cycle. As students apply what they have studied they are better able to see the real meaning of theory. They have a concrete idea of relationships and better understand concepts that are interrelated. They are also better able to understand the reasons why certain practices are called "approved practices." Application is a vital step in problem solving. It is the testing of tentative conclusions. This period of application also provides a meaningful forum wherein to provide feedback to students regarding how well they understood the basic concepts and principles taught in class. The laboratory is one of the places for this application to take place.

FUNCTIONS OF APPLICATION IN LABORATORY LEARNING

Application in laboratory learning provides students with an opportunity to develop manipulative skills. Certainly, this is a crucial function and one that the laboratory serves well. Without the laboratory, students would merely gain an idea of the specific skills they need in an occupation. For example, horticulture students would know that they need to be able to stake a tree properly. However, by being able to go to the laboratory and actually stake trees, students develop specific proficiency. Not only do they have a chance to practice the manipulative skills that the teacher has taught and demonstrated but they also have the opportunity to master (at least at an entry level) the manipulative skills that they need for success in their chosen occupation.

The fact that students have time to practice the important manipulative skills of their specialty until they become proficient is vital. Without a good laboratory for practice, the accomplishments of the local agriculture program would be seriously compromised. This practice must be extensive enough to develop the competence and quality of work required in the industry by those who seek entry-level positions. Therefore, it is essential that the products of students' laboratory work evidence standards of work quality, as identified by the advisory committee when planning the course of study.

Planning for Laboratory Instruction

Planning for laboratory instruction is different from planning for classroom learning. The basic focus of planning for laboratory instruction is to identify the objectives to be accomplished in the laboratory, to identify the work to be done (i.e., the development of the desired skills and abilities), to be sure that all students have an opportunity to develop the desired skills, and to guarantee that all students receive the opportunity to show you they have learned. Thus, the function of the teacher becomes planning how to manage students' practice rather than planning the presentation of basic information.

Planning laboratory instruction involves making decisions. Teachers need to decide which skills and procedures must be practiced and when. They need to decide what work is to be completed in the laboratory, which will serve as the carrier or facilitator through which the desired skills are developed. Once the appropriate work has been identified, teachers need to decide how to be sure that all students have work and that all students have the opportunity to develop the important skills. Invariably, teachers discover that all students cannot practice the same skills at the same time, and thus teachers have to determine how to ensure that each student ultimately gets to develop each skill. Of course, it is essential that the teacher devise a way of teaching and of managing the lab that ensures that students have the necessary classroom instruction, demonstrations, and supplementary aids, such as skill sheets or plans, to be ready to practice the skill or develop the ability once they are given the chance to do so in the laboratory.

Identifying Laboratory Work. In most agriculture laboratories, skills cannot be developed unless materials and projects are available to facilitate such skill development. For example, if I am to learn to be a florist, then I need plants, flowers, containers, foliage, florist supplies, and other materials. The quantity of such supplies available determines the potential learning experience. It is the job of the teacher to provide for these needs. Certainly, for basic skill development the school needs to provide consumable supplies. A laboratory fee will probably need to be assessed. Teachers also need to be resourceful and to be promoters. Industry sometimes donates materials, the Defense Department often sells surplus items inexpensively, and sometimes the FFA alumni affiliate can be helpful. If products that are produced or built in the process of developing skills can be sold, money can be generated to replace consumed supplies.

Unless there is a market for such products, the cost of providing meaningful laboratory experiences becomes prohibitive. The same thing is true in production agriculture. If one is to learn to weld, there must be welders, the ancillary equipment, electrodes, metal, and other supplies available for student use. In the case of learning basic woodworking skills, one must have wood, woodworking equipment, and the necessary hardware. In order for the learning of the woodworking skills to be most meaningful, some sort of useful, relevant project needs to be constructed. Meaningful projects produce items with utility in the real world, such as a gate or trailer, as opposed to an exercise that teaches the skills but has no practical use. If there is no market for the project, the cost of providing the learning experience soars.

Laboratory learning can only be meaningful if the students have received proper instruction preceding their work in lab. The very idea of application via laboratory work implies that one is applying previous learning. Thus, teachers of agriculture must have taught the basic processes and skills in class and be sure students have the basic cognitive and affective understanding that undergirds the development of the psychomotor skill that is to be learned in the lab. For example, prior to students grinding valves in a mechanics laboratory, they must study

- The concept of internal combustion
- The functions of valves
- Specifications
- Parts
- Equipment to be used
- Terminology
- Conditions under which grinding valves is warranted
- How to grind valves

Of course, there must have also been a good demonstration, and the students may well have a skill sheet (see Chapter 7) to follow in attempting the operation in a lab. All of this instruction was preparatory to the students' laboratory assignment to grind valves. Without such thorough prior instruction, the laboratory becomes a meaningless comedy of errors.

Grouping Learning Activities

Generally, there is not sufficient work, materials, or equipment for all students in the laboratory to be doing the same thing at the same time. For example, if students need to practice transplanting shrubs, seldom are there enough shovels, pruners, and other equipment to accommodate everyone in the class at the same time. In the case of an animal care program, a school generally does not have enough aquaria so that each student or even pair of students can simultaneously master the process of setting up an aquarium. Thus, teachers have to design what are called learning centers in order to have all of the students productively involved.

A learning center is an area (physically and in terms of subject matter) within the laboratory where similar (like) work is performed. Study Table 9-2 to get an idea of the probable learning centers in the laboratories of some of the specialty areas of agricultural education.

The specific learning centers in a laboratory during a given part of the school year are determined by the teacher's course of study. Although the

Table 9-2 Example Laboratory Learning Centers for Agricultural Instruction

Specialty Area	Possible Centers
Production agriculture	• Cold metals
	• Hot metals
	• Concrete
	• Woodwork
	• Small gas engines
	• Electricity
Natural resources	• Nature center
	• Monitoring stations
	• Mechanics shop
	• Taxidermy area
	• School park
	• Wildlife habitats
Agricultural mechanics	• Engine overhaul
	• Testing room
	• Welding
	• Electrical systems
	• Equipment set-up
	• Paint room
Horticulture	• Flower shop
	• Greenhouse
	• Equipment maintenance
	• Commercial jobs
	• Arboretum
	• Landscape construction
	• Turf plots
Research	• Biotechnology
	• Computers
	• Library

ideal of presenting the basics in the classroom, followed immediately by practice, is never perfectly attained, there must be a reasonable degree of connection between the development of understandings and the development of skills.

The actual learning centers selected may also be related to the work that is available for sale. For example, if there is no market for various flower arrangements or there are no dogs to be groomed, then the prospective learning center for these labs may never be a reality, at least beyond cursory practice, because of the prohibitive cost of providing such activities simply for practice with no way to recoup costs via marketing.

Once learning centers have been identified or chosen, the teacher must plan the specific learning activities to be completed in each of the learning centers. This should not be construed to mean that every group of students who works in a learning center completes the same activities. For example, the group that works in the flower shop one week might make three planters, but the group the next week may not make any. It all depends on the work that needs to be done during the time students are in the learning center. However, if the teacher is rotating students through a learning center to learn striking an arc and laying a bead in welding (initial skill development), then all students who go through that center will complete the same activities because this is a basic skill development-learning center. This planning is fairly well dictated by the work to be done and the prior instruction that students have had. Before students can engage in meaningful learning, they must have a precise idea of what they are to do, what tools and supplies they need in order to do it, and what procedures they must follow. Once teachers develop assignment sheets (see Chapter 7) for the basic learning center assignments, this part of planning can be used and reused. Table 9–3 illustrates the point that has been presented.

Rotating Students through Learning Centers

If all students cannot do the same tasks at the same time, and must therefore divide into work groups to work at separate learning centers during lab, then the teacher must have a viable system to make sure that, over a period of time, all students get all of the essential experiences. This plan is called a *rotation system*. It ensures that each work group of students has a chance to work through all the learning centers over a period of time. Depending on the nature of the program, this rotation could be completed once during the school year or it could be completed every week or two.

For example, in horticulture a group of students might work in the greenhouse for one day, work in the flower shop the next, care for a school golf green the third day, prune trees the fourth day, and maintain horticulture equipment

Table 9–3 Sample Activities for a Learning Center in Horticulture

Learning Center: Greenhouse	
Activities to be performed	
Daily	Water
	Control temperature
	Remove damaged foliage and spent flowers
Monday	Disbud mums
Tuesday	Space Easter lilies
	Fertilize Easter lilies
Wednesday	Replace aspen pads
Thursday	Start bench for cinereria
Friday	Build bench for cinereria

the final day of the week. However, in an agricultural industrial mechanics program a group of students may work on engine overhaul for an entire grading period. The length of time in a center, the number of centers, and the frequency of rotation from one center to another depends on the nature of the work being done and the scope of the work. It takes longer to assemble an aquaculture system than it does to make a corsage and therein lies the determiner of rotation frequency.

Rotation is not the easiest way to manage laboratory learning. Certainly, the easiest way to manage a laboratory is to have every student practicing the same skill. Rotation is not the preferred way to manage a laboratory based on educational soundness either. Rather it is a necessity. However, rotation is far better than planning enough lab work for each individual student for one day, and then starting to plan from scratch all over again the next day. The primary strength of rotation is that it brings an acceptable degree of systematization, clarity, and structure. It also provides for accountability in that it allows every student to work in every area.

A rotation schedule is planned by grouping work into a number of areas that can accommodate the number of students in the class in mathematically divisible units. For example, if one has twenty-four students, then four learning centers that accommodate six students each might be used; or six centers of four each; or two centers for six students each and four centers for three students each. Another important consideration is that the learning centers require about the same amount of time or divisible portions thereof. For example, if there are four centers of six students each, then all four centers need to require the same amount of time (one week, one month, or whatever is needed). In the case of using two groups of six each and four groups of three each, the groups of three would have to have twice the number of stations as the groups of six, that is, groups of six spend one month (or other unit of time) in a center while the groups of three work in two stations of two weeks each. See the sample charts in Figures 9–1 and 9–2, which illustrate the previous discussion. Figures 9–3 and 9–4 show examples of rotation schedules in selected areas of agricultural instruction.

Developing Students' Leadership Abilities through Involvement in Laboratory Instruction

The agricultural laboratory should approximate the real-world setting to the extent possible. In most businesses (farms and off-farm businesses) the work for the day is identified and the workers are assigned to specific duties. There are general policies regarding getting started, breaks, and how the workplace is to be left at the end of the workday. These same basic notions ought to be built into the agricultural laboratory. Not only will such a scheme create a worklike environment but it will also make the laboratory run smoother and, perhaps most important of all, it will provide an excellent realistic opportunity to help students develop their leadership abilities in a work-related setting.

Four Groups of Six; Each Group Spends an Equal Amount of Time in All Four Learning Centers

Months	Groups			
	A(6)	B(6)	C(6)	D(6)
1	Learning Center 1	Learning Center 2	Learning Center 3	Learning Center 4
2	Learning Center 2	Learning Center 3	Learning Center 4	Learning Center 1
3	Learning Center 3	Learning Center 4	Learning Center 1	Learning Center 2
4	Learning Center 4	Learning Center 1	Learning Center 2	Learning Center 3

Figure 9-1 Rotation schedule for four learning centers.

In order to implement the preceding suggestions, teachers can use students in the following positions of responsibility: laboratory supervisor, leader of a learning center, safety engineer, equipment manager, and demonstration assistant.

The laboratory supervisor's responsibilities include the following:

1. Be sure everyone has the tools, equipment, and materials needed.
2. Be sure everyone starts on time.
3. Record attendance and complete progress charts as students progress through their assigned skills and jobs.
4. Provide assistance for students needing help. This may be personal assistance that the supervisor gives, or it may consist of getting other students or the teacher to provide the needed assistance.
5. Signal time for clean up.
6. Be sure all students adequately complete their clean-up responsibilities.
7. Inform the teacher when the lab is in order and the students are ready to be dismissed.

As students have an opportunity to accept such responsibilities under the teacher's direction they develop confidence in themselves. They also learn how to best get others to perform their duties for the good of the total organization (in this case, the class).

Two Groups of Six and Four Groups of Three, with Varying Times Depending on the Learning Center in Question

Months	A(6)	B(6)	C(3)	D(3)	E(3)	F(3)
			Groups			
1	Learning Center 1	Learning Center 2	Learning Center 3	Learning Center 4	Learning Center 5	Learning Center 6
1	Learning Center 1	Learning Center 2	Learning Center 4	Learning Center 3	Learning Center 6	Learning Center 5
2	Learning Center 2	Learning Center 1	Learning Center 5	Learning Center 6	Learning Center 3	Learning Center 4
2	Learning Center 2	Learning Center 1	Learning Center 6	Learning Center 5	Learning Center 4	Learning Center 3
3	Learning Center 3	Learning Center 4	Learning Center 1	Learning Center 1	Learning Center 2	Learning Center 2
3	Learning Center 4	Learning Center 3	Learning Center 1	Learning Center 1	Learning Center 2	Learning Center 2
4	Learning Center 5	Learning Center 6	Learning Center 2	Learning Center 2	Learning Center 1	Learning Center 1
4	Learning Center 6	Learning Center 5	Learning Center 2	Learning Center 2	Learning Center 1	Learning Center 1

Figure 9-2 Rotation schedule for six learning centers.

The position of responsibility needs to be rotated among the students. There will be some students who do not have the ability to handle this level of responsibility, in which case the teacher may choose to more closely supervise these students. Some teachers let the laboratory act as a club and democratically select this and other positions of responsibility. In such instances teachers work with the students to help them realize that they need to rotate most of the responsibilities among everyone in the class during the year. The position of leader of a learning center requires that the following duties be completed.

1. Know the specific objectives to be accomplished in the learning center each laboratory session.

Sample Rotation Schedule

GROUPS	MONDAY	TUESDAY	WEDNESDAY	THURSDAY	FRIDAY
A	Stick Cuttings	Greenhouse Maintenance	Build Bench	Space Mums; Disbud	Lay Patio Patterns; Record Book
B	Stick Cuttings	Lay Patio Patterns	Greenhouse Maintenance	Sterilize Soil	Space Mums; Disbud; Record Book
C	Stick Cuttings	Replace Windows	Lay Patio Patterns	Greenhouse Maintenance	Sterilize Soil; Record Book
D	Stick Cuttings	Install Mist System	Install Mist System	Lay Patio Patterns	Greenhouse Maintenance; Record Book
E	Flower Shop	Flower Shop	Flower Shop	Flower Shop	Flower Shop; Record Book

Figure 9-3 Sample one-week rotation schedule for horticulture.

2. Be familiar with the skills and procedures to be followed in order to act as an instructional leader. This student may have been given basic demonstrations and practice prior to lab in order to be an effective assistant for the other students in the learning center.
3. Know the location of all tools, equipment, and supplies that are needed to complete the activities assigned for the learning center.
4. Get students in the learning center organized to complete the assigned activities.

The equipment manager keeps an accurate list of what tools are being used and who is using them. He or she checks in tools, checks out tools, and makes sure the tools are clean and put away in their proper places.

The safety engineer is a student who is assigned to guarantee that approved safety practices are used in all laboratory work. To facilitate completing this kind of assignment, the safety engineer needs to be provided with a safety checklist that can be used to guide inspections of safety conditions in the laboratory. Such a checklist includes, but is not limited to, items such as: safety glasses worn by all students in the laboratory; students wearing safe clothing; and when chemicals are used, approved practices (taught in class) used.

Demonstration assistants are students who have special skills. They may have developed these skills through previous employment, or they may have performed exceptionally well when they were in a given learning center. At any

Senior Laboratory Rotation

	A	B	C	D	E	F
	Don Adams Lois Bres Jim Bood Tom Smith	Gerry Crowl Luther Dodson Susan Easter Larry Toth	Carl Muscle Gerri Blake Lloyd Walton Beverly Neyer	Carol Purcel Marilyn Wallace Loomis Linkous Archie Roark	Roscoe Paine Mary Luther Connie Price Richard Meredith	Matthew Brown Jill Musser Tom Bell Judy Milum
	1st 3 weeks	2nd 3 weeks	3rd 3 weeks	4th 3 weeks	5th 3 weeks	6th 3 weeks
A	Orientation	Woodwork; Sawhorse	Electricity; Soldering	Leveling Jack	Project	Small Engines
B	Orientation	Small Engines	Woodwork; Sawhorse	Electricity; Soldering	Leveling Jack	Project
C	Orientation	Project	Small Engines	Woodwork; Sawhorse	Electricity; Soldering	Leveling Jack
D	Orientation	Leveling Jack	Project	Small Engines	Woodwork; Sawhorse	Electricity; Soldering
E	Orientation	Electricity; Soldering	Leveling Jack	Project	Small Engines	Woodwork; Sawhorse

Figure 9-4 Rotation schedule for laboratory work.

rate, these students can help the teacher in giving basic demonstrations and providing assistance to students who are having particular difficulty in an area.

As students have an opportunity to serve in these various roles they develop additional skills in communicating, organizing, and providing direction. They also receive recognition and praise for their work and this provides incentive to continue to strive for excellence in their laboratory endeavors.

The teacher needs to be sure that all students who can cope with the particular responsibility have their opportunity to develop in this way. The teacher also needs to be sure that students are assigned increasing levels of responsibility in the laboratory while still developing their technical skills, so that they are always challenged and so that they continue to grow.

PROVIDING FOR SAFETY INSTRUCTION

An essential element of good laboratory management is providing for safety instruction. Every agricultural laboratory has many dangerous areas and situations. It is crucial that students learn to work in their environment safely not only for their immediate welfare but also so that they develop the essential safety habits needed for their future employment in the industry. From the standpoint of teachers, there are at least three compelling reasons for making sure that their students learn to work safely in the laboratory. First, teachers care about the individual students and do not want anyone to become injured. Second, teachers want to get students totally prepared for a successful future. The third reason teachers must be sure students receive superb safety instruction is to protect their own welfare. The legal climate is such that teachers who do not provide satisfactory safety instruction and supervision may be found liable for failure to do so. Thus every teacher must be sure he or she is adequately covered by reliable professional liability insurance.

The first thing to remember about safety instruction is that safety is largely a question of attitude. Thus, teachers must use instructional practices that impact not only the psychomotor and cognitive domains but also that impact favorably on the affective domain.

The cornerstone for safety instruction is thorough classroom instruction. Students must be taught the specific safety practices that will be expected and required in the laboratory in question. Students must also be taught specific safety practices that pertain to each area of the lab and to specific pieces of equipment and tools.

Structured classroom instruction with complete note taking is essential. Teachers should use a variety of techniques that appeal to all five senses when teaching safety principles and practices. Another effective technique is to bring in people from agribusiness and industry who can relate vivid experiences they have had or seen, and to offer industry specifications on how safety is handled in real-life settings.

Teachers also need to demonstrate and role-model the specifics of safe psychomotor operations. Finally, students must show proficiency on a general safety test, as well as on a specific test for each category of tools and equipment, chemicals and dangerous agents, and specific learning centers. These tests then must be kept on file as evidence that formal instruction has taken place and that each student has demonstrated mastery. There are a number of good commercial sources of safety units, tests, and audiovisual aids. Throughout the duration of the course, students must be reminded of safety. As new problems are discussed, teachers can reiterate previous cautions.

Once in the laboratory there needs to be an effective safety-conscious environment. Teachers must use state and national recommendations for color-coding, display safety posters and exhibits (many good materials are available through each state's safety agencies), and guarantee that all guards and other safety devices on equipment meet state standards. The laboratory must always be neatly arranged, clean, and well lit.

Each day and throughout each laboratory period students need to be reminded of safety. This can be done by the teacher, by the laboratory supervisor, and by the safety engineer if one is used. Another important practice is for teachers to set the example by wearing safe clothing and safety glasses, and following the same rules that they have set forth for their students.

The final element of a good safety program is for the teacher and assigned students to conduct safety inspections daily. As tools and equipment are found to be in anything other than top operating condition, appropriate restorative maintenance must be provided. All in all, the goal of a good program of safety instruction is for students to develop a mind set for safety and practice daily habits of working safely.

SUPERVISING LABORATORY INSTRUCTION

There are a number of tasks that teachers must perform if they are to have an effective laboratory that promotes the goals and objectives of their agricultural instruction program. Teachers have to effectively begin the laboratory period (preferably in the classroom), manage students as they work in the laboratory, evaluate students' laboratory performance, manage tools and equipment use, supervise computer use, and end the laboratory in a desirable way.

Beginning the Laboratory Period

If teachers begin laboratories in a businesslike manner, then this mood is apt to prevail for the rest of the laboratory period. However, if teachers begin laboratories in a disorganized and confused manner, then the entire laboratory period is apt to be tainted by that approach and attitude throughout the session.

The first thing to be accomplished in getting the laboratory started is to achieve the proper mental set. The mental set desired is that "we're here to work, we'll accomplish much, and we'll enjoy doing it." This can be achieved by beginning the laboratory promptly and in a systematic and businesslike fashion, preferably in the classroom where students can sit in the academic environment and check notebooks for project timeliness and daily objectives before reporting to the desired learning centers of the laboratory.

Students need to report for lab promptly and be dressed and ready for work within a set number of minutes after being dismissed from the classroom. The roll should then be checked at each learning center. This can be done by the teacher or by the laboratory supervisor for that day.

Assignments for the day then need to be given. They may be given on paper, on the chalkboard, or orally. Each learning center group must know its assignment. The teacher needs to be sure that the students know the goals or objectives for the lab activities and that they clearly see the relationship between their lab work and their previous classroom learning. The types of forms as shown in Figures 9-5 and 9-6 could be used to distribute assignments to each learning center.

If there is a major new skill or a problem from class or from yesterday's laboratory for which a demonstration is needed in order for the class or groups within the class to be able to complete their assignments for the day, then such a demonstration needs to be given. In fact, a good daily demonstration by the teacher (or a selected student) is a good way to help students continuously accumulate new abilities. It also enriches the period of laboratory instruction. There should be some planned group instruction as a part of the laboratory period almost every day.

Each day students need to be reminded of general safety concerns because of the nature of a given day's assignments. This can be done by the teacher or a student who is prepared for it, such as the safety engineer for that day.

Prior to allowing students to begin work, the teacher or an appointed student, such as the laboratory supervisor for that day, needs to double-check to be sure that every student knows what to do and how to get started. Only then should students be allowed to start working.

If this procedure or a similar one is not followed, then some students begin work, others do not, there is more loafing, there is confusion as to who is supposed to do what, and the general atmosphere of the laboratory is very unsatisfactory and not conducive to learning.

Supervising Students as They Work in the Laboratory

Once students have been properly organized for the day's laboratory and have begun work, then the real job of the teacher begins, that is, supervising students as they work. Teachers must remember that this is a laboratory for learning, not simply for consuming time. It is the teacher's responsibility to manage and direct this important learning experience to be sure that each student's skills and work ethic develop as much as is possible. This means there must be effective teaching taking place.

One of the best ways of helping students to learn during a laboratory is to use "coaching." In this activity the teacher observes a student or group of

LEARNING CENTER: _____

STUDENTS:

a. _____ d. _____
b. _____ e. _____
c. _____ f. _____

TASKS TO ACCOMPLISH:

1. _____
2. _____
3. _____
4. _____
5. _____
6. _____

TOOLS AND MATERIALS NEEDED: _____

GRADE: _____

COMMENTS: _____

Figure 9-5 Task assignment sheet.

AQUARIUM ASSISTANT

NAME: _____

ACTIVITY	DATE _____ MANAGER _____ (Grade)	DATE _____ MANAGER _____ (Grade)	DATE _____ MANAGER _____ (Grade)	DATE _____ MANAGER _____ (Grade)	DATE _____ MANAGER _____ (Grade)
Feed fish					
Fill out temperature form					
Check breeding tanks					
Inspect for and report any disease					
Clean 4 tanks daily — Partial ⅓ of tank — Fill with water of the same temperature — Run diatom filter for five minutes — Clean diatom filter — Clean any corner filters which require it (list tank # of filters cleaned)					
Wipe off front of tanks					
Sweep aisle between sales and breeding tanks and see that all equipment is in good order and working condition and stored properly					
Additional assignments					

Figure 9–6 Form for distributing assignments.

students having problems, and takes time to discuss with the students the skill or operation with which they are having difficulty. Through one-on-one or small-group interaction the teacher offers help by making suggestions, raising questions, performing a troublesome operation for the students, repeating a demonstration, or giving a private demonstration. In essence, the teacher is coaching the students as to how they can improve on their performance. This act of coaching occurs at a very personal level, at the point when the instruction is the most needed and relevant, and in order to help the students achieve success. Thus it is a very potent strategy to use in managing and directing laboratory learning.

On occasion, as teachers are involved with such coaching they find a student who is having an unusual or rare problem that one seldom encounters. However, this is a very important problem and one with which every beginning worker in the industry should be familiar. When such conditions exist, the teacher seizes on what is called the "teachable moment" (that is, a moment when readiness for instruction (see Chapter 2) is high and apt not to be replicable; it is a naturally occurring point in time when instruction has its finest hour). The teacher may stop all laboratory work and have every student report to the learning center in question and there elaborate on the situation, explaining background information, theory, and practice, and teaching every student how to solve this unique problem. Such instruction is invariably exciting and memorable to students.

However, teachers cannot be everywhere in the laboratory at once. Thus teachers need to make effective use of group or learning center leaders if laboratory instruction is to be supervised and guided as well as it must. In order to make the best possible use of students as group leaders, the teacher needs to prepare the leaders for their responsibilities. Sometimes such leaders already have the special skills needed to give special demonstrations or to provide assistance in a specific learning center. These skills may have been developed through the students' supervised agricultural experience programs or because of special interests of the students. However, even if the students do not have predeveloped special skills, the teacher can work with them in advance of when they are to serve as learning center leaders to develop the skills necessary to provide specific help to the rest of the students in the learning center.

Serving as learning center leaders does not always demand that these students possess special manipulative skill. It does mean that such leaders be cognizant of what activities must be performed in their learning center and where the necessary equipment for each day's work is located. This knowledge must be provided prior to laboratory. Teachers can coordinate this pretraining during students' studyhall time, after school, or prior to the start of school. Without this orientation as to what must be done in each lab and a general understanding of how it is to be done, student leaders will not be very effective at reducing the teacher's load and contributing to good laboratory management.

The central problem in providing for a well-managed laboratory is that each student is working individually or in a small group. Whenever teachers divide the number of minutes in the laboratory period by the number of students, they find there is very little time available to provide specific assistance to each student. For example, with a class of twenty-five students in a sixty-minute lab there is no more than two minutes available per student after beginning and closing time is considered. Even if it is a three-hour lab, there is not much more than six minutes per student, and this estimate does not allow for the time it takes a teacher to walk from one learning center to another. Therefore, the teacher must provide as much in the way of supplementary instructional assistance as possible. One excellent way of providing such assistance is by using information sheets, job sheets, or skill sheets, which were discussed in Chapter 7.

Much of the energy of the teacher who wants to do a good job of supervising laboratory learning must be directed toward checking the progress of students. Teachers must be on the move observing the progress of students, ensuring safety, raising questions, offering guidance, showing how, giving praise, answering questions, locating tools and supplies, and being sure that the laboratory environment is conducive to the growth of each student. This is a relentless task, and only determination and a high level of energy get a teacher through each day's laboratory. However, other alternatives to this kind of supervision spell disaster or at least disappointment for all who are concerned.

Evaluating Student Performance

Assessing Work Habits. There are three areas of evaluation of students' performance in the laboratory that are discussed in this text (see Chapter 14 for evaluation of learners). The first area is that of assessing students' work habits. Teachers need to use the laboratory to help students learn to be prompt, attend to their assigned work, and develop efficient work habits and a positive attitude toward working. One way of facilitating this is by using a checklist, such as the one shown in Figure 9-7, daily or weekly.

Teachers can also have the laboratory supervisor or a learning center leader complete such assessments. The checklist needs to be shared and discussed with each student. Suggestions for needed improvement should also be offered. The results of this phase of the students' evaluation could possibly count as 10 percent or more of their laboratory grades.

Assessing Process Skills. Another area of laboratory evaluation is that of evaluating how well students follow prescribed processes or procedures. For many laboratory assignments specific procedures are prescribed. See the sample job sheet in Figure 9-8. Any of these aids that prescribe procedures or steps can be used as the checklist to follow in determining whether or not students followed directions. Of course, point values need to be assigned

Rating Scale: 5 = Excellent
4 = Very Satisfactory
3 = Satisfactory
2 = Needs Improvement
1 = Poor

Dressed appropriately for lab	1 2 3 4 5
Begins work promptly	1 2 3 4 5
Organizes work before beginning	1 2 3 4 5
Completes each task before moving to another	1 2 3 4 5
Cooperates with others	1 2 3 4 5
Follows instructions	1 2 3 4 5
Accepts responsibility	1 2 3 4 5
Is courteous	1 2 3 4 5
Returns tools and materials to proper location	1 2 3 4 5
Helps with clean-up	1 2 3 4 5

Figure 9-7 Work habits assessment sheet.

to the stages based on their importance. This lends some degree of objectivity to the assessment.

Product Assessment. Product assessment is perhaps the most important phase of the assessment of laboratory learning. Well-designed lab activities are created to develop specific abilities. Student assessment must be based on the objectives of the assignment. In order to do a fair job of evaluation of products, a score sheet such as the one shown in Figure 9-9 should be developed based on the objectives of the laboratory activity. This score sheet can serve to provide the criteria to be used in the assessment. Before the students start the lab assignment they should be provided with a copy of such a score sheet so that they can concentrate on the specifications included therein.

To aid in the learning process and the evaluation of learning, teachers should display product examples to which students can compare their products. There are two types of product examples that are necessary. One shows diagnostic examples; the other shows examples representative of various marks (A, B, C, D, and F). For example, in the case of a flat lap weld, a teacher could display examples of welds that were and were not uniform in width and thickness or other characteristics. Likewise, the teacher could display examples of "A" welds, "B" welds, and so on along with diagnostic score sheets. The same is true for corsages. Teachers will want to display corsages which do and do not follow basic design techniques, and then also display "A" corsages, "B"

Activity Sheet IX
Air Cleaners

Objectives
1. To be able to explain why air cleaners are important in the operation of an engine
2. To be able to explain how the different types of air cleaners work
3. To be able to properly service air cleaners

Ways and Means
I. Study Information Sheet IX. If you have difficulty understanding the information, ask your teacher for help.
II. When you believe you understand the information presented, answer the following questions. These questions are designed to help you understand this unit. Be sure to give a complete answer for each question.
   1. Why is it important to have an air cleaner on the air intake of an engine?

   2. Explain how the dry-type air cleaner removes dirt from the air.

   3. Explain how the oil-bath type air cleaner removes dirt from the air.

   4. Your teacher may have other questions related to this unit.

III. Now check your answers with those on Answer Sheet IX. Correct the answers you missed before going to Statement IV. The answers to these questions will be important in later units.
IV. Complete the activities on Project Sheet IX-A.
V. Complete the activities on Project Sheet IX-B.
VI. Secure your teacher's signature to signify your successful completion of this unit.
   Signature: _____
VII. Retain this completed unit for future reference. Go on to the last unit.

Figure 9-8 Sample activity sheet.
Source: Printed with permission of The Ohio Curriculum Materials Services. From Ohio Agricultureal Education Curriculum Materials Service (1970). *Individual Study Guide on Carburetion.* Columbus: Author.

Sawhorse
EVALUATION SCORE SHEET

Name: _____ Date: _____

Criteria	Points Allowed	Earned
1. *Correct Dimension* (50 points)		
a. Height at one end	8	
b. Height at opposite end	8	
c. Spread of legs at one end (width)	8	
d. Spread of legs at opposite end (width)	8	
e. Length of beam	10	
f. Beam extension beyond leg at one end	4	
g. Beam extension beyond leg at opposite end	4	
2. *Correct Leg Angles* (20 points)		
a. 90-degree angle between beam and table top at one end	6	
b. 90-degree angle between beam and table top at opposite end	6	
c. All four legs on floor	4	
d. Correct bevel on the bottom of all four legs	4	
3. *Joint Preparation, Correct Angles* (30 points)		
a. Leg on beam (leg 1)	3	
b. Leg on beam (leg 2)	3	
c. Leg on beam (leg 3)	3	
d. Leg on beam (leg 4)	3	
e. Outside brace at one end	6	
f. Outside brace at opposite end	6	
g. Screws properly countersunk	6	
Points Earned	100	
4. *General Appearance* (minus points)		
a. Free from marks	1 @	
b. Cracked, split, or chipped boards	1 @	
c. Screw placement	1 @	
d. Machine and tool marks	1 @	
e. Excessive glue	1 @	
Penalty Points		
Final Score		

Figure 9-9 Evaluation score sheet.
Source: Courtesy Dr. Joe A. Gliem, Department of Human and Community Resource Development, The Ohio State University.

corsages, or "unacceptable" corsages along with the sheet used to assess those grade categories.

Once the basis for evaluation of lab work is established and the evaluation is completed, there are two types of records that need to be kept. One is the recorded grade for future reporting. The other is a progress report so the teacher knows who has yet to experience a given task. The progress chart shown in Figure 9–10 can easily be used for this latter record.

Managing Tools and Equipment

As if teachers did not already have enough to do in trying to manage laboratories effectively, they also must be sure to manage all tools and equipment, including computers. Without a viable management system for maintaining and upgrading tools and equipment, there will not be a laboratory to manage. Tools, equipment, hardware, and software are very expensive, and no administrator should settle for a teacher who does not do a good job of keeping them in good condition or updated. Also, if tools, equipment, and computers are in disrepair, broken, or missing, the students will be unable to stay productively involved during laboratory, the consequences of which can be readily predicted.

If students do not learn to properly use and care for tools, equipment, hardware, or software as a part of their instruction, then they will encounter serious problems when they enter the workforce. In essence, students need to develop acceptable equipment use habits. The dollars involved in industry demand such an attitude.

It is the preference of the authors to store or locate tools of all kinds on a display board at the learning center where they will be used. In the case of CDs and DVDs appropriate filing systems should be located at the learning center. Larger tools and pieces of portable equipment should be kept in a logical storage area and checked in and checked out using a system that everyone understands. Wherever tools are kept, they should be kept clean and ready for the function for which they were intended.

Every student who uses a tool or piece of equipment should return it cleaned and ready to use to the place where he or she found it. At the close of every laboratory period, the supervisor and teacher must be certain that everything is clean, in its place, and, in the case of computers, certain that no personal files are left on hard drives. Otherwise problems mount.

Teachers must also provide students with instruction on refitting tools, upgrading software packages, and fine-tuning equipment. Students need to learn to sharpen hoes, shovels, chisels, chain saw chains, and other tools. They must learn to replace glass or plastic on the green house, to readjust gate hinges in the school barns, to sharpen clippers used for dog grooming, and readhere PVC pipes on aquaculture tanks. This is an important part of preparation for work and is vital to the successful management of tools and equipment.

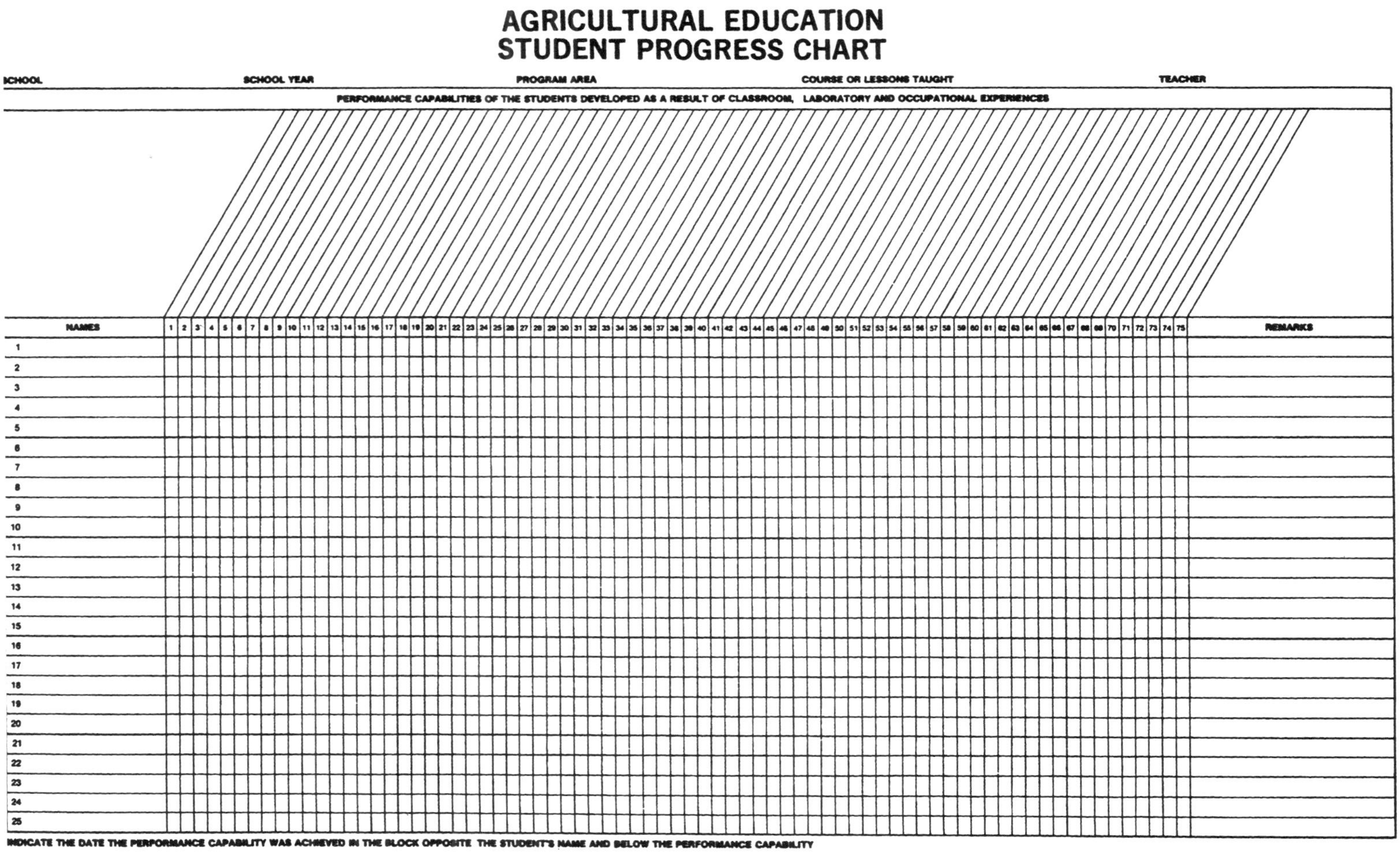

Figure 9-10 Progress chart for agricultural instruction.

Source: Printed with permission of The Ohio Curriculum Materials Service.

Ending the Laboratory Period

Just as the laboratory period needs to be started in a businesslike manner, it is also imperative that it be ended in such a manner. The following procedures are suggested:

1. *Signal a time to stop.* This may be done by blowing a whistle, ringing a bell, or other procedures that can become habitual for students. On giving the stop work signal, everyone must immediately cease work; otherwise, the system does not work. This expectation has to be clearly defined early in the school year, reinforced, and practiced.

2. *Each student puts away his or her work, tools, and equipment.* Students need to clear the laboratory so it can be cleaned daily. Projects cannot be left strewn about. Students cannot be allowed to leave tools out, reasoning that they'll be needed again tomorrow or by students in another laboratory. Such practice leads to a situation in which no one can find anything. If a student is acting as the equipment manager, all items used in the laboratory were checked out by that person and must be checked back in by that person.

3. *Everyone joins in to clean the entire laboratory.* The only way this works is if everyone has a specific duty during clean-up. Teachers should demonstrate how they want each clean-up job performed; otherwise, perfection will not be attained. These duties need to be rotated because some are much less pleasant than others. Many teachers in agriculture mechanics laboratories use a clean-up wheel such as the one shown in Figure 9–11. A similar wheel can be created for all types of laboratories. All students' names in the class are included in the center circle. Then one clean-up duty for each name is listed on the outer circle. The wheel rotates clockwise one name each class period.

Note: In many agricultural laboratories, clean-up may be assigned by learning center rather than in general. For example, in a horticulture laboratory, students in the retail shop may have a list of duties to attend to, whereas those who are working in the greenhouse have an entirely different set of responsibilities.

4. *Once all clean-up duties have been completed, students may clean themselves up and dress for the rest of their schoolday.* If students are allowed to attend to personal grooming before the teacher (or supervisor) checks their clean-up duties and clears them to prepare themselves to leave the laboratory, they loiter and the clean-up never gets done.

5. *Once students are groomed and ready to leave the laboratory they must wait until the teacher is satisfied, primarily via the laboratory supervisor, that everything is in order.* The teacher then dismisses the class. The students do not leave until the teacher has dismissed them; otherwise, laboratory clean-up is never well done.

Chap. 9 Application of Learning: The Laboratory

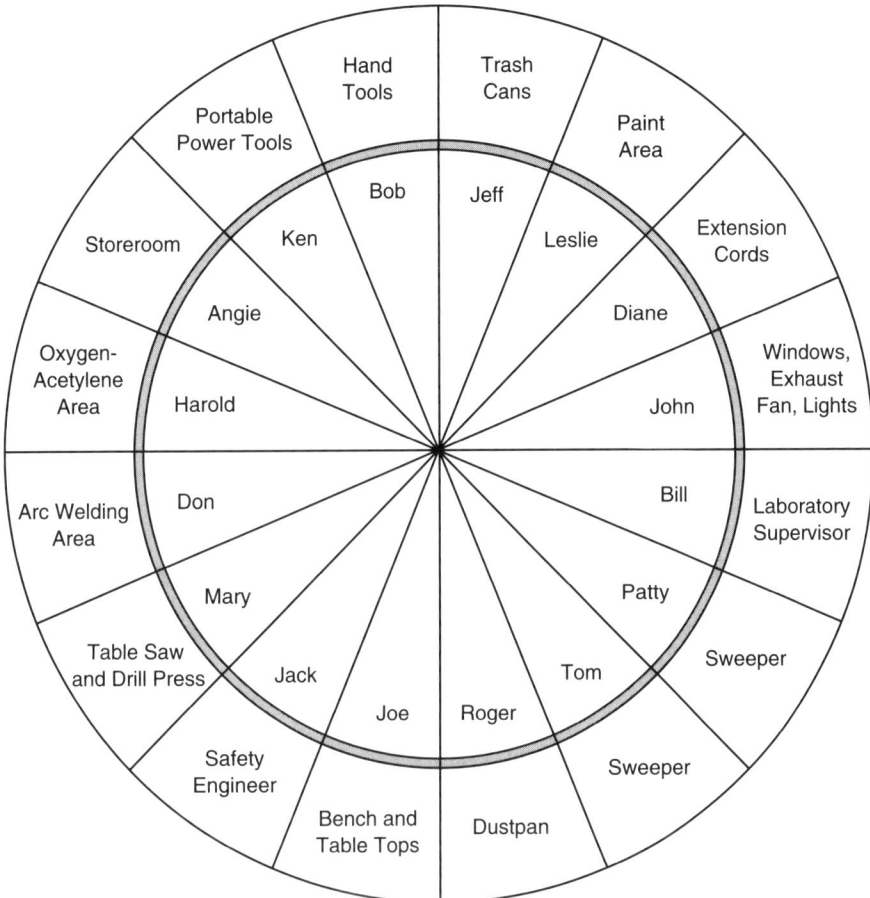

Figure 9-11 Agricultural mechanics laboratory clean-up detail.

If the laboratory is not cleaned after every class period, no system will work. A clean laboratory is an efficient and productive laboratory and must be a priority item with teachers and students alike.

DEVELOPING A SALES POLICY

An important phase of the agricultural laboratory is the completion of work for customers. In agricultural mechanics this might consist of repairing someone's tractor. The horticulture class may sell bedding plants to the public, prepare floral arrangements for a wedding for someone in the community, or landscape a residential site. Students studying food processing may market fresh meat to the public, and students in small animal care may sell pets and pet products, and provide pet services.

In all of these instances relationships with people outside the class must be established and maintained. This is an essential component of training for employment outside of the school setting. The work generated by providing such services and products is essential to the success of the laboratory phase of the instruction program. Without the work created for the sale of products and services, the laboratory would not provide a vibrant learning experience. Instead, it would consist of a series of concocted exercises. Students would not be immersed in reality, and the school could not afford enough materials to allow the students to develop the level of skill needed in order to enter and progress in entry-level jobs after graduation.

The problem is that whenever the school agrees to produce or to serve for a fee, then potential work has been diverted from the private enterprise of the community. If the businesses in the community have not been involved in recommending the need for students to have these experiences and in helping establish an agreeable sales policy, problems can develop. Therefore, teachers must involve retail-level representatives in the agricultural industry in question in establishing such policies. For example, the owner of the local pet store must be asked to be a key consultant in writing the companion pet curriculum so skills can be developed that prepare students to be successful employees of that store. This relationship is important for the agricultural instruction program.

The cornerstone of any successful sales policy rests on the idea that the experience of selling to the public is a vital component of preparing people for entry-level jobs in the industry. Generally, the bulk of the sales of products is made to students, teachers, and workers in the school. In the case of floral work for weddings or funerals, an agreement to do no more than a certain number of each type of wedding or funeral may be needed. Because the horticulture industry encounters more problems of retailers fearing their market being diluted, teachers of horticulture need to help their industry partners realize that by introducing flowers to more people in the community the potential market is increased in future seasons.

It needs to be understood that products and services provided in the agricultural instruction program need to be sold for somewhat less than would be the case in industry. This is because students who are developing proficiency make mistakes and produce work that may be somewhat inferior to commercial grade. By the same token, however, the prices charged by the school should not be so low as to dramatically draw consumers out of the public market place.

In the case of services rendered, such as repairing a tractor, landscaping a home, stocking a pond, or producing a company newsletter, the school needs to charge for actual costs plus a standard percentage that covers some overhead and breakage and maintenance of the equipment used. These fees can be set annually after consulting with the advisory committee of the program.

Unless such a procedure is followed, too many "picky" problems are encountered throughout the year, in which case prevention of such problems is possible and essential.

Given that satisfactory policies have been formulated, the teacher must use sales as an important part of teaching students for their future jobs. Inventories must be kept, sales records and procedures need to be similar to those of industry, pricing procedures need to be taught, and so on. The laboratory becomes a place to learn retailing as it applies to the agricultural specialty in question. This entrepreneurial, or business management, phase of the instruction needs to be experienced and understood by all students as much as the psychomotor skills.

SUMMARY

Good laboratory management is important for any teacher. Nothing less is acceptable. The laboratory must truly be a laboratory for learning, including more than learning skills.

Students must practice what they have learned in the classroom until they are truly skilled workers. Yet laboratory is more than applying what has been learned in the classroom. It is a place where additional instruction is given, competence is developed under the watchful eye of a masterful teacher, and important work attitudes and safety are developed.

Students learn to lead and to follow as a consequence of their laboratory experiences. They must be productive but safe. They must learn to use tools and equipment properly and to care for them as well. Finally, they must learn to leave the laboratory ready for the next users and to follow an established routine. In laboratories that are operated properly learning will take place, and students will become prepared to enter jobs and colleges after graduation.

FOR FURTHER STUDY

1. Visit a local agriculture program and analyze how the laboratory instruction is conducted. Identify the strengths of the operation. Then develop an improved plan for the operation of the laboratory based on what you have learned in this chapter.
2. List the basic steps one needs to follow in planning laboratory instruction.
3. Suppose you have a group of twenty students in your agriculture program (you select the length of time in lab each day and the tasks to be

completed). Plan a rotation schedule for one week, month, or semester (you may select the period of time).
4. Outline the basic strategies you can use as a part of your laboratory management systems to help develop students' leadership abilities.
5. Describe the sequence of events you would follow in operating each laboratory session, beginning with the starting bell for the period and ending with the final bell of the laboratory period.

10

Application of Learning: Supervised Agricultural Experience

> Mike is interested in animals. He raised rabbits and mice for a local pet shop during his first two years in an agriculture program. During his junior year he worked in the pet shop as a part of his supervised agricultural experience program. During his senior year he worked for a dog groomer. He hopes to own his own business or teach agriculture after further education.

Students studying agriculture apply what they learn in the classroom not only in laboratory but also in their supervised agricultural experience programs. These experience programs provide a wealth of relevant problems for study in class. Many useful interest approaches can also be developed from these student experiences.

Supervised agricultural experience consists of all the agricultural activities of educational value conducted by a student outside of class for which systematic instruction and supervision are provided by parents, the agriculture teacher, employers, or other adults. It is *supervised*. It is *agricultural* because it helps the student learn about agriculture. It is *experience*, or "learning by doing," or experiential, because it allows students to apply practices and principles learned in the classroom and to develop new skills and abilities.

OBJECTIVES

After studying this chapter, you will be able to

1. Explain the need for supervised agricultural experiences.
2. Work with students in developing supervised agricultural experiences.
3. Relate instruction and supervised practice.
4. Supervise agricultural experiences of students.

Throughout the rest of this chapter "supervised agricultural experience" is referred to as "supervised experience or SAE."

NEED FOR SUPERVISED EXPERIENCE

The need for supervised experience in agricultural instruction is established because of improved learning, student personal development, and career and technical development.

Supervised Experience and Learning

Students with programs of supervised experience learn more in agricultural courses. This greater learning might be caused by several factors. Students with programs of supervised experience have a greater need to learn. They also have the opportunity to practice what is taught in the classroom.

> An agriculture student had limited space for a project but desired to raise hogs. She built a portable hog house in the agricultural mechanics laboratory and used an existing concrete area at her home to assemble the portable unit. She fenced in a small lot and raised three groups of feeder pigs per year. The project enabled her to learn much about swine production and marketing, as well as related mechanical skills. The profit from the project was a major factor in the student attaining the state FFA degree.

Personal Development

Different types of supervised experience may develop different abilities. Most types contribute to the personal development of students. Supervised experience provides opportunities for students to work, earn money, achieve a degree of financial independence, and assume increasingly greater responsibility. Students who participate in supervised practice are provided with an opportunity to develop their managerial ability. Those with entrepreneurship or production projects plan work, develop budgets, work with financial agreements, make decisions, solve problems, put plans into action, and keep records. A student develops his or her program by displaying initiative, a trait essential in management of one's self and others.

Work habits are developed and practiced in supervised experience. Students on work experience programs work under the supervision of employers, the teacher, and parents in establishing desirable work habits. For all projects managed by a student, he or she must set work goals and organize the available time to accomplish the goals. The result is often improved work habits.

Abilities in cooperation are developed by participation in supervised experience. The teacher, parents, work supervisor and students cooperate in planning and evaluating programs of supervised experience. During work experience, students must cooperate with employers and employees because goals may be accomplished in relationship to others rather than in isolation. Thus, human relations skills are also learned and practiced.

As discussed previously student self-concept development is critical in teaching and learning. During supervised experience students are expected to budget the financial aspects of supervised experience. They set goals and work to attain them. This ability to follow through on what has been planned is important in the personal development of individuals.

In summary, supervised practice is an integral component of instruction in agriculture. Traits developed include

1. Ability to assume responsibility
2. Willingness to follow good work habits
3. Readiness to show initiative
4. Ability to get along with others
5. Willingness to learn on the job
6. Flexibility[1]

Career Development

Programs of supervised experience are important in the career development of students. Opportunity is provided for a student to grow gradually into a career. Various occupational choices can be explored by supervised experiences. Students can then make intelligent career decisions.

Earning money is a part of most supervised experience programs. Students learn to develop budgets and to spend money wisely. They can accumulate savings as they invest toward future goals. Students are supervised in financial planning as they earn money either through work experience or ownership programs.

Work experience is advantageous to students as they seek employment. They have a personal job reference. Many employers desire individuals who have had a previous work record. Students participating in supervised experience

---

[1]Amberson, Max L., and Anderson, B. Harold. *Learning through Experience in Agricultural Industry.* New York: McGraw-Hill, 1978, pp. 18–19. Out of Print.

have been making productive use of their time. They have a real advantage over others when competing for placement and advancement.

Supervised experience provides students with an opportunity to develop a special area of expertise. Some students may learn to do experiments in agricultural science, others may specialize in landscaping, others may specialize in machinery reconditioning, others may choose sales, and still others may select swine production, to name only a few. Supervised experience is a form of individualized instruction in which students can develop competence in areas of interest to them. As they gain technical skills, they prepare for occupations for which their specialized expertise is desirable.

SUPERVISED EXPERIENCE AND TEACHING

The effectiveness of programs of supervised experience varies with the extent to which they are used in teaching and learning. The interrelationship of classroom, laboratory, supervised agricultural experience (SAE), and FFA activities enhances learning. The teacher needs to orchestrate student involvement in these essential aspects of the agricultural instruction program.

Relating Instruction and Practice

Teachers of agriculture have the responsibility of demonstrating to students that supervised experience is important for learning to take place. Integration of supervised experience can be accomplished by using the situations of students as interest approaches in introducing topics of study, as sources of problems for study, and in applying what is learned.

Interest Approaches. Programs of current and former students can be effectively used to stimulate interest in a topic. For example, a teacher introducing a unit in wildlife management might use slides of a project conducted by a former student to provide cover for pheasants. A unit on merchandise display might be introduced by showing PowerPoint images produced from digital images of displays built by former students at their places of employment. A wood construction unit could be introduced by describing a home improvement project of a current student that won an FFA proficiency award. A sales skills unit could be introduced by having a panel of students describe their roles in selling products to customers. Problems and achievements of current and former students are effective in stimulating student interest.

Sources of Problems. The problem-solving approach to teaching assumes realistic problems are used as a basis for instruction. In agricultural instruction, a major source of problems is the supervised experience of students. Teachers of agriculture must draw on the experiences of their students in formulating problems for classroom instruction. In order for this to be possible, students must be encouraged to develop supervised experience relating to the course of study of the program in which they are enrolling. This assumes, of course, that the curriculum has been planned in accordance with community and student needs.

As the teacher visits students in their homes, work places, and school labs, pictures should be taken and notes kept concerning problems the students encounter and solutions to those problems. These can then be filed for later use as the teacher plans units of instruction. A written plan on wiring a charging circuit in agricultural mechanics can be based on resolving the problem of a class member who needed to accomplish this task as a part of a supervised experience program. Instruction in the control of greenhouse pests can be based on solving the problem of a student who is placed in a greenhouse environment for a supervised experience. The problem of how to fell a tree can be studied in relation to a student's timberstand improvement project. In order to be effective in basing instruction on student problems, a teacher must first have students develop realistic projects and then be thoroughly familiar with the work of the students during their programs of supervised experience.

Application of Learning. Students who have learned should show a change of behavior (see Chapter 2). Learning that is applied in practice is learning that has been assimilated. Many teachers close a unit of instruction by having students develop a plan of practice. The plan of practice specifies the way students will apply what was learned in the classroom in their supervised experience, FFA participation, or laboratory experience.

A unit of instruction on weed control may result in a plan for controlling weeds in a row crop production project, recommendations for applying agricultural chemicals for use in sales work in an agricultural business, or a plan for controlling weeds on lawns or golf courses. A plan for applying instruction in supervised experience should be a part of most instruction in agriculture.

TYPES OF SUPERVISED EXPERIENCE

Project work can usually be classified into the three general areas of ownership programs, placement or cooperative programs, and improvement or skill development projects.

Ownership Programs

Students studying agriculture may select three types of ownership programs. Ownership programs relate to the career interests of the students and to the course of study. The three types of ownership programs are production projects, group enterprise projects, and entrepreneurship projects.

Production Projects. A production project is a business venture for experience and profit involving the production of some type of crop or animal. A student may produce a greenhouse crop of mums, a field of cotton, feeder pigs, chinchillas, or a truck crop of tomatoes. Rather than producing the product only once, the student usually is encouraged to carry the project through multiple production cycles on a continuing basis. The teacher assists the student through supervisory visits on a year-round basis if needed, depending on the production cycle of the enterprise.

The student is expected to have a financial interest in the project, either as a full owner or as a partner with parents or others. The production project provides opportunity for the student to learn gradually about agricultural production and also accumulate assets allowing him or her to enter farming or an agricultural business, or to further his or her education.

> An agriculture student became interested in gardening. He raised vegetables, including tomatoes, sweet corn, potatoes, peppers, peas, and cabbage. His specialty, however, became strawberry production. The size of his plot required that labor be hired to assist with picking the crop. He sold his produce at a roadside stand. He annually sold several thousand quarts of strawberries and placed high in the state FFA proficiency competition.

Group Enterprise Projects. Group enterprises often provide opportunities for students to have ownership projects. A group enterprise is similar to a production project; however, a group of students share in the management decisions and work related to the project.

> A farm management class started a cooperative to provide experience in livestock production and management and to learn more about agricultural cooperatives. A governing board and a manager were elected. The students bought shares. They raised broilers and feeder pigs. They worked according to a schedule developed by the manager so that each student did an equal amount of feeding and cleaning. The class made decisions about the projects, with members sharing equally in the profits.

Group enterprises may also be used when students are unable to have a project at home. For example, a group of students can raise bedding plants in the greenhouse that can be marketed by wholesale or retail means. Any project that can be managed by an individual student can also be adapted for use as a group enterprise.

> In one school, four students leased land to raise a cash crop of soybeans. The students each put an equal amount of money into a joint bank account. The business was managed according to a formal agreement among the involved students and their parents.

Entrepreneurship Projects. Entrepreneurship projects are those in which students own, organize, and manage businesses for the sake of profit. This type of project is well suited to most off-farm areas of agriculture. A student who is interested in and is studying agricultural mechanics may start a lawn mower, chain saw, or tool repair and reconditioning business. A student interested in landscape and turf work may develop a home lawn and landscape service business. A student desiring to work as a park naturalist can profit from carpentry skills by building and selling picnic tables and wood signs similar to those used in parks. A student interested in forestry can start a business cutting and selling firewood. A floriculture student can develop a business in servicing weddings. Entrepreneurship provides opportunities for students to learn needed technical and managerial skills. All ownership projects allow students to make decisions and apply the instruction received in class.

Placement or Cooperative Programs

Students may be placed on farms or in agricultural businesses or industries for supervised practice. The primary purpose of the placement is the learning and application of needed agricultural knowledge and skills. An important secondary purpose is that students are able to learn to earn and manage money. A placement program is often referred to as cooperative education. This is because the business, represented by the employer, and the school, represented by the teacher, cooperate in providing the needed learning experiences for the student. Some use the term *placement* to refer to experiences gained outside of school time and the term *cooperative education* to refer to supervised experiences which involve some release time for the student during the schoolday. In either case, the teacher should ensure that the experience is supervised. This supervision includes making sure a placement agreement has been developed and signed. It also includes assisting the employer and the student in creating an education development plan for the work experience program.[2] Just as with other types of supervised experience, all work of students should be related to classroom instruction.

> Sarah has been placed on a dairy farm near her home for two years. She works six nights a week feeding and bedding the young calves, steers, yearling heifers, and forty-five milk cows. Other tasks include giving medical injections, grooming cattle, and delivering calves. She assisted in crop production—raising corn, soybeans, wheat, and oats. She helped bale hay and straw, and exhibited cattle at two county fairs. Sarah hopes to become a farm manager or possibly work for a breed association after further education in dairy science.

---

[2]For examples of cooperating center agreements and supervised agricultural education plans, refer to Amberson and Anderson, *op. cit.* pp. 116–119, 129–131, and to Figures 10-4 and 10-5 in this book.

Improvement and Skill Development Projects

Improvement and skill development projects are designed to provide opportunities for students to learn and practice skills, and improve their surroundings in home or business.

Improvement Projects. An improvement project may be developed by a student to improve the efficiency of an enterprise or of an entire business; the appearance or real estate value of the farm or place of business; or the appearance, value, comfort, and convenience of the home of the student.

Parents or an employer usually finance this type of project, with no degree of ownership by the student. Some examples are

- Garden improvement
- Lawn improvement
- Home landscaping improvement
- Interior landscaping of the home or business
- Home shop improvement
- Nature trail development
- Sheep enterprise improvement
- Home painting
- Home library development
- Business product display improvement
- Records analysis

Most students in applying instruction can use similar improvement projects. For example, following a unit on gardening, students could be encouraged to apply the instruction by having an improvement project on gardening.

Skill Development Projects. Skill development projects are often called supplementary practices. Students learn specific skills by completing these projects or practices. They enable students to broaden their experiences beyond ownership, placement, and improvement projects. The skills, or competencies, are usually performed to learn tasks needed for agricultural careers. Table 10-1 is a list of skills that students can perform for skill development or supplementary practice.

DEVELOPING INDIVIDUAL PROGRAM PLANS

As students enroll in agriculture courses, the teacher works with them and their parents in developing a plan for supervised experience. Much of the instructional program in agriculture is designed for teaching approved practices that are widely accepted by the industry. Most students, therefore, should be encouraged to develop programs that relate to instruction and also to their unique interests and abilities.

Chap. 10 Application of Learning: Supervised Agricultural Experience

Table 10-1 Example Skills for Supplementary Practice

A. Agricultural Production
  1. Take a soil sample
  2. Inseminate cows
  3. Operate a tractor
  4. Prepare seedbed before planting
  5. Calibrate planting equipment
  6. Castrate pigs
  7. Probe for back fat
  8. Dock lambs
  9. Worm sheep
  10. Keep feed records

B. Agricultural Mechanics
  1. Operate arc welder
  2. Pour concrete
  3. Lay concrete block
  4. Tune up a multicylinder engine
  5. Overhaul a small gas engine
  6. Set up a plow
  7. Adjust clutch linkage
  8. Change oil and filters
  9. Cut metal using oxyacetylene
  10. Wire an electrical circuit

C. Horticulture
  1. Prune trees
  2. Identify common shrubs
  3. Make a corsage
  4. Propagate using cuttings
  5. Transplant seedlings
  6. Water greenhouse plants
  7. Prepare a landscape plan
  8. Grade, rake, and level a lawn
  9. Wrap and stake trees
  10. Read blueprints

D. Agricultural Products Processing
  1. Operate churn
  2. Test milk for butterfat content
  3. Slaughter lambs
  4. Trim hams
  5. Grind beef
  6. Use scales
  7. Sanitize facilities
  8. Cut fish
  9. Candle eggs
  10. Set machine

E. Forestry
  1. Plant trees
  2. Operate chain saw
  3. Operate bulldozer
  4. Operate skidder
  5. Identify lumber
  6. Grade lumber
  7. Fell trees
  8. Set backfire
  9. Delimb trees
  10. Operate winch

F. Agricultural Business, Supplies, and Service
  1. Complete sales slip
  2. Use telephone
  3. Balance rations
  4. Mix feed
  5. Test grain moisture content
  6. Operate scales
  7. Bill customers
  8. Develop sales displays
  9. Clean seed
  10. Bag fertilizer

G. Renewable Natural Resources
  1. Inventory fish populations
  2. Evaluate fish habitat
  3. Take water samples
  4. Analyze water samples
  5. Mow lawns
  6. Clean swimming pool
  7. Paint buildings and equipment
  8. Maintain nature trails
  9. Analyze effluent
  10. Identify wildlife species

H. Small Animal Care
  1. Feed animals
  2. Groom dogs
  3. Restrain animals
  4. Order supplies
  5. Take blood samples
  6. Perform euthanasia
  7. Identify fish
  8. Clean and disinfect pens
  9. Identify parasites
  10. Identify animal heat signs

List on this page the careers in horticulture in which you would like to be employed upon graduation. Be sure to talk your plans over with your teacher and your parents. Your plans should cover the period from the present until you enter your chosen occupation. After you have identified a horticultural area as an occupational goal, determine the experience, education, and capital required to be successful in at least two job classifications under each area.

OCCUPATIONAL GOALS:

After completing my education I hope to work in one of the following horticultural areas. (If you have more than one choice indicate by 1, 2, or 3.)

____ Retail Floriculture   ____ Turf   ____ Garden Center
____ Floriculture Production   ____ Nursery   ____ Vegetables and Fruit
____ Landscaping   ____ Equipment and Mechanics   ____ Floriculture Wholesale

First Occupational Goal: _____
Job Classification: _____ Job Classification: _____

EXPERIENCE NEEDED:	EXPERIENCE NEEDED:
EDUCATION NEEDED:	EDUCATION NEEDED:
CAPITAL REQUIRED:	CAPITAL REQUIRED:
OTHER COSTS:	OTHER COSTS:

Second Occupational Goal: _____
Job Classification: _____ Job Classification: _____

EXPERIENCE NEEDED:	EXPERIENCE NEEDED:
EDUCATION NEEDED:	EDUCATION NEEDED:
CAPITAL REQUIRED:	CAPITAL REQUIRED:
OTHER COSTS:	OTHER COSTS:

Approved by Teacher (Date) _____

Figure 10-1 Career goals in horticulture.
Source: Printed with permission of The Ohio Curriculum Materials Services. From Waldman, Dennis. *Horticulture Record Book.* Columbus. The Ohio Curriculum Materials Service, The Ohio State University, p. 1.

Career Goals

A first step in planning supervised experience is the selection of one or more career objectives by the student. Students should be encouraged to make tentative career choices, at least concerning the areas in agriculture of most interest to them. Figure 10-1 provides students the opportunity to specify two career objectives in horticulture and realistically examine the needed experience, education, and capital to reach their goals. There is space for two job classifications that relate to each career objective. The planning of supervised experience in relation to career objectives makes the program meaningful and relevant for students.

Yearly Plan

A yearly plan for supervised experience also needs to be developed by each student. Figure 10-2 is a planning form for use in a two-year horticulture program. The use of such a form encourages realistic planning toward career objectives. An example of a soundly constructed plan using another format is shown in Figure 10-3. The teacher, to encourage supervised experience that relates to the instructional program, can use planning of this nature.

Involving Others

Many programs must be conducted at the home or farm of the parents. Others may be conducted elsewhere with approval of the teacher. The parents should always understand and support the student as she or he becomes involved in supervised experience. The agriculture teacher, through supervisory visits and parent conferences, assists in the development of this understanding and support.

Employers in agricultural business and industry and agricultural agencies should also be made aware of the needs of students for supervised experience. Their support is needed for programs through which students learn about the world of work in agriculture through observation and placement experiences. The teacher must nurture the support of the business community and ensure that the provided experiences are educationally sound.

Some teachers have found it useful to involve lending institutions. These agencies can help students develop meaningful programs because of sufficient financial backing.

Students with Limited Resources

Some students need special assistance in developing programs. They may be in urban or inner-city areas, or they may be students in rural areas without access to land, buildings, equipment, capital, or parental cooperation. Students in this category require extra supervision by the teacher.

Placement in a business, agency, or on a farm can be of great value to a limited-opportunity student. The student gets firsthand experience in performing skills in a real working situation. It is important that transportation be available for the student. The employer must understand the need for the student to learn and not expect an experienced employee.

CAREER GOAL IN AGRICULTURAL EDUCATION

There is much more to job satisfaction than "how much money the job pays" or "how hard one has to work." Each job consists of many characteristics that eventually determine how satisfying the job will be.

Completing the following information will help you to develop a procedure for studying a career choice. Your agricultural education department, school library, guidance services, local employment offices, and especially employers and employees in a prospective occupation are sources to survey.

CAREER GOAL _____
JOB CLASSIFICATION _____
JOB DESCRIPTION
Nature of the work _____

_____
_____
_____
_____
_____

Industry visits made to learn more about the characteristics of my career goal:

Date	Person or Business Visited	Hours Spent	Major Information Acquired about Characteristics of Occupation or Occupational Skills

What are the current local, regional, and national employment opportunities for this career?

_____
_____
_____

What are the estimated local, regional, and national employment opportunities for this career in the next 5-10 years?

_____
_____
_____
_____

Figure 10-2 Planning supervised agricultural experience program.
Source: Printed with permission of THe Ohio Curriculum Materials Services. From *General Record Book (1991).* Columbus. The Ohio Curriculum Materials Service, p. 1.

	9th GRADE	10th GRADE	11th GRADE	12th GRADE
\multicolumn{5}{c}{Production Enterprises}					

	9th GRADE	10th GRADE	11th GRADE	12th GRADE
		Production Enterprises		
Sugar beets	½ acre [0.2 hectare]	5 acres [2 hectares]	15 acres [6 hectares]	20 acres [8 hectares]
Pinto beans		2 acres [0.8 hectare]	10 acres [4 hectares]	30 acres [12 hectares]
Corn	1 acre [0.4 hectare]	10 acres [4 hectares]	15 acres [6 hectares] (ensilage)	25 acres [10 hectares]
Alfalfa		5 acres [2 hectares]	12 acres [4.8 hectares]	25 acres [10 hectares]
Swine	1 gilt (bred)	2 gilts, 1 sow and litter, 8 market hogs	5 gilts, 10 sows and litters, 50 market hogs	25 sows and litters, 100 feeder pigs, 200 market hogs
		Improvement Projects		
	Repaired hog house. Built mail box. Built new steps for home. Painted garage. Cleaned up farm.	Rebuilt hog fence. Painted woodshed. Seeded lawn and shrubs. Painted farm machinery. Graveled driveway. Built windbreak.	Controlled 2 acres of weeds. Put new windows and door in granary. Repaired electric wiring system. Painted tractor. Put new roof on barn. Built fence around lawn.	Eradicated pests. Tested soil. Controlled pests (rats). Built septic tank. Put kitchen sink and bath in home. Improved dairy herd by use of artificial insemination.
		Supplementary Practices		
	Treated seed. Sharpened tools. Dehorned calves (15). Tested milk. Prepared seedbed.	Castrated calves (20). Castrated sheep (100). Treated barley seed. Sprayed fruit trees. Tagged ewes.	Judged livestock. Mixed dairy feed. Docked lambs. Wormed sheep. Mixed concrete.	Trimmed cattle feet. Showed sheep. Showed grain. Made farm gates. Vaccinated sheep.
		Agricultural Mechanics Projects		
	Built hog feeder. Built tool box. Built 20-ft [6.1-m] ladder. Built show box. Soldered radiator.	Built sawhorse. Made chisels for shop. Built farrowing house. Built rack for oil drums. Built homemade concrete mixer. Hard-surfaced cultivator attachments.	Built stock racks. Timed tractor engine. Built feed bunk. Built automatic watering trough. Surveyed for ditch. Built squeeze chute. Built trailer. Built welding table.	Built hog feeder. Built hog house. Painted tractor. Built manure loader. Built weed burner. Built grain trailer. Built grain elevator. Built weed sprayer.

Figure 10-3 Types of supervised agricultural experience projects.
Source: From Amberson, Max L., and Anderson, B. Harold. *Learning through Experience in Agricultural Industry.* New York: McGraw-Hill, p. 39.

Cooperatives can be organized and participated in by limited-opportunity students as a class project. Facilities in the community or at the school are necessary. The teacher often assumes a major supervisory responsibility.

Students having limited opportunities may often conduct gardening projects. The scope of such a project should be sufficient to produce a salable product for marketing.

> One young student began with a halfacre plot growing tomatoes, peas, peppers, green beans, and other vegetables. He built a cold frame to start the plants. He later built a small greenhouse from corrugated fiberglass. He sold his produce to local customers. He showed his produce at the state fair, receiving high placings.

Small animal projects may be used with students having limited resources. They require less housing space, less expensive feeding equipment, and fewer financial resources. Small animals, though, provide students with the opportunity to learn about animal reproduction, nutrition, management, and marketing. Students can raise rabbits, chinchillas, rats, mice, gerbils, chickens, or other species, but may need assistance in developing a profitable market.

Many students having limited opportunity at home may need to make greater use of the school laboratory in applying instruction (see Chapter 9). Teachers can also provide opportunities for students to conduct scientific experiments relating to agriculture in the school facilities.

CONDUCTING SUPERVISORY VISITS

Effective supervision is required for students to develop relevant programs of supervised experience. By definition, supervised agricultural experience programs are to be supervised. The supervision is provided by parents, the teacher, and, in some cases, an employer. Supervision is provided by the teacher in visits to the home of a student, in visits with the employer and student at the place of employment, and through working individually with the student to set goals and resolve problems. The major purpose of visits is to provide individualized instruction. A secondary purpose is to develop essential relationships between parents, employers, and teachers, who are the partners in the instruction of the student.

Orientation Home Visit

An orientation visit to the home of each entering student prior to the beginning of the school year is highly recommended. Such a visit enables the teacher to become familiar with the environment of the student. The teacher can orient the student and family to the nature of agricultural instruction and the need for supervised experience. The visit can be used to motivate the student to plan an outstanding program. Many teachers prepare a notebook containing photographs to illustrate various aspects of the agricultural instruction program, including student participation in supervised experience. Such a notebook can be effective in communicating with students, parents, and employers. Tentative selection of projects should be encouraged during the orientation visit.

Program Planning Visit

The agriculture teacher should, at a minimum, annually visit each student to assist in planning and further developing the program of supervised experience. Student long- and short-term goals should be developed and revised. Based on resources available to the student, a program should be planned to assist the student in reaching short- and long-term goals. The plan should set forth how the program will develop from year to year. Program planning visits provide excellent opportunities for career guidance and counseling.

Instructional Visits

Instructional visits are conducted periodically to provide individual instruction and supervision. Visits of this nature can be used to help the student overcome obstacles. They may also be used to encourage the use of approved practices and to assist in implementing plans of practice developed by the student in the classroom. The teacher can gain information from instructional visits that can be used to make classroom instruction more relevant.

Instructional visits usually provide an opportunity for the teacher and the student to discuss the records relating to the program. The teacher can assist in advising how records might be improved.

Some teachers schedule a minimum of one instructional visit to each student per grading period. This enables an accurate assessment of progress for evaluation purposes and encourages students to have a year-round program of supervised experience. It is recommended that teachers set aside a specific time or times each week for visits. Many schools provide the teacher with one period per day for this purpose. A schedule of visits can be prepared and shared with students so they know when they will be visited, and so that they see that all students are being treated equally. More efficient use of time and travel can usually be accomplished by visiting more than one student each trip. Teachers should keep records of observed projects, including the scope, the amount of effort put forth by the student, and the condition of each project. These records will provide support for the points awarded on the evaluation form in Figure 14–7 (see Chapter 14) and the supervised agricultural experience portion of the grade determined using the form in Figure 14–9 (see Chapter 14).

Employer/Employee Visits

Teachers must conduct visits to orient employers to the agricultural instruction program and to solicit their involvement in providing work experiences for students. A notebook similar to that used in home visits is helpful in describing the program and showing how it might be beneficial to the business as well as the school.

Once a business has agreed to participate in the program and has accepted a student, certain forms need to be completed. A placement agreement and an educational development plan are essential. Figure 10-4 is a placement and educational development agreement. This agreement is designed to protect the interests of the student, the employer, the parents, and the school (represented by the teacher). Various forms are available for this purpose; however, the teacher should ensure that all parties understand all essential aspects of the agreement.

The student, the employer, and the teacher should cooperatively develop an educational development plan. It specifies the skills or competencies the student will achieve as a result of the employment experience. There are various forms for educational development plans. One that is satisfactory is shown in Figure 10-5. This plan specifies the areas in which the student will work, the jobs to be accomplished (skills to be developed), estimated time to develop competence, dates, and an evaluation of the work. The teacher will need to work with the student and the employer to revise and update the plan as needed.

The teacher must visit the employer and the student employee to maintain relationships, ensure the work experience is in accordance with the agreement and educational development plan, obtain problems for class discussion, assist the student with work-related problems, and evaluate the student's learning.

In placing students with employers, the teacher may need to discuss subminimum wage provisions, work permit requirements, and restrictions concerning hazardous jobs. The local or state employment office can help the teacher learn about such legal requirements.

One important aspect of the employer visit is student evaluation. Employers must be taught how to evaluate students. Otherwise, one employer may rate a student worker high when another employer is rating a better-performing student low. Evaluation of supervised experience is discussed in greater detail in Chapter 14.

RECORDS INSTRUCTION

Students are responsible for budgeting and keeping financial records of income and expenses. By analyzing the records, they can devise ways of improving their programs.

Selecting, planning, and conducting programs of supervised experience provides a means to teach agricultural records in a realistic way. Students who

Chap. 10 Application of Learning: Supervised Agricultural Experience

---

_____ PROGRAM

This agreement is to (1) define clearly the conditions and schedule of experience whereby student

Name: _____

is to receive experience in _____

and (2) serve as a guide to the cooperating parties: the _____
                                                           (company or agency)

and the _____
                              (name of school)

Public Schools, in providing the student with opportunities for education and training in the basic skills of the occupation and the technical information related to it. In order that a systematic plan that provides for well-rounded education can be followed, a schedule of work experiences and a course of study paralleling it have been agreed upon by the employer and representative of the school.

The student agrees to perform diligently the work experiences assigned by the employer according to the same company policies and regulations as apply to regular employees. The student also agrees to pursue faithfully the prescribed course of study and to take advantage of every opportunity to improve his efficiency, knowledge, and personal traits so that the student may enter a chosen occupation as a desirable employee at the termination of the training period.

In addition to providing practical instruction, the employer agrees to pay the student for the useful work done while undergoing training according to the following plan:

1. The beginning wage will be $ _____ per _____ for _____ hours per week, which amount is approximately _____ percent of that paid competent full-time employees in the same occupation in the community.
2. A review of the wages paid the student will be made jointly and periodically by the employer and coordinator at least once each semester for the purpose of determining a fair and equitable wage adjustment consistent with the student's increased ability and prevailing economic conditions.

The training period begins the _____ day of _____
20_____, and extends through 20_____. There will be a probationary period of _____ days during which the interested parties may determine if the student has made a wise choice of an occupation, and if the training should be continued.

This plan has been reviewed and recommended by the Local Advisory Committee. It may be terminated for just cause by either party.

Approvals:

(Student)	(Name of Employer)
(Parent or Guardian)	(Name of Company or Agency)
(Chairperson, Local Advisory Committee)	(Teacher-Coordinator)

Figure 10–4A Sample cooperative education agreement.

Source: From Amberson, Max L., and Anderson, B. Harold. *Learning through Experience in Agricultural Industry.* New York: McGraw-Hill, pp. 116–119; and Adapted from the form developed by the Department of Vocational Education, Texas Education Agency.

To whom it may concern: Date _____
The enclosed training agreement is for the following student-trainee:

Student trainee _____ Date of Birth _____

Soc. Sec. No. _____ Grade _____ Available Work Hours _____

Occupational Objective _____ Training Period: Beginning _____

Ending _____

Training Agency _____

   Address _____ Phone No. _____

   Department in which employed _____

   Sponsor _____ Phone No. _____

Parents or Guardian _____

   Home Address _____ Phone No. _____

   Place of Business _____ Phone No. _____

Teacher-Coordinator _____

   Home Address _____ Phone No. _____

   School Address _____ Phone No. _____

The attached is a training agreement for career technical education programs at _____
_____ School. This agreement includes the definite responsibilities to be accepted by the employer, the school (represented by the teacher-coordinator), the parents, and the student.
Copies of the agreement will be on file with the employer and in the teacher-coordinator's office.

Signed _____
Teacher-Coordinator

STUDENT AGREEMENT

The student-trainee agrees to:

- Do an honest day's work, understanding that the employer must profit from the labor to justify hiring a student and providing this cooperative training experience.
- Accept the training station as procured by the teacher-coordinator.
- Do all jobs assigned to the best of his or her ability.
- Be punctual, dependable, and loyal to the school and business where employed.
- Follow instructions, avoid unsafe acts, and be alert to unsafe conditions.
- Be courteous and considerate of the employer, his family, customers, and others.
- Keep the records of the cooperative training program and make the reports that the teacher-coordinator and the employer require.
- Be alert to perform unassigned tasks which promote the welfare of the business.
- Assume responsibility for getting to and from work according to the work schedule.
- Notify the training sponsor *and* the school by 10:00 a.m. if absence is unavoidably necessary.
- Present himself or herself to the training station only on those days when classes were attended all that day.
- Work at least 15 hours (minimum) during the school week.

Figure 10.4B Continued

- Maintain good work habits both in school and on the job.
- Maintain a neat and clean appearance at all times.
- Contact the teacher-coordinator before resigning and under no circumstances resign the position at this training station without first discussing this action with the teacher-coordinator.

I have read and understand the purpose and intent of this training agreement.
I UNDERSTAND THAT ANY VIOLATION OF ANY PART OF THE ABOVE AGREEMENT MAY RESULT IN MY BEING DROPPED FROM THE PROGRAM, AT THE DISCRETION OF THE TEACHER-COORDINATOR.

Signed _____
                Student

TRAINING STATION OWNER-MANAGER AND/OR SPONSOR AGREEMENT

The employer agrees to:

- Teach the student-trainee the trade or business as circumstances permit, and insofar as possible, route the student through the different activities of the business.
- Assist the teacher-coordinator in planning the training program for the student.
- Train the student to do the jobs in the safest and most efficient manner.
- Avoid subjecting the student-trainee to unnecessary hazards.
- Assist the teacher-coordinator in making an honest appraisal of the student's performance at least once a quarter (each 9 weeks).
- Report student-trainee progress intermittently to the teacher-coordinator.
- Require the student to be on the job regularly except:
  a. On regular "no school" days such as school vacations and legal holidays.
  b. When unusual circumstances such as illness make it necessary for the student to be absent.
- Supervise the student-trainee for a minimum of 15 hours per week.
- Confer with the teacher-coordinator before dismissing the student-trainee for unsatisfactory performance.
- Pay wages which are comparable to wages paid apprentices and other beginning employees for the occupation in which the student-trainee is to receive training.

We have read and understand the purpose and intent of this training agreement.
Signed_____   _____
        Owner-Manager   Training Sponsor

PARENT AGREEMENT

The parent agrees to:

- Allow the student-trainee to work in the training station during hours and days as specified in the work schedule.
- Assist the student-trainee in getting to and from work according to the work schedule.
- Assist in promoting the value of the student's experience by cooperating with the employer and teacher when needed.
- Assume full responsibility for any action or happening pertaining to the student-trainee from the time of leaving school until reporting to the training station.

I have read and understand the purpose and intent of this training agreement.

Signed _____
                Parent

Figure 10-4C Continued

TEACHER-COORDINATOR AGREEMENT

The teacher-coordinator, on behalf of the school, agrees to:

- Give related instruction at the school enabling the student-trainee to better understand and carry out duties and responsibilities in the training station.
- Visit the student-trainee on the job periodically for the purpose of supervising the person to ensure that the person gets the most out of the cooperative training experience.
- Work with the employer, student-trainee, and parents to provide the best possible training for the student-trainee.
- Visit the employer at any time the employer deems it desirable.

I have read and understand the purpose and intent of this training agreement.

Signed _____
Teacher-Coordinator

Figure 10-4D Continued

Name _____ School _____

Career Objective _____ *Small Engine Mechanic* _____

DEPARTMENT OR AREA OF WORK	JOBS TO BE ACCOMPLISHED	TIME TO DEVELOP	DATES	EVALUATION COMPLETED	
				Satisfactory	Repeat
Servicing small engines	Identify types of small engines Clean small engines Service carburetor air cleaners Service fuel strainers Service crankcase breathers Lubricate small engines Refuel small engines Service spark plugs Check and adjust carburetors Check compression Check and service batteries				
Starting and operating engines	Start engines procedures Operate and adjust engine speed and load				
Storing engines	Eliminate storage hazards				
Maintaining starters	Repair rope-wind starters Repair rope-rewind starters Repair windup starters Repair 120 volts ac starters Repair direct-current starting and generating systems				
Maintaining and repairing ignition systems	Maintain and repair magneto and solid-state ignition systems Maintain and repair battery-ignition systems				
Repairing fuel systems	Repair fuel tank assemblies Repair fuel pumps Repair carburetors				
Repairing governors					
Repairing valves	Repair valves on 2- and 4-cycle engines				
Repairing cylinders, pistons, and rod assemblies	Check the cylinder and piston and rod assembly for proper operation Repair pistons, rods, rings Check and repair cylinders				
Repairing lubricating mechanisms in 4-cycle engines					
Repairing crankshaft assemblies					

Figure 10-5 Supervised agricultural experience plan for a small engine mechanic.
Source: From Amberson, Max L., and Anderson, B. Harold. *Learning through Experience in Agricultural Industry.* New York: McGraw-Hill, 1978, p. 129.

were surveyed concerning the value of record keeping rated the following abilities highest:

- Determine profit and loss
- Analyze production costs
- Keep useful records
- Maintain up-to-date records
- Maintain accurate records
- Keep neat records
- Appreciate the value of records
- Cultivate initiative in record keeping[3]

The preceding list suggests that students believe they are developing ability in keeping records but do not perceive records as useful in decision making. Teachers should use information from student records in class instruction to emphasize the usefulness of such information.

Records should be kept on all aspects of supervised experience. Record books in agricultural courses are useful for ascertaining ownership, cooperative placement, improvement, and skill development projects.

Ownership Project Records

Ownership record books vary widely from state to state, but there are common elements. The common elements include a budget for each enterprise, an agreement among interested parties, beginning and ending inventories, a record of receipts, a record of expenses, a labor record, and a summary and evaluation.

A teacher will find it effective to use a practice record book and have students record all the items in a hypothetical situation. The teacher may prepare a diary of events containing essential information to be recorded in the record book. The teacher can then teach record keeping as a unit of instruction, keeping the class together. On completion of the practice book each student can then enter items from his or her project in the regular record book.

Ownership records might be taught in seven problem areas

1. Budgeting
2. Developing business agreements
3. Taking inventory
4. Recording receipts, expenses, and major events
5. Determining overhead and operating costs
6. Summarizing records
7. Analyzing and evaluating records

---

[3]Davis, Duane L., and Williams, David L. *The Agricultural Education Magazine*. September 1974, 64.

These problem areas can be spread over two years of agriculture instruction. They should be taught just ahead of students making similar entries in their own books.

Cooperative Placement Records

The primary record to be kept on cooperative placement is the monthly work record. Here the student records daily the work performed on the various types of projects. Wages are recorded by pay period. The major records to be kept include skills or jobs performed, type of project for which the work was performed, and wages, if any. Students should be encouraged to list skills or jobs performed as specifically as possible.

Additionally, students should be taught to summarize their annual earnings and prepare a financial statement.

Improvement Project Records

Keeping improvement project records can be taught in a manner similar to that of keeping ownership records. The teacher can have all students complete a set of practice records on the same project. Then each student can complete a form for each project being conducted. The record should include an agreement between interested parties, a budget, a diary or work record, and an evaluation.

Skill development or supplementary practice records can be kept simply by recording the skill on a monthly work record similar to that used for cooperative placement. Some teachers add a column for the student and teacher to evaluate the quality with which the skill was performed.

SUMMARY

Supervised experience as an integral part of agricultural instruction is essential. Supervision requires teacher time and effort, but the payoff is worthwhile in terms of student development. Teachers believe in supervised experience because of its value in teaching and learning.

Types of supervised experience include ownership programs (production projects, group enterprises, and entrepreneurship projects), placement or cooperative programs, and improvement and skill development projects. Students should be encouraged to develop career goals, develop yearly plans, and involve parents and employers. Students with limited resources may require extra attention on the part of the teacher. Supervisory visits need to be conducted to orient students and parents to the program, assist in program planning, and provide instruction. Record keeping is an integral and essential component of supervised experience.

FOR FURTHER STUDY

1. Conduct a supervisory visit with a teacher of agriculture. Observe whether the visit was used for orientation, program planning, instruction, or evaluation. Record your impressions. List problems that might

be used for later class instruction. List possible interest approaches for later use in the class or laboratory. What laboratory projects might be needed to improve the students' supervised experience program?

2. Observe an agriculture class. List approved practices that could be used by students to improve their supervised experience programs.

3. Examine some student record books. From the books, make a list of ownership projects, improvement projects, and skill development projects.

4. Examine a local course of study. From the topics taught, suggest projects that should be encouraged. The projects should be relevant in applying instruction.

11

Application of Learning: FFA

The superintendent of schools calls you to his office and says, "I've never been associated with agriculture and the FFA before. I know you say it is 'intracurricular,' but I'm not sure what you mean when you use that word. It seems to me it is simply another club like the Spanish club. I want you to explain to me what FFA is for, how it ought to be used, how it is a part of the agriculture curriculum, and what student abilities it is designed to develop." If you encountered this situation, how would you respond?

This chapter presents the FFA the way it is best viewed and used to improve learning: as a laboratory for learning. The FFA, like the agricultural mechanics shop, greenhouse, or other facility, can be viewed as a laboratory. This laboratory is an integral part of the curriculum and its purposely designed learning experiences are necessary if students are to achieve the objectives of the agricultural instruction program.

OBJECTIVES

After studying this chapter, you will be able to

1. Explain how and why the FFA is a laboratory.
2. Discuss how to use the FFA as a place for applying leadership and personal development abilities.
3. Have students apply technical agriculture skills through the FFA.
4. Design ways to have students use FFA activities to apply leadership and personal development skills.

THE FFA AS A LABORATORY

Instruction in agriculture demands that students be taught the basic knowledge, skills, and understandings necessary for successful entry into the workforce. It is essential that this instruction be followed by application of what has been learned. In the area of psychomotor learning the basic knowledge is taught, the skills are explained and demonstrated, and then the learners apply their skills in a functional setting through directed laboratory experiences.

A Vehicle for Structured Learning

An essential component of the agricultural curriculum is the development of leadership and personal development abilities. These abilities must be taught in the classroom as well as in less structured settings and then applied in a functional setting. One of the laboratories for such purposeful practice is the FFA. Although FFA is analogous to the more conventional agricultural laboratories, it is more. The FFA has a built-in wellspring of motivation and rewards through its many incentive programs. These dimensions of FFA positively influence classroom studies, production or entrepreneurial projects, and placement programs. Of course, the same is true of laboratory and out-of-school occupational experiences. Teachers must remember that all parts of the agricultural instruction program are interdependent.

The effective agriculture teacher uses the FFA as a laboratory where leadership, career, and personal development abilities can be practiced. The teacher also uses the FFA for other purposes for which it is suited. One of these, for example, is using FFA activities to gain additional practice in applying the technical agriculture skills that have been studied.

If the FFA is truly to be used as a laboratory, it needs to be planned and organized. In other words, there needs to be structure. Classroom instruction is carefully planned and structured via a course of study and units of instruction. The conventional laboratory is planned as a part of the total course of study and structured through learning centers and specific activities within learning centers. The FFA is no exception.

The vehicle for planning and structuring the FFA as a laboratory for learning is the FFA program of activities. The planned activities in this program need to facilitate students practicing the leadership, career, and personal development abilities that they must possess if they are to enter and satisfactorily progress in the jobs for which they are preparing. The program of activities is then scheduled by month to be sure that all activities are spread across the school calendar. These FFA activities are the laboratory learning centers where leadership, career, and personal development abilities are practiced.

By bringing structure to the FFA as a laboratory, important learning experiences are likely to be acquired in a meaningful way. FFA activities provide real experiences at the site of directed practice. These readily available real experiences are superior to concocted simulations.

Types of Content for Which FFA Can Be a Laboratory for the Application of Learning

Certainly, teachers of agriculture who want students to learn basic leadership and personal development skills could set up mock situations wherein such skills could be practiced. This would be better than no practice at all. However, when real events can facilitate such practice, learning takes on added meaning and the results are better.

For example, agriculture teachers can teach their students how to greet others on social occasions, how to make introductions, and how to apply social graces in general. The students could then role-play these skills. However, if this is the end of the opportunity for planned and directed practice, the learning will be far too superficial. The FFA program of activities will undoubtedly include a parent–member banquet. This banquet is a perfect situation during which to teach the previously mentioned skills and at which to practice using them. These abilities could also be practiced at regular chapter meetings, at career development events (CDEs), on field trips, and at state and national conventions.

Agriculture students need to learn to speak in public. Classroom instruction must be given, and each student should practice giving a short speech in class. The FFA public speaking career development events can provide incentive to learn through the awards and recognition they offer. Of course, the public speaking career development events themselves (creed, prepared speaking, and extemporaneous speaking) are laboratories for practice. The ability to speak can also be practiced in chapter meetings, when giving oral reasons during a livestock judging career development event, when serving as a chapter officer, and when serving on committees in the FFA, as well as in the agriculture class and laboratory.

Another important area for leadership development involves learning how to conduct meetings. Most of the organized classroom instruction for this occurs through learning parliamentary procedure. This is practiced in class. However, the more meaningful practice occurs in chapter meetings, in committee meetings, and through the parliamentary procedure career development event.

Leaders must be able to plan and organize effectively as well as to evaluate the success of projects and programs. Although some rudimentary skills can be formally taught, most of these abilities need to be developed experientially. The FFA provides many opportunities for students to truly learn by doing. Students get practice at planning the FFA program of activities, planning and conducting specific projects, and evaluating the effectiveness of all FFA activities. Such learning is fostered through involvement in FFA and contributes significantly to the total development of the students.

Planning for Practice through the FFA

It must be pointed out that this ideal of using the FFA as a laboratory where leadership, career and personal development abilities, and technical agriculture skills are learned and applied does not automatically happen. The FFA can fit into the total agriculture program only if the teacher sees it in this context and then purposely plans for the FFA to be used as such. It is the responsibility of the teacher to ensure that this planning and follow through occur.

The teacher's first task in moving toward the ideal is to identify the appropriate leadership, career, and personal development abilities to be taught and to be applied through FFA as a laboratory. The teacher also needs to be alert to opportunities to apply technical agriculture skills. The basic abilities that employers believe are needed have been identified by Hampson, Newcomb, and McCracken.[1] These competencies are presented later in this chapter.

Using such a source listing, agriculture teachers must determine which abilities they desire their students to possess on exiting the agricultural instruction program. Perhaps a diagram will help in visualizing how teachers can best conceptualize the process they need to follow in planning for the development of these abilities (see Table 11-1). Teachers and students will first assess, using questionnaires and purchased instruments, the students' current skill package and level of skill development. Notice that all students enter agricultural courses with some leadership and personal development abilities. These abilities have been acquired through previous schooling, parental instruction, religious instruction, and participation in other organizations. However, it is not often that students enter agricultural instruction with nearly the total array of abilities or the level of mastery that will be needed by the time they enter post high school opportunities. If one subtracts the abilities possessed on entering agricultural instruction from abilities needed on graduation, the difference represents the abilities that must be developed in agricultural courses *and* through other experiences in and out of school prior to graduation. Expressed algebraically, the formula for determining what leadership, career, and personal development skills need to be developed in agricultural instruction is $Y = Z - X$,

Y = the leadership, career, and personal development abilities to be learned in studying agriculture

Z = the abilities and level of mastery students need to possess on graduation

X = the level of ability with which students enter the agricultural instruction program

---

[1] Hampson, M. N., Newcomb, L. H., and McCracken, J. D. *Essential Leadership and Personal Development Competencies Needed in Agricultural Leaders in Ohio* (Summary of Research Series No. SR12). Columbus: The Ohio State University, 1977.

Chap. 11 Application of Learning: FFA

Table 11-1 Conceptualization of How Teachers Determine Which Leadership, Career, and Personal Development Abilities to Include in Agricultural Instruction Programs

X	Y	Z
Leadership, Career, and Personal Development Abilities Possessed by Students as They Enter Agricultural Instruction	Leadership, Career, and Personal Development Abilities to Be Learned in Agricultural Instruction	Leadership and Personal Development Abilities Needed Upon Graduation
- -	- - -	- - - -
- -	- - -	- - - -
- -	- - -	- - - -
- -	- - -	- - - -
- -	- - -	- - - -
- -	- - -	- - - -
- -	- - -	- - - -
- -	- - -	- - - -
- -	- - -	- - - -
-	- -	- - - -
-	- -	- - - -
-	- -	- - - -
-	- -	- - - -
-	- -	- - - -

The same concept expressed algebraically:

$$X + Y = Z$$

Once teachers determine Y, or the leadership, career, and personal development abilities they wish students to develop through agricultural education, they are well on their way to planning successfully for the development of such abilities.

Agriculture teachers then need to examine the abilities they wish their students to develop and determine which abilities can be taught in formal units of instruction in the agriculture classroom. For example, you can teach parliamentary procedure, social graces, good grooming, duties of officers, and public speaking. These abilities should be grouped into units of instruction and arranged in the course outline. However, there are other important abilities, such as being consistently dependable, demonstrating good judgment, being punctual, cooperating with others in group activities, providing service to the community, and developing self-initiative, that cannot readily be taught as formal lessons. These abilities have to be provided for in more informal ways through FFA activities, supervised experiences, observations of the teacher as a role model, and laboratory experiences. However, any of the abilities that can be developed through participation in FFA, laboratory, or experience programs must be planned. In the case of FFA, the teacher identifies appropriate activities which, when participated in by students, aid in the development of given abilities. The teacher then ensures that such activities are in the FFA program of activities. The teacher may also use laboratory or supervised experience programs in order to facilitate the development of these important abilities.

Thus the teacher identifies the work to be done (abilities to be developed) and develops a means of accomplishing it. The result is that $Y = \text{FFA} + \text{Lab} + \text{SAE} + \text{Other Learning}$, where:

Y = the leadership, career, and personal development abilities to be developed by the time the students graduate

FFA = the abilities included in the FFA program of activities

Lab = the abilities developed through participation in the conventional lab

SAE = the abilities developed through experience in the student's supervised agricultural experience program

Other Learning = the abilities developed in various other ways as a part of agricultural instruction and through instruction in courses other than agriculture (i.e., in the rest of the school curriculum and other experiences)

Teachers of agriculture must involve students in planning for the application of leadership, career, and personal development abilities. As students work together to plan the program of activities, the teacher will provide them with a copy of the course outline and a copy of the leadership, career, and personal development abilities that graduates need, and will encourage them to devise activities that will help students develop and practice these essential abilities.

Much of the remaining planning for practicing leadership, career, and personal development abilities must be of a sixth-sense nature. That is, teachers have to be alert to identify every situation possible in which students can apply these abilities or have experiences that further the overall goal. For example, the teacher uses group leaders in laboratory; elects class officers for each class; sends students to leadership conferences; has students speak to civic clubs or at a board of education meeting; and allows students to sponsor, plan, and implement impromptu events.

The Teacher's Responsibility for Guiding the Practice of Leadership, Career, and Personal Development Abilities in the FFA

An important concept for teachers to accept is that they are the directors for achieving all sought-after learning outcomes. This means that teachers must accept responsibility for directing formal and informal learning. Formal learning is learning that is planned for in very precise terms, such as learning that results from the study of specific units of instruction in the classroom. Informal learning is learning that takes place through participating in an activity that may have been planned but was not specifically designed solely to produce a given learning outcome. For example, a teacher takes a group of FFA members to a career development event, they win the competition , and the teacher decides to treat them to a meal in an exclusive restaurant where two students enjoy their first experience with formal dining. Informal, yet very valuable learning takes place in this scenario. The teacher who recognizes and accepts the responsibility for directing such learning will help students to get more from every experience than a teacher who is indifferent to or unaware of such opportunities for learning.

Now, given that leadership, career, and personal development skills can be applied through FFA activities, and that teachers need to be alert to opportunities for promoting and guiding such application, one needs to consider the kinds of skills that may be applied. The only way for teachers to use FFA as a laboratory for the application of leadership, career, and personal development skills is to be ever mindful that FFA affords this needed opportunity and to search constantly for situations that lend themselves to this kind of practice.

For example, teachers should realize that a community service project is not merely an FFA project that helps their community. They should also realize it is a way to:

- Learn and practice the community development process
- Develop and practice cooperation skills
- Develop and practice citizenship values
- Develop and practice committee skills
- Develop and practice technical agriculture skills

For example, if the project is restoring a trout stream, students develop skill in chain saw operation, dredging, laying riprap, and establishing erosion cover. Also, they will

- Learn how to work with community leaders
- Develop pride in their community and environment

The same notion applies to most career development events. There are opportunities to practice technical abilities and also to cooperate with teammates, express one's self orally, develop social skills, foster school pride, be dependable, and much more.

Using the Program of Activities to Provide the Structure for Practice

Because of the breadth and depth the FFA provides in teaching agriculture, it is easy to miss many opportunities, or fail to fully exploit the available opportunities the FFA provides to students. The best way to combat this problem is to plan the program of activities with these added dimensions and goals of the FFA in mind. The program of activities needs to be correlated with the course of study so as to provide the optimum chance to use FFA events to apply previous learning.

As teachers guide agriculture students in planning their FFA program of activities, they should encourage the students to consider activities that allow the students to practice and develop essential leadership, career, and personal development abilities. Every teacher needs to work diligently to ensure that goals are stated in clear and measurable terms and that the ways and means for accomplishing the goals are so specific, sequential, and clear that students can follow the methods in a step-by-step fashion without needing very much specific direction from the teacher. Thus, the methods for completing planned activities in the FFA are quite analogous to the step-by-step directions provided to a group of students when they are assigned a job to complete in the conventional agricultural laboratory.

APPLICATION OF TECHNICAL AGRICULTURE SKILLS THROUGH THE FFA

Let's explore the many ways in which an agriculture teacher can use the FFA to provide students with additional opportunities to apply technical agriculture skills. We examine three areas within the structure of the FFA that provide a majority of these opportunities.

Applying Technical Agriculture Skills through FFA Career Development Events

Many of the FFA career development events are specifically designed to encourage agriculture students to develop specific technical agriculture skills. When students prepare for these CDEs, they are practicing or applying previous learning from the classroom and laboratory. Obviously, when students compete in career development events at all levels (local, district, area, state, regional, national), they are further applying previous learning as well as developing additional proficiency.

Examine the national FFA career development events in Table 11–2 to get an idea of the specific ones that are available in the various specialty areas of agriculture through the national FFA career development events program.

Of course, at the state and local level there are additional career development events such as forestry, game and waterfowl, tractor troubleshooting, animal care, wool, crops, and others. In addition, there are activities that focus on product marketing and sales and service.

Hence, the FFA provides motivation to master many areas of the agricultural curriculum through its use of awards and recognition. The career development events offer structure for the application of this knowledge and skill, and by preparing for and participating in the career development events,

Table 11-2 National FFA Organization Career Development Events (CDEs)

Agricultural communications	Floriculture
Agricultural issues forum	Food science and technology
Agricultural mechanics	Forestry
Agricultural sales	Horse evaluation
Agronomy	Job interview
Creed speaking	Livestock evaluation
Dairy cattle evaluation	Marketing plan
Dairy cattle handlers	Meats evaluation and technology
Dairy foods	Nursery and landscape
Environmental and natural resources	Parliamentary procedure
Extemporaneous public speaking	Poultry evaluation
Farm business management	Prepared public speaking

students have additional opportunities to master the skills that are associated with each competition.

Using the Proficiency Awards Program to Foster the Application of Technical Agriculture Skills

The proficiency awards program was designed to motivate students to become occupationally proficient in the areas aligned with their career choice. This program is directly connected to the students' supervised agricultural experience programs and the classroom and laboratory learning that the students use as a basis for developing proficiency in an area. The proficiency awards program recognizes excellence in numerous specific areas.

Teachers need to be familiar with the wide range of awards available to their students. The proficiency awards listed in Table 11-3 are currently available nationally.

Teachers must study the specifics of each proficiency award application in order to best facilitate their students' success in a given area. The following categories are represented in the various proficiency award applications: (1) scope of supervised agricultural experience program, (2) earned income, (3) inventory, (4) financial statement, (5) related activities, (6) marketing experience, (7) proficiencies attained, (8) skills and competencies, (9) improved practices, (10) achievements and accomplishments, (11) experiences gained, (12) safety, (13) leadership development, (14) other information, and (15) supporting evidence.

Using the FFA Degree Program to Promote Application of Technical Agriculture Skills

An FFA program of long standing designed to promote the application of technical agriculture skills (as well as leadership, career, and personal development skills) is the degree program. From the very outset FFA members are

Table 11-3 FFA Proficiency Awards

- Agricultural Communications—Entrepreneurship/Placement*
- Agricultural Mechanics Design and Fabrication—Entrepreneurship/Placement*
- Agricultural Mechanics Energy Systems—Entrepreneurship/Placement*
- Agricultural Mechanics Repair and Maintenance—Entrepreneurship/Placement*
- Agricultural Processing—Entrepreneurship/Placement*
- Agricultural Sales—Entrepreneurship
- Agricultural Sales—Placement
- Agricultural Services—Entrepreneurship/Placement*
- Beef Production—Entrepreneurship
- Beef Production—Placement
- Dairy Production—Entrepreneurship
- Dairy Production—Placement
- Diversified Agricultural Production—Entrepreneurship/Placement*
- Diversified Crop Production—Entrepreneurship
- Diversified Crop Production—Placement
- Diversified Horticulture—Entrepreneurship
- Diversified Horticulture—Placement
- Diversified Livestock Production—Entrepreneurship
- Diversified Livestock Production—Placement
- Emerging Agricultural Technology—Entrepreneurship/Placement*
- Environmental Science and Natural Resources Management—Entrepreneurship/Placement*
- Equine Science—Entrepreneurship
- Equine Science—Placement
- Fiber and/or Oil Crop Production—Entrepreneurship/Placement*
- Floriculture—Entrepreneurship/Placement*
- Food Science and Technology—Entrepreneurship/Placement*
- Forage Production—
- Entrepreneurship/Placement*
- Forest Management and Products—Entrepreneurship/Placement*
- Fruit Production—Entrepreneurship/Placement
- Grain Production—Entrepreneurship/Placement
- Grain Production—Entrepreneurship
- Grain Production—Placement
- Home and/or Community Development—Entrepreneurship/Placement*
- Landscape Management—Entrepreneurship/Placement*
- Nursery Operations—Entrepreneurship/Placement*
- Outdoor Recreation—Entrepreneurship/Placement*
- Poultry Production—Entrepreneurship/Placement*
- Sheep Production—Entrepreneurship/Placement*
- Small Animal Production and Care—Entrepreneurship
- Small Animal Production and Care—Placement
- Specialty Animal Production—Entrepreneurship
- Specialty Animal Production—Placement
- Specialty Crop Production—Entrepreneurship/Placement*
- Swine Production—Entrepreneurship
- Swine Production—Placement
- Turf Grass Management—Entrepreneurship
- Turf Grass Management—Placement
- Vegetable Production—Entrepreneurship/Placement*
- Wildlife Production and Management—Entrepreneurship
- Wildlife Production and Management—Placement

* Entrepreneurship SAE programs to compete with placement SAE programs in these categories.

challenged to attain the greenhand FFA degree, chapter FFA degree, state FFA degree, and American FFA degree.

An analysis of the requirements for these degrees reveals that they are firmly grounded in students' supervised agricultural experience programs, where students must properly apply the knowledge and skill they have learned in order to meet the minimum requirements for the degrees.

As students seek to attain the higher degrees, the competition becomes more keen and the scope and quality of their performance must increase; as it does, so does the amount and quality of application of their knowledge of and skill in their particular area of technical agriculture.

APPLICATION OF LEADERSHIP, CAREER, AND PERSONAL DEVELOPMENT ACTIVITIES THROUGH THE FFA

If FFA is to serve its role as a laboratory where students gain important practice in using the leadership, career, and personal development abilities needed for success in life, it must offer many opportunities for many students. One important key to using the FFA successfully in this manner is to be sure that opportunities are available to and used by many students. If the FFA is only for the chosen few, it will not adequately serve as a learning laboratory. Perhaps an examination of a number of facets of the FFA that can be used as learning centers where leadership, career, and personal development abilities can be applied (and developed) will help the reader grasp the full extent of possibilities for using the FFA as a laboratory.

In order to gain an overall perspective of the capacity of the FFA to provide an opportunity for students to apply (and develop) leadership, career, and personal development abilities, study Table 11-4. This table consists of the leadership, career, and personal development abilities Hampson and colleagues recommended that agriculture students develop. For each ability there is an indication of whether that ability can be practiced (or developed) through the following aspects of the FFA: chapter meetings, committee work, career development events, chapter banquet, parliamentary procedure, public speaking, community service, or participation in state and national activities. Following is a general explanation of how some of these types of specific applications are secured.

EXAMPLES OF HOW SELECTED ABILITIES ARE APPLIED THROUGH VARIOUS ASPECTS OF FFA

Chapter Meetings

By participating in chapter meetings, students have the opportunity to learn by doing what they have studied in the classroom. The chapter meeting is a somewhat realistic replica of a democratic society and its government. At a

Table 11-4 Aspects of FFA That May Provide for Practice and Development of Given Abilities

Desired Leadership, Career, and Personal Development Abilities by Area of Classification	Chapter Meetings	Committee Work	Career Development Events	Chapter Banquet	Parliamentary Procedure	Public Speaking	Community Service	State and National Activities
Area 1: Leading individuals and groups								
Follow democratic procedures	X	X	X		X		X	X
Keep group progressing toward goals and objectives	X	X			X		X	X
Demonstrate tact and diplomacy	X	X	X	X	X		X	X
Involve others in group decisions and actions	X	X					X	X
Be consistently dependable	X	X	X	X	X	X	X	X
Make and substantiate decisions	X	X	X		X	X	X	
Collect and evaluate necessary information		X	X			X	X	
Set meetings, date, and place	X	X			X		X	
Develop meeting agenda	X	X			X		X	
Demonstrate good judgment	X	X	X	X	X	X	X	X
Provide constructive criticism	X	X	X	X	X		X	X
Inform individuals of their roles and responsibilities		X		X			X	
Area 2: Developing good work habits								
Attend work regularly		X	X				X	
Complete assigned work to best of one's ability		X	X	X	X	X	X	
Work cooperatively with others	X	X	X		X		X	X
Follow business rules and policies		X	X		X		X	X
Demonstrate speed and accuracy in work		X	X	X	X	X		
Provide work instructions to others		X		X			X	
Identify unsafe and inadequate work habits		X					X	
Work under pressure		X	X	X	X	X	X	
Area 3: Participating in social activities								
Extend courtesies to others	X		X	X	X	X	X	X
Participate in conversation appropriate for the occasion		X	X	X			X	X
Meet and greet people	X	X	X	X		X	X	X

Table 11-4 Continued

Desired Leadership, Career, and Personal Development Abilities by Area of Classification	Chapter Meetings	Committee Work	Career Development Events	Chapter Banquet	Parliamentary Procedure	Public Speaking	Community Service	State and National Activities
Use proper manners in a restaurant			X	X				X
Demonstrate correct eating etiquette with various types of food in various situations			X	X				X
Dress appropriately for various occasions	X		X	X	X	X		X
Maintain good posture	X		X	X	X	X	X	X
Be punctual for social events	X	X	X	X			X	X
Area 4: Participating in committees and groups								
Serve as committee chairperson		X		X			X	
Participate as a committee member		X		X			X	
Select members of a committee		X		X			X	
Present a committee report		X		X	X		X	
Identify committee objectives		X		X			X	
Delegate responsibilities to other committee members		X		X	X		X	
Give recognition and thanks for work done	X	X		X			X	X
Serve as an officer	X	X					X	
Use proper parliamentary procedure	X	X			X		X	
Maintain satisfactory group membership	X	X	X				X	X
Area 5: Participating in professional, business, and civic organizations								
Participate as a member of an organization at the local, state, or national level	X	X	X	X	X			X
Assume responsibility for the operation of the organization	X	X		X	X		X	
Identify the principles and purposes of the organization	X	X					X	X
Interpret the constitution and bylaws of the organization	X	X						X
Vote on organizational concerns	X	X			X		X	X
Area 6: Managing financial resources								
Prepare a personal budget								
Set financial goals for the future								

(continued)

Table 11-4 Continued

Desired Leadership, Career, and Personal Development Abilities by Area of Classification	Chapter Meetings	Committee Work	Career Development Events	Chapter Banquet	Parliamentary Procedure	Public Speaking	Community Service	State and National Activities
Write checks and maintain checkbook register								
Calculate interest on savings account								
Identify the cost of ownership of a car								
Identify the value of real estate in the local community							X	
Area 7: Developing communication skills								
Present information to group	X	X	X	X	X	X	X	
Communicate clearly in written form		X	X	X			X	
Function as a spokesperson for a group	X	X	X	X	X	X	X	
Introduce a speaker at a meeting	X	X		X			X	
Participate in conversations and discussions	X	X	X	X			X	
Use correct telephone procedures		X	X				X	X
Write letters correctly when appropriate		X		X			X	X
Area 8: Developing citizenship skills								
Cooperate with others in group activities	X	X	X	X	X	X	X	X
Respect national symbols	X		X	X	X		X	X
Respect, maintain, and improve the environment			X	X			X	X
Stay well informed of state, national, and local issues						X	X	X
Provide service to community (i.e., local, state, and national)		X					X	
Vote on issues and in elections	X	X			X		X	
Help authorities in specific cases when needed							X	X
Stay informed about the law							X	X
Area 9: Developing personal skills								
Complete a personal inventory of strengths and weaknesses								
Demonstrate personal integrity	X	X	X	X	X	X	X	X
Determine future courses (lifestyles)								X
Maintain positive attitude	X	X	X	X	X	X	X	
Develop self-initiative		X	X			X	X	

Table 11-4 Continued

Desired Leadership, Career, and Personal Development Abilities by Area of Classification	Chapter Meetings	Committee Work	Career Development Events	Chapter Banquet	Parliamentary Procedure	Public Speaking	Community Service	State and National Activities
Manage use of time		X	X	X	X	X	X	X
Respect the rights of others	X	X	X	X	X		X	X
Demonstrate sincerity	X	X	X	X	X	X	X	X
Demonstrate enthusiasm	X	X	X	X	X	X	X	X
Demonstrate confidence	X	X	X	X	X	X	X	X
Exhibit receptiveness to suggestions	X	X	X	X	X		X	X
Demonstrate the ability to work with others	X	X	X	X	X	X	X	X
Demonstrate patience	X	X	X	X	X	X	X	X
Be a good sport	X	X	X	X	X	X	X	X
Be responsible for personal actions	X	X	X	X	X	X	X	X

chapter meeting students follow democratic procedures, work cooperatively with others, extend courtesies to others, meet and greet poor students and guests, serve as officers, and respect the rights of others.

In preparation for a good chapter meeting, students need to complete assigned committee work to the best of their ability, develop a meeting agenda, and be informed of the issues.

Of course, if all the abilities indicated in Table 11-4 are to be developed, such development will not be left to mere chance. Rather, the agriculture teacher will point out opportunities, encourage students to participate, and offer the instruction needed for students to be successful.

Committee Work

Committee work provides opportunities for students to apply and develop specific leadership, career, and personal development abilities. In many ways committee work is a learning center for FFA as a laboratory. Not only are there standing committees provided in the program of activities but there are also opportunities for students to acquire abilities through many special committee assignments.

Through committee work a student learns how to be patient, progress toward goals and objectives, serve as a committee chairperson, participate as a committee member, participate in conversations and discussions, and manage the use of time, among other things.

Although every student cannot be a chapter officer, every student can serve on a committee, and a majority of students can, in fact, chair a committee or subcommittee. Take the case of a chapter that has 100 members. Through standing committees, activity committees, and special projects, such as a community service or an agricultural literacy project, plus additional ad hoc committees, all 100 members are needed to properly run the organization. Teachers can also effectively use committees as a part of their classroom structure, and for all practical purposes group work in the conventional laboratory is committee work. Thus, teachers who sincerely want their students to develop the leadership, career, and personal development abilities needed for success in life will make good use of committees in developing such abilities. By doing so every student can actively serve in several distinct ways during each year he or she is enrolled in agricultural education.

Career Development Events

While most career development events are primarily concerned with testing students' mastery of technical skills, the very nature of participating in them contributes to the development and application of leadership, career, and personal development skills. For example, students who try out for the soils team must show that they are dependable, can make and substantiate decisions, can work under pressure and cooperate with others in group activities, and can be good sports, to name only a few abilities.

When students travel to a forestry competition, horse judging CDE, or horticulture CDE, they have to practice social skills and human relations skills in addition to the technical skills. Thus, prudent career development event participation can add to the development of agriculture students in a variety of dimensions.

Chapter Banquet

The chapter banquet is the social event of the FFA year. It is also the place where much student recognition is received and where students and guests alike are able to see the cumulative work of the FFA chapter for the year. Because it is a student activity, it is planned by students and conducted by students. Thus the chapter banquet is a real problem to be solved that gives students a felt need to learn some rather specific skills.

For example, the students who are to introduce guests or award winners need to learn how it is properly done, and then practice the skill, so they can successfully make such introductions and presentations at the banquet.

Some students are not sure about their social graces, including proper etiquette. Therefore they want to know how to handle themselves properly. This is a great opportunity for the teacher to teach a unit of instruction on social graces.

Many students will have to serve on committees. They will have to arrange for facilities, food, and decorations; secure awards; generate interest in attendance; and invite guests—they will even have to learn to keep secrets because many of the awards are surprise announcements during the banquet. All of these duties provide additional opportunities for students to practice leadership, career, and personal development abilities. The teacher must direct this informal laboratory learning.

Parliamentary Procedure

In studying parliamentary procedure, students develop many important leadership abilities, such as developing agendas, serving as officers, making decisions, giving reports, contributing to discussions, involving others in group decisions, assuming the responsibility for the operation of the organization, and following democratic procedures. Then students apply these abilities as they serve on committees, attend chapter meetings, compete in the parliamentary procedure CDE, and attend and participate in state and national activities.

Public Speaking

The basic ability of public speaking is taught in the classroom. The ability to speak in public is also practiced in the classroom. However, it is also practiced in the public speaking CDE and when oral reasons are given in various other career development events. Public speaking skills are also used in chapter meetings, committee meetings, class discussions, and at state and national activities.

One's public speaking also leads to the opportunity to practice other leadership, career, and personal development abilities, such as collecting and evaluating information, completing assigned work to the best of one's ability, extending courtesies to others, meeting and greeting people, and functioning as a spokesperson for a group.

The real challenge for the teacher is to ensure that the great majority of agriculture students develop the aforementioned abilities associated with public speaking. If only those who compete in the public speaking career development events develop these abilities, much will have been lost. Teachers of agriculture can foster these abilities by selecting teaching techniques that lend themselves to using such abilities and by causing all students to serve in roles and complete tasks for which they use these kinds of skills.

Community Service

A comprehensive community service project, such as restoring a historic structure like a one-room school or a covered bridge, can provide numerous opportunities for students to apply basic leadership, career, and personal development skills.

In order to complete such a project successfully, students study the community development process, cooperate with civic and government leaders, cooperate with other clubs, plan technical work, raise money, work as a team, share the recognition, and improve the environment. The result is that students practice technical and leadership, career, and personal development skills. Thus, another laboratory learning center will have significantly contributed to the total development of the agriculture students' lives.

State and National Activities

When students participate in state and national activities, they almost always have to travel, eat in restaurants, and stay in hotels. Traveling provides a great opportunity for contributing to the personal development of students. They also learn that they are a very small part of a much bigger organization, but they can develop much pride in themselves and their local chapter while at the same time being challenged to improve themselves. Participation at this level provides many opportunities to be courteous, to cooperate, to observe rules and regulations, and to lose and win graciously.

Thus, FFA as a laboratory provides many ways to help students apply the leadership, career, and personal development abilities the teacher deems important. The real situation is not as clear-cut as the preceding categories suggested because all of those activities are interrelated, not only one with the other but also with the agriculture classroom and laboratory experiences and with the students' supervised agricultural experience programs as well.

SUMMARY

Too many agriculture teachers view the FFA as the purpose for the existence of agricultural instruction. Whenever such a perspective governs their use of FFA, the results are invariably disappointing.

When the FFA is viewed as an integral component of the curriculum, a tool for teaching, and a laboratory wherein students can both develop and apply leadership, career, and personal development abilities needed by all citizens and agricultural employees, it will make its maximum contribution to the agricultural instruction program. For FFA to be used in this manner, teachers must purposely use the organization to bring about the kinds of change in learners that are described in this chapter.

Teachers should not simply insist that members go to career development events to stroke their own egos and strive to win every award available whether it fits the course of study or not. Rather, they carefully select activities and avenues for using FFA to aid in the accomplishment of predetermined educational goals.

FOR FURTHER STUDY

1. Study the program of activities of a local FFA chapter and determine how many of the leadership, career, and personal development activities in Table 11–4 are included.

2. Visit with a local FFA advisor and discuss how he or she monitors which students have experienced various activities.
3. Accompany local FFA members on an activity and determine how many leadership, career, and personal development abilities and technical agriculture skills they are able to practice.
4. Examine an agriculture teacher's course of study and determine how many formal leadership, career, and personal development topics are included.

PART IV *Teaching Special Populations*

The emphasis in preparing teachers of agriculture is normally on secondary and postsecondary instruction. Within the secondary and postsecondary setting, learners with special needs often present a challenge to the teacher. In addition to teaching learners with special needs, many beginning teachers look forward to teaching youth but are somewhat apprehensive about teaching adults. Part IV of this book is designed to provide help with modified methods to successfully teach the client groups of (1) learners with special needs and (2) adults.

Chapter 12 provides definitions of learners with special needs with various types of disabilities. It presents the importance of proper attitudes on the part of the teacher. Special learning needs and teaching methods are provided for each learner. Practices are suggested for student attitude development, personal development, and supervised experience. Instructional enrichment methods are presented that teachers will also find of value in teaching all students.

Adult education is described in Chapter 13 as an important function of the public schools. Various types of adult education in agriculture are described, and a rationale for adult education is presented. Adult learner characteristics are included to enable better understanding on the part of teachers. Methods for course planning and techniques for class and individual instruction are also discussed.

12

Teaching Learners with Special Needs

Ronald liked to work with plants. He also enjoyed visiting with people. He enrolled in the floriculture program at the area career and technical center near his home. He impressed his teacher as being a student who had a natural feeling for plants, who was polite and courteous in his relationships with others, and who accomplished assigned tasks according to instructions. At graduation he received a vocational certificate certifying his competence in horticulture. He was employed driving a horticulture delivery truck for a large wholesale firm in the nearby metropolitan area. He achieved an excellent work record because of his ability to work with people, to care for plant materials, to follow instructions, to read road maps, and for his dedication to his work. He also became active in community affairs, serving as the youth sponsor for his church.

Ronald was disabled. He had a learning disorder resulting from early brain damage. He had been diagnosed as *special needs* and was in special education prior to enrolling in the area career and technical center. Because of his disability, Ronald had some difficulty making decisions. He was unable to complete narrative problems in mathematics but could calculate numerical problems. He could handle money and make change without error. Ronald's participation in an agricultural instruction program helped him to enter a useful and rewarding career.

The strongest rationale that can be given for advocating the active participation of special needs learners in agricultural instruction is their performance in the world of work. During the early stages of education reform, the President's Committee on Employment of the Handicapped[1]

---

[1] President's Committee on Employment of the Handicapped. *Hiring the Handicapped: Facts and Myths.* Washington, DC.

reported that the work record of people with disabilities compared favorably with the nondisabled. This is true of their ability to perform tasks, safety records, attendance, and advancement.

Every individual should be afforded the chance to lead a full and rewarding life with the personal dignity that comes from possessing useful skills and having a chance to apply them in productive work.[2] Agriculture teachers must keep an open mind about the occupational areas a learner with special needs can best pursue and assist him or her in developing for employability. However, teachers are sometimes accused of focusing much of their attention on the advantaged students in order to show greater FFA and program achievement. Teachers should keep in mind the value added by agricultural instruction. The disadvantaged students are sometimes those who progress most in programs where teachers have been prepared to teach to their special needs. Legislation requires that, insofar as possible, learners with special needs be educated in regular rather than special education classes. If teachers are to meet the needs of all students, they must be prepared to do so. Learners with special needs often require attention from the teacher in the form of an awareness of special learning needs, provision of individualized instruction, adjusting of teaching methods, or giving extra attention to specific learning requirements.

OBJECTIVES

After studying this chapter, you will be able to

1. Explain the school's responsibility to serve learners with special needs.
2. Describe the motivational forces influencing learners with special needs.
3. Assist in preparing an individualized educational plan for a student.
4. Use teaching methods that are modified to meet specific learning needs.
5. Describe how you would teach students having specific special needs.

LEARNERS WITH SPECIAL NEEDS

There are two legal mandates that protect students from discrimination and ensure that they have equal access to all aspects of education. These laws include Section 504 of the Rehabilitation Act of 1973 and the Americans with Disabilities Act (ADA) of 1990.[3]

---

[2] Weisgerber, Robert, Ed. *Vocational Education: Teaching the Handicapped in Regular Classes.* Reston, VA: The Council for Exceptional Children, 1978, p. 3.
[3] *Teaching Students with Disabilities: Instructor Handbook.* The Ohio State University. Office for Disability Services. Printed with permission from Patty Carlton, Assistant Director, Office for Disability Services, The Ohio State University, 2001.

Section 504 states that, "no otherwise qualified individual with a disability in the United States . . . shall, solely by reason of his or her disability, be excluded from participation in, be denied the benefits of, or be subjected to discrimination under any program or activity receiving Federal financial assistance. . . ."

Title II of the ADA states that, "a public entity shall make reasonable modifications in policies or procedures when the modifications are necessary to avoid discrimination on the basis of disability, unless the public entity can demonstrate that making the modifications would fundamentally alter the nature of the service, program, or activity."

Section 504 and the ADA specify that students are considered to have a disability if they meet any one of the following criteria:

- He or she has a physical or mental impairment that substantially limits one or more of the major life activities, or
- Has a record of such an impairment, or
- Is regarded as having such an impairment.

A major life activity is considered a basic human function, such as seeing, hearing, walking, breathing, speaking, caring for self, learning, performing manual tasks, and working.

Disability Conditions

Teachers of agriculture will not necessarily encounter every type of disability condition in their classrooms during their careers. However, teachers need to be aware of the conditions that exist and be able to find professional guidance in identifying various disability conditions. The following classifications of learners with special needs offer insight and suggestions for teaching these students. Public Law 94-142 identifies and defines the following disability conditions:

1. *Autism.* A developmental disability significantly affecting verbal and nonverbal communication and social interaction, generally evident before age three, that adversely affects a child's educational performance.
2. *Deaf-blindness.* Simultaneous hearing and visual impairments, the combination of which causes such severe communication and other developmental and educational problems that they cannot be accommodated in special education programs solely for children with deafness or with blindness.
3. *Deafness.* A hearing impairment that is so severe that the child is impaired in processing linguistic information through hearing, with or without amplification, that adversely affects a child's educational performance.
4. *Hearing impairment.* Impairment in hearing, whether permanent or fluctuating, that adversely affects a child's educational performance, but that is not included under the definition of deafness.

5. *Mental retardation (developmental handicap).* Significantly subaverage general intellectual functioning existing concurrently with deficits in adaptive behavior and manifested during the developmental period that adversely affects a child's educational performance.

6. *Multiple disabilities (multihandicapped).* Simultaneous impairments (such as mental retardation-blindness, mental-retardation-orthopedic impairment, etc.), the combination of which causes such severe educational problems that they cannot be accommodated in special education programs solely for one of the impairments. The term does not include deaf-blindness.

7. *Orthopedically impaired.* A severe orthopedic impairment that adversely affects a child's educational performance. The term includes impairments present at birth abnormally (e.g., clubfoot, absence of some member, etc.) or impairment from other causes (e.g., cerebral palsy, amputations, and fractures or burns which cause contractures).

8. *Other health impaired.* Limited strength, vitality, or alertness caused by chronic or acute health problems, such as a heart condition, tuberculosis, rheumatic fever, nephritis, asthma, leukemia, or diabetes, that adversely affects a child's educational performance.

9. *Serious emotional disturbance.* A condition exhibiting one or more of the following characteristics over a long period of time and to a marked degree that adversely affects a child's educational performance:

    a. An inability to learn that cannot be explained by intellectual, sensory, or health factors;

    b. An inability to build or maintain satisfactory interpersonal relationships with peers and teachers;

    c. Inappropriate types of behavior or feelings under normal circumstances;

    d. A general pervasive mood of unhappiness or depression; or

    e. A tendency to develop physical symptoms or fears associated with personal or school problems.

    The term includes schizophrenia. The term does not apply to children who are socially maladjusted unless it is determined they have a serious emotional disturbance.

10. *Special learning disability.* A disorder in one or more of the basic psychological processes involved in understanding or in using language, spoken or written, that may manifest itself in an imperfect ability to listen, think, speak, read, write, spell, or do mathematical calculations.

11. *Speech or language impairment.* A communication disorder, such as stuttering, impaired articulation, language impairment, or a voice impairment, that adversely affects a child's educational performance.

12. *Visually disabled.* A visual impairment that, even with correction, adversely affects the child's educational performance. The term includes both partially seeing and blind children.

Students with Learning Disabilities[4]

Students with learning disabilities often learn differently from their peers. Although they have average or above average intelligence, there is frequently a discrepancy between their ability and their achievement in specific areas. Learning disabilities are presumably caused by a central nervous system dysfunction. Learning disabilities are permanent disorders that interfere with integrating, acquiring, or demonstrating verbal or nonverbal abilities and skills. There are generally some processing or memory deficits.

Each student with a learning disability has his or her own set of characteristics; one is not necessarily like another. Teachers of agricultural education can help identify some learners with special needs by noticing students' difficulties in the following areas:

Reading comprehension	Oral expression	Abstract reasoning
Written expression	Auditory processing	Visual spatial skills
Mathematics	Visual processing	Processing speed

Keep in mind that one student does not have difficulty with all of these areas but generally only a few of these areas. Also, it is not unusual for a learner with special needs to be gifted in some areas.

SUGGESTIONS FOR TEACHING MODIFICATIONS

Exam accommodations: Assist students in arranging for appropriate exam accommodations whether you arrange these accommodations yourself or coordinate them with the special education staff at your school.

Multimodality instruction: A multimodality approach to instruction assists students in finding a modality that is consistent with their learning strength. Providing important information and assignments in both oral and written formats helps avoid confusion.

Alternative format: Some students need print material in an alternative format.

Study aids: Study questions, study guides, opportunities for questions and answers, and review sessions help learners with special needs who require a lot of repetition.

Exam aids: Permit students to use simple calculators, portable spell-checkers, and scratch paper during exams.

Flexible exam format: Students who have language-based or writing disabilities may need more time on essay exams. Others may want to tape record answers, use a scribe, or use a computer. Be open to a flexible exam format as long as the student is able to demonstrate his or her knowledge.

---

[4] *Teaching Students with Disabilities: Instructor Handbook, op. cit.*

Students Who Are Visually Impaired[5]

Students with visual impairments are constantly challenged by classroom instructional strategies. Although they can easily hear lectures and discussions, it can be difficult for them to access class syllabi, textbooks, overhead projector transparencies, PowerPoint presentations, the chalkboard, maps, videos, written exams, demonstrations, and library materials. A large part of traditional learning is visual; fortunately, many students with visual disabilities have developed strategies to compensate.

These students' abilities vary considerably. Some have no vision; others are able to see large forms; others can see print if magnified; and still others have tunnel vision with no peripheral vision or the reverse. Furthermore, some students with visual impairments use Braille, and some have little or no knowledge of Braille. They use a variety of accommodations, equipment, and compensatory strategies based on their widely varying needs. Many make use of adaptive technology, especially print to voice conversions using scanners and voice production software. Textbooks are often converted and put on disks for later use. Others use taped textbooks or equipment to enlarge print (closed circuit television [CCTV]) or actual enlargements.

Students Who Are Deaf or Hard of Hearing[6]

Individuals who are deaf or hard of hearing rely on visual input rather than auditory input when communicating. Using visual aspects of communication (body language, gestures, and facial expression) often feels awkward to people who are accustomed to the auditory; however, it is essential that instructors learn to effectively communicate with students who are deaf or hard of hearing.

Students who are deaf or hard of hearing do not all have the same characteristics. Some have a measure of usable residual hearing and use a device to amplify sounds (FM system). Some choose to speak; others use very little or no oral communication. Some students are extremely adept at speech reading, whereas others have a very limited ability to read lips. For some, sign language or finger spelling are the preferred means of communication; other communication choices include gestures and writing. Most students who are deaf or hard of hearing have experience communicating with the hearing population. Let them be the guide on how best to communicate.

Students with Attention Deficit Hyperactivity Disorder[7]

Attention deficit hyperactivity disorder (ADHD) is characterized by a persistent pattern of inattention and/or hyperactivity that is more frequent and severe than is typically observed in individuals at a comparable level of development.

---

[5] *Teaching Students with Disabilities: Instructor Handbook, op. cit.*
[6] *Teaching Students with Disabilities: Instructor Handbook, op. cit.*
[7] *Teaching Students with Disabilities: Instructor Handbook, op. cit.*

SUGGESTIONS FOR TEACHING MODIFICATIONS

Preferential seating: Students with visual impairments may need preferential seating because they depend upon listening. Because they may want the same anonymity as other students, it is important that you avoid pointing out the student or the alternative arrangement to others in the class.

Exam accommodations: Exam accommodations—which may include adaptive technology, a reader/scribe, extra time, a computer, closed circuit TV, Braille, enlargements, tapes, or image-enhanced materials—may be needed.

Arranging for accommodations: A meeting with the student is essential to facilitate the arrangements of accommodations and auxiliary aids, which may include, in addition to exam accommodations, access to class notes or the taping of lectures; print material in alternative format; or a script with verbal descriptions of videos or slides, charts, and graphs, or other such visual depictions converted to tactile representations.

Orientation to classroom: You may also ask the student if he or she would like an orientation to the physical layout of the room identifying the locations of steps, furniture, lecture position, low-hanging objects, or any other obstacles.

Use of language: Although it is unnecessary to rewrite the entire course, you can help a visually impaired student by avoiding phrases such as "Look at this" and "Examine that" while pointing to an overhead projection. Use descriptive language. Repeat aloud what is written on an overhead or chalkboard.

Lab assistance: Students may need an assistant or lab partner in lab classes. Help the student find an assistant.

Print material in alternative format: Have reading assignments ready three to five weeks prior to the beginning of classes. Students with visual impairments will likely need all print material in alternative format which means that they need print material converted to audio tapes, scanned onto disks, Braille, enlarged, or image enhanced. Conversion of materials takes time. It is important that they have access to class materials at the same time as others in your class.

Guide dogs: Keep in mind that guide dogs are working animals. They must be allowed in all classes. Do not feed or pet a guide dog. Because they are working, they should not be distracted.

SUGGESTIONS FOR TEACHING MODIFICATIONS

Gaining attention: Make sure you have a deaf student's attention before speaking. A light touch on the shoulder, a wave, or other visual signals will help.

Preferential seating: Offer the student preferential seating near the front of the classroom so that he or she can get as much from visual and auditory clues as possible or clearly see a sign language interpreter if one is used.

Effective communication: Do not talk with your back to the class (for example, when writing on the chalkboard). It destroys any chance of the student getting facial or speech reading cues. Your face and mouth need to be clearly visible at all times. Avoid sitting with your back to a window, chewing gum, biting on a pencil, and other similar obstructions.

Videos and slides: Provide videos and slides with captioning. If captioning is not available, supply an outline or summary of the materials covered. If an interpreter is in the classroom, make sure that he or she is visible.

Class discussion: When students make comments in class or ask questions, repeat the questions before answering, or phrase your answers in such a way that the questions are obvious.

Class notes: Students may need your assistance in getting class notes. When a student is using a sign language interpreter, captioning, or lip reading, it is difficult to take good notes simultaneously.

Sign language or captioning services: When a student uses a sign language interpreter, discuss with both the student and interpreter(s) where the interpreter(s) should be located to provide the greatest benefit for the student without distracting other class members. When a student uses a captioning service, discuss with the student and captioner the appropriate location.

Role of the interpreter: The interpreter is in the classroom only to facilitate communication. He or she should not be asked to run errands, proctor exams, or discuss the student's personal issues. The interpreter should not participate in the class in any way or express personal opinions.

Interpreter classroom etiquette: The interpreter is in the classroom to facilitate communication for both the student and the instructor. Speak directly to the student, even though it may be the interpreter who clarifies information for you. Likewise, the interpreter may request clarification from you to ensure accuracy of the information conveyed.

English as a second language: For many deaf students, English is a second language. When grading written assignments or essay tests, look for accurate and comprehensive content rather than writing style. Students should be encouraged to go to the writing center for assistance if necessary.

Students with ADHD or ADD (without hyperactivity) may have difficulty with one or more of the following areas:

Concentration	Following directions
Distractibility	Listening
Organization	Sitting for lengthy periods
Completing tasks	Transitioning
Sedentary tasks like reading	Planning

Some students with ADHD take medication for their condition. This medication may be a stimulant, which actually calms them and helps them focus on tasks. Antidepressants may also be used.

SUGGESTIONS FOR TEACHING MODIFICATIONS

Assistance with structure: Handouts with clearly delineated expectations and due dates and frequent opportunities for feedback provide these students with assistance with organization and structure. Study guides and review sheets are also helpful in providing structure.

Exam accommodations: Many students with ADHD use exam accommodations including extended time and a distraction-reduced exam space.

Access to class notes: Some of these students have difficulty focusing and concentrating and for this reason may need access to classroom notes. Your assistance may be needed to ensure that they get notes.

Classroom distractions: If a student appears extremely distracted, it may be appropriate to encourage the student to sit near the front of the class, away from doors, air conditioning units, windows, or any other possible sources of distraction.

Students with Mobility or Medical Impairments[8]

Mobility impairments are often caused by conditions such as cerebral palsy, multiple sclerosis, muscular dystrophy, or spinal cord injury. Students may use crutches, braces, or a wheelchair, and in a few instances, may be accompanied to class by a round-the-clock nurse. Medical impairments are often hidden disabilities, caused by such conditions as arthritis, asthma, cancer, orthopedic limitations, postsurgery, chronic fatigue syndrome, or seizure disorder. The student may have limited energy; difficulty walking, standing, or sitting for a long time; or other disabling characteristics.

Functional limitations may be episodic for some students who may experience dizziness, disorientation, and difficulty breathing during a recurrence. For example, with asthma or a seizure disorder, students may have

---

[8] *Teaching Students with Disabilities: Instructor Handbook, op. cit.*

periods when they function without any accommodations, but at other times their functional limitations are quite severe.

Even with the same disability, students with mobility or medical impairments may have a wide variety of characteristics. For example, persons who have experienced a spinal cord injury are likely to show differing degrees of limitation. They may require different types of class accommodations or may not need accommodations, depending on their functional limitations.

SUGGESTIONS FOR TEACHING MODIFICATIONS

Exam accommodations: Students who have upper body limitations and are unable to use their hands will likely need exam accommodations, which may include extended time, a scribe, or voice recognition software.

Access to class notes: Students who are unable to use their hands may need assistance in finding a note taker, or they may elect to tape record lectures.

Tardiness: Some students are unable to quickly get from one location to another because of architectural barriers or difficulty in using adaptive transportation. The transportation system is influenced by traffic, weather, and scheduling problems. For these reasons, a student may be late getting to class. Please be patient when this happens.

Seating arrangements: In a few situations, a student may be unable to use the type of chair provided in a particular classroom. Check with your school about making special seating arrangements.

Inaccessible classroom: If your classroom is inaccessible and a student is unable to get into your classroom, your class location must be moved to an accessible location.

Laboratory courses: Some students may need assistance for laboratory courses. These students may need to be paired with an able-bodied student or a teaching assistant. A student using a wheelchair may need a lower lab table to accommodate the wheelchair.

Missed exams or classes: Some students experience recurrence of a chronic condition requiring bed rest or hospitalization. These students need extra time to complete work and the opportunity to make up tests. Other arrangements may be necessary if a student misses a class excessively because of a disability and is unable to make up the essential requirements of the class. In either situation, it is essential not to penalize a student for a disability and at the same time maintain the integrity of the requirements of the class.

Field trips: Make arrangements for field trips or other out-of-classroom experiences as soon as possible so that all students are able to experience all class teaming opportunities.

Students with Psychiatric Disabilities[9]

Students with psychiatric disabilities exhibit a persistent psychological disorder that adversely affects their educational access, their academic performance, and daily functioning. They frequently require medication. Psychiatric disorders include but are not limited to the following:

> Depression is a major disorder that can begin at any age. Major depression may be characterized by a depressed mood most of each day, a lack of pleasure in most activities, thoughts of suicide, insomnia, and feelings of worthlessness or guilt.
>
> Bipolar disorder causes a person to experience periods of mania and depression. In the manic phase, a person might experience inflated self-esteem and a decreased need to sleep; however, in the depressive phase, a person may experience lack of energy, lowered self-esteem, and lack of interest in family, friends, and school.
>
> Anxiety disorders can disrupt a person's ability to concentrate and cause hyperventilation, a racing heart, chest pains, dizziness, panic, and extreme fear.
>
> Schizophrenia can cause a person to experience, at some point in the illness, delusions and hallucinations. *(Source: University of Minnesota Disability Services web site: http://disserv3.stu.umn.edu/AG-S/3-5.html)*

In most situations you will not be aware that you have a student with a psychiatric disability in your classroom. Because students do not show any outward signs of the disability, that does not mean that their disability is any less disabling than a more visible disability. Many of these students are fearful of and have faced stigmatization because of their disability. Some do not need or request any accommodations, and some require a variety of accommodations. For some, the disability is temporary. With medication or therapy, they recover. However, some students face a constant or a recurring battle to keep their disability under control.

SUGGESTIONS FOR TEACHING MODIFICATIONS

Exam accommodations: Assist these students in arranging for exam accommodations when requested. The exam accommodations most likely used are a distraction-reduced exam space and extra time.

Make-up work: During periods of serious psychiatric episodes, these students may miss class. Collaborate with students about arrangements to make up tests and other assignments allowing them extra time.

Welcoming and supportive environment: Many students with psychiatric disabilities fear stigmatization because of their disability. If a student shares his or her disability with you, be supportive and welcoming when a student requests assistance in arranging for accommodations.

---

[9] *Teaching Students with Disabilities: Instructor Handbook, op. cit.*

TEACHER ATTITUDE

It is especially important that teachers exhibit positive attitudes toward learners with special needs. A positive attitude on the part of the teacher will encourage positive attitudes on the part of students. Positive student attitudes enhance learning.

Some teachers claim that time spent working with learners with special needs detracts from the educational experience of other students. This is sometimes true. However, a teacher who undertakes the thorough planning needed to effectively present instruction for special needs learners will find that instruction for all students will have been more thoroughly planned. Often the modified materials used for presentations to learners with special needs are equally effective with other students.

There is a legislative mandate to provide educational opportunity for learners with special needs within the regular school program whenever possible. But regardless of whether there was a legislative mandate, teachers have a responsibility to every student being served. Teachers must not attempt to exclude someone from the educational process simply because he or she may be more inconvenient to teach.

Students with mild disabilities in regular classes have been shown to progress as well as or better than those in special education classes. Schools, therefore, provide educational opportunity in the *least restrictive environment*. Whenever possible, a student with a disability is to be educated with his or her nondisabled peers.

CONCEPTS AND PRACTICES

There are some general approaches to teaching learners with special needs that work. These concepts and practices stress the value of meeting the individual needs of students. Although these practices are recommended for all students and were discussed earlier in the text, they are emphasized here because of their modified application with learners with special needs.

Student Attitude Development

Many learners with special needs have had their attitudes shaped by failure and rejection. The challenge to the teacher is to develop a positive atmosphere in which all students can learn.

Self-Concept. Self-concept may be the most important single factor in determining what a person is able to do under any given circumstance. It is how a person feels he or she will do in a given area that determines his or her success. A student may not perceive himself or herself as a studious individual, and hence not play the role of a student. There are specific ways a teacher can enhance the self-concept of all students. They include:

1. Praise achievements, no matter how small. Students react positively to praise. If a student does a good job of sweeping the pet shop floor, praise

that student. If a student is more attentive than normal during a class period, praise him or her. For some students, praise may need to be given for five minutes of attentive work rather than expecting them to be attentive through an entire class period before receiving your positive reinforcement.

2. Avoid labeling students. In conversations, never refer to students as retarded, special education students, or slow learners. Such labels tend to lower a student's self-concept.

3. Ignore negative behavior, when possible. Students receiving criticism of negative behavior are reinforced in their belief of low self-worth. Instead, if the behavior is a minor problem, learn to live with it. For example, if a student cannot seem to sit still, it might be better to let the behavior continue than to call attention to the student for the behavior. Channel the behavior in a contributory way.

4. Treat students with dignity. This can be done by valuing students' opinions and by treating them with respect.

5. Encourage each student's assets and interests. If a student draws well, encourage him or her to develop plans for a class project. If a student is a good welder, have that student give a class demonstration.

6. Compare a student's performance with his or her previous work, not that of other students. A student may recognize that his or her work does not compete with others in the class. A teacher, by stressing how the student's floral arrangement is improved over a previous attempt, will help the student develop a sense of progress.

7. Point out areas of accomplishment rather than focusing on mistakes. If a student creates a sign for the outdoor nature center at the school, but not perfectly, a teacher should recognize that the student has finished a project that he or she had never been able to do previously. In this way, the student is not discouraged from trying again.

8. Listen intelligently to each student. A student who wants to tell the teacher about the playful antics of a lamb at home wants to feel that the teacher is interested in what is said. You must take time to listen.

9. Support students in their attempts to express themselves. Learners with special needs may have some difficulty communicating ideas. Encourage them when they offer to share with the class or the teacher.

10. Provide each student with honest success experiences. Look for experiences in which students may succeed. Do this by giving them responsibilities that match their capabilities and by practicing small increments of the responsibility until it builds to successful accomplishment. Keep in mind that a very small task can be a success for many students.

11. Provide specific help for improvement rather than generally negative comments about work. Rather than telling a student where he or she erred, give specific suggestions. For example, instead of saying, "Don't

hold the parrot that way," suggest, "It is easier to control the parrot when you hold it like this."

Recognition and Reward. Being rewarded for doing something correctly usually causes that behavior to be repeated. The rewards should be appropriate to the situation and meaningful to the student. Rewards should be consistent for all students.

The FFA organization provides an excellent vehicle for recognizing achievements of students. The number of awards provided should be sufficient so that all students who have achieved in some area are recognized. The FFA Achievement Awards program provides a system through which FFA members are recognized for their accomplishments in both the instructional program and in their FFA leadership activities. Students participating in the achievement awards program compete against a standard set of performance objectives rather than against other chapter members.[10] The awards are intended to involve all students studying agriculture in the awards system of the FFA and to provide a broad array of opportunities for recognition. An individual contract containing a checklist of skills to be achieved can be developed with each student. Achievement award certificates are available to teachers from the National FFA Organization.

Learners with special needs often respond to immediate rewards rather than the promise of later recognition. Therefore, teachers must devise a system for giving immediate recognition to desired behavior. An FFA merit point system is an example of providing both immediate and delayed gratification. The student receives points for displaying desired behavior. As points accumulate, the student again receives recognition for the total effort.

Fear of Failure. Fear of failure may result in avoiding an activity in which students perceive they will fail. Students have been known to be absent from school because of a planned activity (such as an examination) they believe they might fail. Some students will simply refuse to try tasks they believe they cannot successfully accomplish.

A person may have potential for a certain job, but it is only after that person believes that he or she can do it that it will be performed successfully. Once a task or job has been accomplished, proficiency and confidence result, which facilitate success in additional work.

In the classroom or laboratory, learners with special needs may not feel capable of performing certain tasks. A teacher may need to provide a meaningful related learning activity. Confidence can then be developed to do work that previously was thought to be too difficult. Learning activities should be designed so students believe they can succeed in doing them but at the same time believe they have been appropriately challenged. Learning activities designed with this range of challenge will enhance student satisfaction and self-concept and help them overcome fear of failure.

---

[10] *FFA Advisor's Handbook.* Indianapolis, IN: National FFA Center, 2000.

Motivation. Students are motivated at all times, but often they are motivated along lines that are not educational. A student may be motivated to study in the classroom, to be disruptive, or to daydream. It is the responsibility of the teacher to channel students' motivations along positive, constructive lines. Because disadvantaged students have sometimes experienced academic failure and frustrations, they will have to be provided with positive experiences before a consistent motivation to learn is established.[11]

Feedback and Reinforcement. Students need to be shown that their actions do make a difference. Feedback or knowledge of results is a motivational technique based on theories of behavioral change. Feedback of a positive nature will provide reinforcement and increase the probability of repeated behavior.

Reinforcement can be verbal (written or oral), nonverbal (physical gestures or appropriate contact), or material (money, food, special privileges). Feedback must be appropriate to the individual. Stephens[12] suggests some tactics for use by teachers:

- *Direct reinforcement.* When students perform or respond correctly, the teacher rewards them. Direct reinforcement can be positive or negative and can be used to develop appropriate or inappropriate behavior.
- *Shaping.* Behavior is shaped through differential reinforcement of gradually improving responses on the part of the student. In this procedure the teacher may recognize and compliment a student attempting to drive a nail for holding the hammer correctly, and later for completing a nail-driving project correctly.
- *Cuing.* Students are coached using a step-by-step process in which they are signaled concerning the appropriate performance for each step. A teacher making a corsage would tell a student to watch and do what the teacher does; the teacher would perform only one step at a time; and the student performs the step while observing and listening to the teacher.
- *Contingency contracting.* Performances and rewards are specified in advance. Interim rewards may be used for progress toward desirable performance.

Student Personal Development

Learners with special needs must develop skills in citizenship if they are to successfully enter adult life as productive workers. There are several techniques that can be used by teachers to develop students in this area. Some are best used in conjunction with the National FFA Organization (see Chapter 11).

---

[11] Thomas, Hollie B., Jr. *Motivating Disadvantaged Students to Learn.* U.S. Department of Health, Education, and Welfare, National Institute of Education, Task Force on Dissemination, 1973, p. 9.

[12] Stephens, Thomas M. *Teaching Skills to Children with Hearing and Behavior Disorders.* Columbus, Ohio: Charles E. Merrill Publishing Co, 1977, p. 216.

Learners with special needs should have the opportunity, through the FFA, to conduct group planning, accept responsibility, develop awareness of the rights of others, and develop work habits and social skills.

Supervised Agricultural Experience

Supervised agricultural experience programs (see Chapter 10) are a desirable part of agricultural instruction for all students, including those classified as learners with special needs. It is especially important that learners with special needs be encouraged to apply instruction because, for these students, abstractions are less meaningful than real hands-on experience.

Instruction and assistance should be provided to students at the time of practical application. Demonstrations are nearly always needed, and close supervision of the work is required.

Student placement with cooperating employers is especially important for special needs learners. The successful work experience gained in supervised agricultural experience will enable students to gain the needed confidence and work record for further employment.

Supervised agricultural experience also teaches students to manage money. Basic personal accounting should be considered as a part of the intracurricular course of study.

Student Safety

Teachers of agriculture are responsible for the safety of their students (see Chapter 9). Special needs learners may require extra instruction and supervision in this area. Therefore, teachers must devote time to safety instruction and supervise a safety test (in some states this is the law) for each student on every potentially hazardous piece of equipment, technique, or process.

Guards on machines must be kept in place. Demonstrations must be given for each step of each job. The teacher must provide supervision. It is helpful to establish a checking procedure whereby students must demonstrate what they are going to do before the task is performed.

Individual Instruction

The teacher is responsible for preparing instruction that meets the individual needs of all students, whether they are classified as disadvantaged, disabled, regular, or gifted. The technique of teaching using individualized instruction has been discussed previously (see Chapter 7). Special needs learners must have an individualized educational plan (IEP). The plan is usually prepared by a team but chaired by the special education teacher. The IEP involves assessment of need or diagnosis, specification of learning objectives, planning and providing learning activities, and assessment or evaluation of learning. Agriculture teachers need to be proactive in providing objectives and activities to the special education teacher for inclusion in the IEPs of learners with special needs enrolled in agriculture classes. Given the hands-on, intracurricular nature of agricultural education, the agriculture teacher is equipped to suggest techniques and strategies that are contextual to the agriculture curriculum. By working together, both teachers are proactive, rather than reactive, to the needs of the learners.

The special needs teacher will lead the team of professionals that perform the following duties:[13]

- *Assessment*
  - Psychological testing
  - Organize and facilitate assessment
  - Assess students' needs and make program recommendations
- *Consultation and Program Planning*
  - Facilitate the development of behavior management programs
  - Facilitate the development of IEPs
  - Discuss modified programs and teaching strategies with teachers; offer recommendations and provide support in program planning for students who are encountering academic or emotional difficulties
  - Consult with administrators, teachers, teaching assistants, and parents regarding appropriate special education programs for students
  - Provide information regarding interpretation of assessments, medical documents, and special needs
  - Assign students to district special education programs
- *Provision of Resources*
  - Recommend materials and resources for special education students
  - Order materials and resources for special education
  - Consult in-service teachers, teacher assistants, and parents regarding special education issues

ENRICHING INSTRUCTION

Instructional techniques that teachers of agriculture might use with learners with special needs are not unlike those that might be used with any group of students. General techniques have been discussed in previous chapters (see chapters 6 and 7). However, with learners with special needs, techniques that

---

[13] The Alberta Teacher's Association. Battle River Regional Division 31. 5402-48A Avenue, Camrose, Alberta, Canada.

appeal to multiple senses and those that reinforce and reemphasize learning are needed. The techniques that follow, although useful with all students, may have special value with students who are disadvantaged or disabled.

Using Instructional Materials

Learners with special needs may have limited vocabularies and be behind grade level in reading or other core subjects. Their orientation may be toward verbal rather than written communication. Standard materials will not meet the needs of students with these characteristics.

To develop the ability of students to read and understand literature in agriculture, the use of glossaries, word lists, or vocabulary lists is helpful. Students should be challenged to learn new words and their meanings. New words can be taught by connecting the word with its visual image, when possible. Either a picture or a real object can be used to illustrate the word. The word can be used in simple sentences to develop further understanding.

The reading level of instructional materials is an important consideration. References for student use should represent a variety of reading levels. During supervised study, materials should be assigned according to the reading capability of students.

Because of the difficulty some students have in dealing with abstractions, real objects should be used for demonstrations when possible. Visual aids may be used to illustrate concepts. With visual aids, use questioning to assess understanding. Students may have trouble visualizing two-dimensional illustrations of three-dimensional objects.

Multimedia presentations are useful and desirable. They appeal to multiple senses and can be replayed for repetition and overlearning. Tape recordings may be especially useful for this purpose.

A camera and the resulting photographs can stimulate interest and discussion. Impromptu photos of individuals involved in field trips or special events may be posted on a bulletin board. The result is often a better group spirit among students.

Students with specific disabilities such as blindness or deafness may need specialized aids. Special education teachers can assist teachers of agriculture in selecting appropriate materials.

Sensory-Rich Classroom

A sensory-rich classroom can arouse student interest, stimulate thinking, and provoke questions. There should be a variety of things to see, touch, hear, smell, and taste.

Developing learning centers of various varieties and grades of hay or silage can enrich a unit of instruction on feeding ruminants. You can also enrich the unit of instruction by conducting experiments on rate gains of animals fed different protein levels. A natural resources unit of instruction on recognizing game birds can be illustrated with color photographs of various species. The ingenuity and imagination of the teacher serves to develop and maintain a climate of anticipation about learning.

Peer Instruction

A student who has mastered a particular skill or some knowledge is often able to work effectively in teaching other students. This method of peer instruction is especially effective in providing individualized or small-group instruction for learners with special needs. Peer instruction using students from another classroom has been found to be more effective when:

- The assistance program was completely voluntary.
- Learning to help was regarded as an educational activity for the assistants.
- Assistants were provided with feedback.
- Helping became a status activity.[14]

When students from the same classroom help other students, care should be taken to avoid the stigma that might be attached to being helped. Vary the procedure so that occasionally a learner with special needs works with another learner with special needs. Take advantage of possible opportunities for learners with special needs to demonstrate skills to regular students. This opportunity may come when a regular student is making up work because of an absence or when a student with special needs possesses a specific skill. For example, a student with special needs may have practiced to become proficient as a dairy holder. The student could then help others learn showmanship skills. Peer instructors should be taught, among many things, to be patient, to relate with others, and to express themselves. They must practice demonstrations for small groups of students to develop their abilities in these areas.

SUMMARY

The goal of every school must be to provide educational opportunities for learners with special needs within the regular school program whenever possible. In regular classes, students who are mildly handicapped can progress as well as or better than they do in special classes. By placing learners with special needs in regular classes, they avoid the stigma that is often associated with special needs education.

---

[14] Lippitt, Peggy. *Students Teach Students.* Fastback 65. Bloomington, IN: The Phi Delta Kappa Educational Foundation, pp. 14–15.

All agriculture teachers are likely to teach learners with special needs in their classes. Assistance can usually be obtained when needed from the special needs teacher from whose class a student is being placed in your classroom. Teachers continually serving students who are disabled may desire further in-service education to better prepare them for the task. The contribution to individuals and to society because of effective instruction of these students is great. Teachers must enthusiastically accept the challenge of teaching every student enrolled in their classes.

FOR FURTHER STUDY

1. Read the following case problem and answer the question in the last sentence.

CASE PROBLEM

Although Mary's legs were severely crippled, she could use her crutches to walk. During the first week of school she went with the rest of the horticulture class on a field trip to a large commercial greenhouse. Shortly after the tour of the facilities began, she became tired from walking. Especially difficult for her were the gravel walkways. Ms. Brown, the class's teacher, gave her a chance to rest briefly before continuing on the tour. As the field trip continued, Mary's rest stops became more frequent and her slow walking put the trip farther and farther behind schedule. Mary sensed that she was spoiling the trip for the other students, so she made up a story that she did not feel well and wanted to go back to the bus to wait until the others had finished the tour. What should the teacher have done?

---

2. Visit a school that offers an agriculture program. Observe the students in one or more classes. Discuss with the teacher which students might be classified as learners with special needs. Find out how the teacher modifies instruction to meet their needs.

    a. List the number of students with each classification of special needs.

    b. What challenge does the teacher face in teaching these students?

    c. What special techniques does the teacher use?

3. Visit with a special education teacher. What recommendations does he or she have for teaching various learners with special needs in his or her classes?

4. Observe a special education teacher during class. Describe techniques being used, and rate the effectiveness of those techniques with the various classifications of learners with special needs.

13

Teaching Adults

> Mr. Miller was interviewing for a position as a teacher of agriculture. During his visit with the superintendent of schools, she emphasized that an important part of being an agriculture teacher in that community involved conducting an adult education program. The community had an active young farmer chapter and had become accustomed to course offerings for adults each year relating to some phase of agricultural financial planning. Mr. Miller was confident. He was pleased that an important part of his preparation to teach had prepared him to instruct adults.
>
> Agriculture teachers should be prepared to teach adults. Confidence in teaching adult students will enable the teacher to better serve the community in many ways.

OBJECTIVES

After studying this chapter, you will be able to

1. Explain the need for adult education in agriculture.
2. Describe adult learning characteristics.
3. Plan and teach adults in group settings, in individual instruction situations, and by using resource people.

TYPES OF EDUCATIONAL PROGRAMS

What types of adult education programs have been offered by teachers of agriculture? What programs might a teacher be expected to conduct? There traditionally have been several forms of adult education offered by teachers of agriculture.

Young Farmer Education

Young farmer education has developed as a distinct type of adult education in many states. The young farmer educational program is designed to help young men and women recognize and solve their problems in becoming established in farming and the community. The idea is to help young farmers advance in efficiency and status in agriculture. Those enrolled are usually out-of-school young farmers who are provided systematic instruction designed to be helpful in establishing them in farming. A student may be living at home with parents, working as a farm laborer, as a partner in a farm business, as a farm renter or owner, as a farm manager, as a part-time farmer, or in an agriculturally related occupation. Teachers provide on-farm instruction for the agricultural enterprises.

Adult Farmer Education

Adult farmer programs have traditionally been offered to help farmers and others increase their proficiency in agriculture through systematic instruction. Classes normally are held in the evenings, especially during nonpeak farming seasons, such as during the winter months. Complete programs include year-round individualized on-farm instruction by the agriculture teacher. Adult farmer instruction has evolved in different forms. A brief description of three of the forms follows.

Adult Farmer Seminars. Adult farmer seminars are provided as a series of classes (about fifteen to twenty), usually during the winter months. Some classes dealing with seasonal topics are held at other times. The topics that are studied are planned jointly by the enrollees and the teacher. The topics may appear to be somewhat unrelated to each other. The purpose of the class is to study current problems faced by the farmers. Examples of class topics might include techniques and equipment for improved pesticide application, value-added enterprises, genetic improvement in cattle, or hazard waste identification and disposal.

Adult Farmer Enterprise Classes. Enterprise classes usually involve a series of related sessions. Examples of agricultural enterprises are a corn production enterprise, a soybean production enterprise, a Christmas tree production enterprise, and a dairy herd enterprise. An enterprise class might consist of a series on cotton production or a series on beef herd management.

Even though agricultural mechanics topics are usually not considered enterprises, agricultural mechanics classes might be offered to adults to improve the proficiency of enrollees in a specific area of agriculture. In agricultural mechanics, a series might be offered in global positioning systems, agricultural waste management systems, or other similar areas.

Enrollees in enterprise classes include farmers and others interested in learning more about the enterprise. The specialized nature of these classes, with more time devoted to each topic, enables study in greater depth than is possible in seminar classes.

Adult Farm Business Management Instruction. Farm business management programs are more intensive than the other types of adult farmer instruction. Instructors of adult farm business management programs are usually either full-time or at least part-time adult instructors. These programs emphasize records analysis and management decision making. Instructors provide in-depth help to farmers with individualized on-farm instruction in record keeping and analysis. Formal classes are held year-round.

Adult Education for Agricultural Business

A type of adult instruction with potential for further development is programming for agribusiness professionals and their employees. Such instruction may take various forms but is generally designed to prepare individuals for entry or advancement, or to update employees in the technology of their specialization. Multiple-teacher departments having teachers with specialized expertise might offer courses on such topics as nursery plant propagation or turf management. The turf management course could be planned to serve the needs of golf course greenskeepers. Courses of this nature have more potential in urban areas because of the greater number of potential students interested in highly specialized topics.

Avocational Adult Education

Agriculture teachers in many communities offer adult instruction of an avocational nature. Such classes may deal with gardening, floriculture, home landscaping, beekeeping, garden tractor repair, or other topics of interest to participants. With many people living on a little land there is much interest in communities for these programs.

REASONS FOR ADULT EDUCATION

What is the case for adult education in agriculture? Why should it be offered in the public schools? The concept of continuous or lifelong learning is generally accepted in U.S society. One must, however, have more than an intellectual acceptance of the importance of adult education. The teacher of agriculture should exhibit an operational commitment to the further development of continuing education in agriculture. Adult education is of benefit to those who enroll and to the community in which they reside. Teachers who instruct adults will find that even the instructor benefits.

Benefits to Adult Learners

Agricultural technology is rapidly changing. The education gained in one's youth needs to be updated. Records analysis has switched from a paper-and-pencil operation to computers. Computer programs are used to balance rations, calculate the proper size of farm machinery, and apply fertilizer at varied rates in the same field. Energy conservation is now more important than in earlier years, and rural/urban interface issues are highly prevalent. These examples illustrate the rapid accumulation of new knowledge that is currently taking place. The adult who continues to study and learn in agriculture will continue to succeed.

Adults desire continuing education to make midcareer changes in occupations or to become more proficient in their current practices. Farmers may attend classes to learn how to better control weeds in their crops. Fertilizer salespersons may attend classes so they can better advise their clients on a broader range of agricultural problems. A basic need of young adults is getting started in an occupation; a basic need of middle-age adults is establishing and maintaining an economic standard of living. Adult classes can be a means of adults achieving their goals.

Adult education fills a social need for the participants. Adults desire to interact with others about their problems and successes. They desire to jointly seek solutions to common problems. Adult education in agriculture can meet social needs. Some adults may form car pools to travel to and from class. Some instructors have an informal dinner to conclude a class and award certificates to participants. These activities fill a social need for some individuals.

Benefits to the Community

The community benefits when its citizens achieve higher levels of education, income, and social consciousness. Adult education has led to early adoption of many agricultural practices. These practices have improved the efficiency of the agricultural enterprise, allowing a small percentage of the population to produce the necessary food and fiber. The remaining portion of the population has been released to produce products and services that have improved the standard of living of the community at large.

Some adult education programs have had a direct impact on the social needs of a community by studying about and acting on such topics as community health, zoning regulations, school financing, and wildlife conservation.

Benefits to the Teacher

The teacher usually finds adult classes to be stimulating and rewarding. Adults attend classes and participate in programs because they want to rather than because they are required to. Motivation is easier.

Teachers who have adult classes often indicate that such classes improve their high school or postsecondary instructional program. They find that the planning required to provide relevant instruction for adults helps also in providing more practical instruction in classes for youth. The wealth of agricultural knowledge learned by a teacher in classes where information is shared among adults can be effectively used at other levels of instruction.

Effective adult instruction builds the reputation of the teacher in the community. This reputation pays off in relationships with other teachers, administrators, and students.

The teacher develops with the adult student a feeling of mutual respect. The teacher needs to respect the adult for the practical experience he or she has gained from years of practice. The adult learner develops respect for the teacher as he or she can relate information to practice. As the teacher visits the adult class members in their places of work, the instructor can share information learned from observing the workplaces of others and from reading literature.

A beginning teacher should consider taking advantage of opportunities to teach adult classes. Adults are supportive of their instructors and will help a class succeed in accomplishing its purposes. The teacher will find adults willing to share the knowledge they possess. The beginning teacher must avoid appearing as a know-it-all and should instead treat the adult students as fellow learners.

CHARACTERISTICS OF ADULT LEARNERS

Do adults learn differently from youth? What are the unique characteristics of adult learners? How should a teacher make use of adult learning characteristics in planning for instruction? It should be emphasized that the principles of learning presented in Chapter 2 are applicable for teaching both youth and adults. The basic ability to learn changes little, if at all, with age. Studies of the same individuals over long periods of time have shown they continue to learn and improve in intellectual ability, at least until they approach the age of sixty. The changes that do occur are changes in health and physical condition, attention, motivation, and ways of viewing experiences.

Physical Characteristics

The speed of performance and reaction time tends to decline with age for adult learners. If one were potting bedding plants in the greenhouse, for example, one should not expect older adults to work at the same speed as younger adults.

Adults also experience physical changes with aging. Some of these changes important in teaching include failing eyesight, loss of hearing, and more frequent illness. Health affects performance on intelligence tests. Therefore, the adult in ill health could be expected to be somewhat impaired in his or her ability to learn. Older adults become increasingly less interested in learning physical skills but often become more interested in intellectual pursuits.

Psychological Characteristics

Adults are a less captive audience than are youth. Adults may elect to remain in or leave a class or program based on their perception of whether it meets their interests. The teacher needs to develop skills in gaining an understanding of what each adult wants and needs, and in creating opportunities for satisfying those desires. Adults may be goal oriented, activity oriented, or learning oriented.

Goal Oriented. A goal-oriented person might be motivated to enroll in adult education because he or she perceives the class as helping to meet a goal. The instruction might help in solving a problem the person wants to solve, or it might help him or her learn to do something needed to accomplish a goal. For example, an adult may have a goal of repairing personal farm machinery. An adult class in welding may be a necessary step in accomplishing the goal. A goal of improving farm income might stimulate participation in a farm business management course. A desire for a promotion at work may lead one to seek additional skills through adult education.

Awards can be motivating factors. They may be given for achievement, attendance, participation in planning committees, and participation in field trials or demonstration plots. The Young Farmer Association (YFA) has many awards that encourage goal-oriented individuals to participate and achieve. A goal-oriented adult usually has the ability and desire to use what is learned by immediately applying the knowledge in practice.

Activity Oriented. An activity-oriented person wants to participate. He or she enjoys working on activities, often with others. This type of person would be motivated to participate in a floral design course primarily to enjoy the activity. A winter swine show might be planned as a part of a series of classes. The show provides opportunity for competition, sociable activity, and additional knowledge.

Learning Oriented. A learning-oriented person would be motivated to participate in adult education because of the desire to learn. More highly educated individuals desire to continue to learn. As more people obtain higher levels of education, the desire of the population to participate in adult education classes will increase. This type of motivation will cause adults to seek avocational and enrichment courses.

Different adults may attend the same class for different reasons. A grain farmer might attend a course on hedging of grain futures to increase income.

Others, however, might attend for the social interaction and still others because they simply want to learn about the topic. An awareness of these motivating forces can help a teacher meet the needs of enrollees.

Attitudinal Characteristics

The older the adult, the more stable are social skills, values, and attitudes. Years of experience may make it more difficult for a person to learn a new skill related to a familiar job. Old habits and attitudes may interfere with new learning. It is easier for an experienced person to learn a completely new task than to learn to do a familiar task in a new way.

Previous opinions or biases can affect the conclusions adults reach. An adult may have great difficulty accepting logical conclusions from evidence if the conclusions seem to be incongruent with a deeply held value or belief. If one believes in organic gardening, the value of using chemicals may be rejected, even though their worth may have been proven in given situations.

Adults bring greater maturity of judgment to learning situations than do youth. It has been shown that adults age forty or older have an advantage in learning information that is related to experience and that calls for sound judgment. Because of more experiences, adults can relate theory to practice more effectively than high school learners. They can evaluate the potential of improved practices and make judgments concerning feasibility in specific situations. Because of the maturity of adults, they are more likely to be discerning and are more likely to reject the irrelevant.

PLANNING FOR ADULT COURSES

The principles discussed in Chapter 3 relating to selection and sequencing of course content also apply to adult instruction. The hours of class time available to teach adults is often less. Adults often like to be involved in course planning decisions. Adults want to study problem areas they perceive as practical and related to their needs. These problem areas may be identified in various ways. The teacher should involve the learners and also research other sources in planning the program of study.

Observing Adult Learning Needs

The best information concerning the needs of adult learners can be obtained by visiting their farms or places of employment. The teacher must visit adult farmers and businesspersons in their specialized areas of teaching in order to keep abreast of developments and plan relevant instruction.

> A teacher, through visits to farmers in the community, discovered several who were planning to build swine confinement facilities. The teacher then worked with the farmers to plan a series of classes on the topic. His

series enabled mistakes to be avoided by those developing the facilities. Other farmers also participated, and some decided to also develop confinement facilities, whereas others determined that they should not build that type of facility at that time.

Involving Adult Learners

Participants need to share in diagnosing needs, in formulating objectives, and in planning. There can be a sense of shared responsibility for planning and learning.

A teacher in a community with many glass greenhouses observed the inefficient use of energy to heat them. She organized an adult course for the growers relating to energy conservation in glass greenhouses. As a result of the course, many growers installed insulating plastic layers over their greenhouses.

Using Adult Education Planning Committees. Adult education planning committees are useful in identifying topics and planning a program of study. Committee members serve as eyes and ears for the teacher as they observe and listen to others. They can then relate the observed needs of potential participants to the teacher as a program of study is being developed. The teacher can also test with the committee topics and ideas that have been identified through conversation with other agricultural experts or by researching literature.

One teacher has a nine-member adult education planning committee. Class members elect three members of the committee each year for three-year terms. An effort is made to include committee members representative of the various types of agricultural enterprises and the various geographic areas of the community. This committee meets once each year to evaluate the adult education program. It also meets once to plan the program of study for the next year.

Using the Class in Planning. A committee of the whole class can be used in much the same way as an adult planning committee. During a final class session for the year, a discussion can be held concerning problem areas of study for the next year. The teacher can further study the identified topics during the period the class is in recess. Problems relating to the topics can be identified and observed. Information dealing with the problem areas can be accumulated. Then, during the first class session of the new school year, the entire class can review a tentative program of study. A committee of the whole class is normally more effective for relatively small classes. The teacher may come to the class with a tentative course outline and have the class review and make suggestions for improving the course during the first session. This approach can be used with all types of adult classes in agriculture.

The teacher must involve more than current class members in planning an adult education program. There may be adults with unmet needs who have not participated because nothing had been planned for them. By involving such individuals in planning, the community is better served.

Involving Others

Indirect input concerning potential problem areas can be obtained from community leaders and those who do business with the adult learners. Community leaders can identify problem areas related to political decisions. Example problem areas that might be identified by them include:

- Effects of rural zoning
- Impact of industrial plant closing
- Development of an industrial park
- Development of a community recreational area
- Taxation for community services

Those who do business with adult learners often observe problem areas for potential study. They have the opportunity to observe when improved practices are needed. They also hear quickly from their clients when one of their products has failed to solve a problem.

Estimating Future Needs

Research and development information may indicate problem areas that should be discussed in adult classes. The information might identify either an emerging problem or a potential solution to an existing problem.

New laws and regulations often influence what should be taught in adult classes. For example, the law requiring certification of chemical applicators resulted in many adults participating in classes to prepare them to take the certifying examination. Regulations of the Environmental Protection Agency have resulted in adult classes being held to discuss ways of handling agricultural waste products. Permitting regulations for large-scale animal agriculture require adults to continue to further their knowledge.

Census data can be used to detect shifts in agricultural enterprises in a community, or trends in age of farmers and farm workers. These data can serve as a topic of study, or the data might target related topics in need of study. For example, rapid increases in rural/urban interface problems might indicate a need to work with farmers and nonfarmers on topics of mutual concern.

Societal problems and trends can also be important problem areas for study. The plight of migrant farm workers, youth unemployment in the community, or the lack of skilled agricultural workers can serve as the basis of a class for seeking possible solutions to one or more specific problems.

Inform Potential Students of Course Plans

Adults need to be made aware of instructional programs available to them. The teacher will find it frustrating to have a well-planned class arranged and few students in attendance. Good attendance is ensured when a sizeable number of potential students are involved in planning and when there is adequate publicity. Radio, television, newspapers, e-mail listservers, and community bulletin boards should be used to inform potential members of classes. Advisory committee members could commit to each inviting five participants. Some teachers have a telephone network to inform students of classes. Information sheets can be sent home with high school students. Teaching adults involves planning the program of study, informing potential class members, visiting farms and businesses of likely enrollees, lesson planning, conducting the class, and follow-up instruction.

TECHNIQUES FOR ADULT INSTRUCTION

The teacher of agriculture is in a unique position in the community. He or she can be an expert concerning the subject matter of the course to be offered. The teacher also can be thoroughly acquainted with the problems and needs of class members related to the course. In addition, he or she is versed in the use of teaching techniques. All of this information can be orchestrated to provide exemplary programs in local communities.

All adult instruction should be thoroughly planned. Teachers will find they want to be even more ready to teach adult classes than they are for younger students. Preparation may involve preparing visual aids, reading, visits to farms and businesses, securing information, and developing lesson plans. Community resources should be used to the extent possible on each topic of study. The time is well spent as the teacher becomes something of an expert in the area. He or she will be able to use much of the same information prepared for adults in teaching high school or postsecondary classes. The preparation of lesson plans is discussed in Chapter 5. The same planning procedures are recommended for adult classes.

Group Techniques for Teacher Use

Group techniques are presented in Chapter 6. Some, however, are especially suited for use with adults and receive additional emphasis here. Those especially suited for adult instruction include the discussion and panel discussion.

Discussion. The discussion method is highly recommended for use with adults. Discussion provides an opportunity for participation. It involves class members mentally. It brings together the knowledge of the students, with each contribution adding to the total information. Discussion must be kept focused and within the planned time frame. It is the responsibility of the discussion leader to:

1. Encourage the expression of ideas by members of the group.
2. Ensure that facts are available as needed.
3. Ask questions from time to time to give direction to the discussion.
4. Make occasional summaries without injecting personal ideas.
5. Serve as an efficiency expert to smooth out rough spots without using autocratic tactics.[1]

Discussion involves a general interaction among students and the teacher. In discussions, all students should be seated so they can be seen and heard by all other students. Informality is essential. The leader must keep the class on the subject with focusing comments and questions. The discussion technique is limited in that the information must be already known by discussion members. It needs to be used with other techniques when new information is needed.

Panel Discussion. A panel is usually composed of three to five members. The members should have different expertise, experience, or viewpoints on the topic being presented. A discussion of the topic is usually held among panel members. Participation may also be invited from the class members. The members of the panel should be oriented to the teaching objectives for the class and the problems and concerns of class members.

A variation of panel discussion is the symposium. In the symposium, short presentations are made by each member of the panel. Following the presentations either a panel or general discussion is held.

These techniques are best suited for local experts, especially those with limited teaching ability but with the ability to respond well to questions and interact in an informal way.

Use of Resource People

The use of resource people is common in adult education classes in agriculture. Resource people bring special expertise to the topic of study. In this way, the class can learn information above and beyond that held within the group.

Generally, a teacher who knows the students and their home situations and who has worked with a class is preferred by the students when the teacher

---

[1] Kahler, Alan A., Morgan, Barton, Holmes, Glenn E., and Bundy, Clarence E. *Methods in Adult Education.* Danville, IL: Interstate, 1985, pp. 91–92.

has the competence to teach the class. However, there are times when the use of resource people is recommended because of the new insights they can bring to the group.

Resource people should be spaced properly throughout the course. They should not be used as a substitute for the teacher. When resource people are overused, the teacher loses the leadership of the class. One must be especially careful to properly use resource people who represent commercial companies desiring to sell products or services to class members. Teachers must protect the integrity of the school and the class. Classes may fail to accomplish their objectives when a resource person takes over the class and presents a canned program that does not meet the needs of the students.

How then should resource people be properly used? The teacher, and perhaps the planning committee, should review the course of study months ahead of the classes and decide for which sessions resource persons might appropriately be used. A list of possible resource people for each chosen session should be prepared. Other professionals in agriculture may be helpful in making suggestions. At least three months ahead of the class, appropriate arrangements should be made to secure the resource people. A confirming letter should be immediately sent. A reminder letter should follow three to four weeks prior to the session. The teacher should know the exact costs that are involved, including whether there is a fee and who will pay expenses such as meals and travel.

Resource people should have the necessary information to properly prepare for the class. They should know the start and end times, the place of meeting, the expected number of students, the teaching objectives, the questions to be answered, and they should have some information about class members.

Prior to the session, the class should be prepared for the resource person. The topic might be introduced and questions and concerns identified. The class then is prepared to effectively use the resource person. During the session, the resource person should be properly presented to the class. The teacher then has the responsibility of keeping the discussion on topic and either stimulating or discouraging participation, whichever is needed. The teacher also has responsibility for the time schedule. At the end of the session, the resource person should be thanked for participating. The teacher should summarize the discussion, perhaps through use of skillful questioning of class members. Class members should be challenged to think about possible application of what was learned. This summary is sometimes more effective when the beginning of the next class session is devoted to this purpose. The important principle to keep in mind is that the responsibility for the class rests with the teacher, regardless of whether a resource person is used.

Use of Individual Instruction

Individual instruction encourages behavior change by students and speeds adoption of recommended practices. Individual instruction can occur before or after class meetings, at the place of business, on farms, or in the agriculture classroom. Adults desire the advice of the teacher and generally appreciate his or her interest in their problems.

Visits to class members can be used to gain information for later adult class sessions. Visits also develop rapport and open up communications so class members will be more willing to share problems and concerns with the teacher.

The instructor should teach on these visits. A procedure should be adopted to diagnose needs, specify problems, discuss alternative solutions, seek additional information, and make tentative choices concerning the proper procedure to follow. A friendly but businesslike approach should be adopted as the mode of operation. A schedule for visits to enrollees and potential enrollees should be developed and followed. Without such a schedule, these important teaching situations will not occur

EVALUATING ADULT INSTRUCTION

Evaluation of adult learning is usually more informal than for high school students. Grades need not be given. It is desirable to evaluate adult courses in order to assess whether the course was worthwhile and to gain information that would be helpful in making future improvements.

Student Reaction

A good informal method of getting student input for class improvement is simply to have students write on a sheet of paper two or three things they like about the course and two or three things they would like to see improved. The results can be discussed with the class if appropriate, and changes made, when feasible.

Outcome Measures

For each class session, teachers can develop with students a list of practices that might be adopted so that the instruction is applied. These lists can be duplicated. Students can be asked to indicate those they already were using before the session, those they plan to adopt, and those that are not feasible for them. This is a good test of the relevance of the instruction. It also provides information helpful to the teacher for individual instruction in future visits.

Applying Instruction

The payoff from adult instruction results when approved practices are applied on the farms, in the homes, and in the businesses of enrollees. One also desires that students use their new-found knowledge and skills in creative ways to improve their lives. The teacher should, therefore, emphasize the use by students of knowledge gained in adult education in agriculture. There are several effective ways to encourage application of learning. One effective method is to have the adult class plan demonstration plots or field trials of approved practices. Various rates of fertilization, types of weed control, varieties of seed, or types of insecticide control can be evaluated for the situations that exist in the community. Often high school classes can cooperate with adult classes in managing and caring for demonstration plots or field trials. Care must be used in ensuring proper research procedures are followed. The plots provide opportunity for photographs to be taken and used in high school and adult instruction the ensuing year. Adults are more willing to adopt practices that can be demonstrated as being effective.

Follow-up visits by the teacher of agriculture to the farms, homes, or businesses of adult class students are effective in securing adoption of improved practices. Problems related to a specific situation of an individual person can be discussed. In many cases, obstacles that seem to be preventing adoption can be overcome when the teacher discusses possibilities with the adult student.

Follow-up meetings may be important to discuss problems and evaluate progress related to implementing approved practices. Such meetings provide opportunities for adult students with similar problems to share experiences and work out solutions.

The adult education program is evaluated according to the relevance of the topics discussed, the effectiveness of instruction, and the extent to which instruction is useful in the lives of the students.

SUMMARY

Teachers will find working with adults to be stimulating, challenging, and rewarding. Types of programs include young farmer education, adult farmer education, adult education for agricultural business, and avocational adult education. Adult students, the agriculture teacher, and the community benefit from adult education. Adult students are somewhat different from youth in physical, psychological, and attitudinal characteristics. Planning for adult courses involves observing learner needs, involving the learners and others, estimating future needs, and informing potential students of course plans. With adults, discussion, panel discussion, resource people, and individual instruction may be more heavily used than with youth. Evaluation of learning may be more informal but should be oriented to application of learning in the lives and workplaces of the students.

FOR FURTHER STUDY

1. Interview one or more teachers of adult courses. Determine
    a. How course content is determined.
    b. What teaching methods are used.

Chap. 13 Teaching Adults

    c. How adult students differ from high school students.

    d. How visits are conducted.

2. Develop a list of the major characteristics of adult learners.
3. Develop a list of benefits enjoyed by a teacher who teaches adults.
4. Using the information in Chapters 5, 6, and 13, prepare a daily plan to teach an adult class.
5. Accompany an agriculture teacher on an adult visit. List the instructional content discussed and explain the teaching technique used by the instructor.
6. Observe an adult class. Describe the teaching methods used. Contrast how this class differs from secondary classes.
7. Interview three adults. Probe their interests in continuing education. Write a case study based on each interview.

PART V Evaluation of Learning

Student evaluation is an essential task of every teacher. Suggestions for evaluation have been presented throughout the book in order to assess specific learning. In Chapter 14, evaluation is discussed in a comprehensive sense. Student performance in the cognitive, affective, and psychomotor domains must be developed and assessed by teachers. Presented in the chapter are reasons for evaluation of learning, test planning, procedures for development of various types of tests, and test administration. Methods of rating affective development and performance skills, including supervised experience assessment, are included. You will also find help in this chapter on reporting student grades.

14

Evaluation of Learning

> Sara was a good student in the agriculture program. She received the highest score on nearly every test given in class. Her agriculture teacher awarded her an A on her report card. Becky had more difficulty in school. She received a D in her agriculture course because of her test scores.
>
> The agriculture teacher placed both Becky and Sara on an SAE "Coop" placement with a local nursery. After a short period of time, the nursery manager telephoned the teacher. He praised Becky's work, indicating that she was an excellent worker. However, he expressed displeasure with Sara's know-it-all attitude, her "bossiness" with other employees, and her lack of work ethic. The manager awarded Sara a D on her report card while awarding Becky an A on her report card.

Evaluation is assessing what students have learned. Evaluation provides useful information to students. It also helps teachers determine whether students are learning what is being taught, and in so doing assists the teacher in self-assessing effectiveness of content delivery.

Were the grades the students received in the preceding example appropriate? It depends on how they were evaluated and the basis for the evaluation. For example, was the evaluation based on course objectives? Were the course objectives related to developing knowledge, attitudes, and performance of skills?

OBJECTIVES

After studying this chapter, you will be able to

1. Explain the reasons for evaluating students.
2. Evaluate achievement based on objectives.
3. Plan a test.
4. Construct various types of paper-and-pencil tests.
5. Develop various types of rating scales and performance assessment instruments.
6. Develop a grading procedure.
7. List alternative evaluation strategies.

REASONS FOR EVALUATING LEARNING

The evaluation of learning is a significant part of the job of a teacher. Accurate assessments provide a clear picture of a student's performance in relation to the objectives of the curriculum and in relation to other students. Completing various types of assessment prepares individuals for evaluations throughout their careers. They are learning that performance is a major determinant of their status in life and that their status is often a result of assessments of their performance as perceived by others. Therefore, teachers have a responsibility to use evaluation instruments and procedures that are unbiased, justifiable, and realistic.

Needs Assessment

Knowledge of the level to which students have developed enables the instructor to effectively plan for teaching. Thus, baselines of the students' educational levels are needed, along with a set of measurements and tools for benchmarking their progress. For example, a diagnosis of learning difficulties provides information helpful in designing individualized instruction. Pretesting is a tool that can be used to suggest when remedial instruction is in order, or whether students have already been taught the unit that is planned. Because it is a waste of time to present content the students have already mastered or material that the students lack the prerequisite skills to grasp, a pretest can help determine knowledge level and, therefore, a starting point for teaching the content. Agriculture teachers have been challenged to take students from where they are to where they want to be. In order to accomplish this task, it is necessary to determine where they are. One important reason then for evaluation of students is to determine what students know or can do prior to teaching a unit of instruction.

Instructional Improvement

Evaluation of student learning provides valuable feedback to the teacher concerning the effectiveness of instruction. Tests can aid teachers in assessing the effectiveness of instructional materials used. Tests can also be helpful in evaluating the usefulness of teaching methods or techniques.

Evaluation can indicate that an entire class failed to grasp an essential concept and can further help the teacher ascertain where additional emphasis is needed. Exam results are useful in determining whether the material is above or below the ability of the students. Teacher-developed evaluations assess the teacher's instruction as much as they do the students' achievements. From evaluations, a teacher can see where better examples or other adjustments are needed. Even the process of planning and constructing an evaluation will encourage the teacher to effectively present important knowledge and skills on which the students will later be assessed. If a few individual students are finding the material either too easy or too difficult, the teacher may need to further develop individualized instruction that is challenging but attainable.

Motivation

Evaluation can be used to encourage some students to study and learn. Students who want good grades tend to respond favorably to the challenge of a test. Fear of failure, desire for academic success, and avoidance of guilt and anxiety may positively persuade students to learn material that is not otherwise rewarding. It has been shown that frequent testing with teacher-made tests, compared with infrequent testing, has generally resulted in improved student performance. This may be because the teacher is better at using feedback to improve instruction, or it may be because students are motivated to learn to do well on an examination.

However, students who experience test anxiety or who have a history of performing poorly on written exams must be provided with alternate approaches to evaluation other than traditional paper-and-pencil tests if their motivation is to remain at its highest levels.

Self-Appraisal

Students may use evaluations to rate their own performance and correct errors on their own. The use of evaluation instruments for this purpose is especially helpful in developing performance skills. Students can self-rate their progress, their project, or laboratory work to determine if it meets a preestablished standard of mastery. They then can choose to improve their performance by further study and practice. For evaluation to be useful for self-appraisal, the standards for performance must be clearly communicated to students.

Instruction

A test is often the climax of a unit of instruction. It functions as a vehicle to encourage students to synthesize previous material as they review for the exam. It also reemphasizes important objectives. To be most effective, the teacher should give corrective and constructive feedback as soon as possible after the students have completed an exam. This process will enable students to learn material they earlier failed to grasp. Content on which students have been tested is more likely to be remembered.

Grades

Schools are accountable to students, parents, and the community for teaching. An important measure of what has been taught is what students have learned. Most schools express the extent to which each student has learned in the form of grades. Grades are the means of communicating with others about the degree to which a student has achieved the objectives of a course. Teachers of agriculture must be able to justify the grades they award. One important input as decisions about grades are being made is the scores on teacher-developed evaluations of learning.

Evaluation Based on Objectives

Well-written objectives describe the learning expected of students. Well-designed evaluation results describe whether the objectives were reached at a satisfactory level. Thus, evaluations of student achievement should measure achievement of objectives. Evaluations should not be designed to measure obscure bits of knowledge but should measure the central focus of instruction. Objectives, therefore, are not only important in planning for teaching but are also essential for use in constructing evaluations.

An evaluation is more likely to be valid if it is written to measure whether students have achieved specified objectives. Evaluations that are not valid have no value other than as a measure of general knowledge, unrelated to the instruction that has been given. Teachers normally would want the content of the evaluation to be:

- The same as that presented in the course
- As comprehensive as the objectives on which the students are being evaluated
- Specific to the competencies that were identified in the objectives

Planning a Test

Tests can have motivational and instructional value if properly constructed. The objectives for the course serve as guides to the students about what is important to be learned and, therefore, serve as the guide to test construction. A haphazardly designed test that emphasizes trivial knowledge frustrates students. They may become discouraged rather than being motivated to learn more.

Specification charts (see Table 14-1) help to ensure a valid and comprehensive test. These charts can help teachers make decisions about test content. Exams that have been planned with the help of specification charts are more likely to cover the important knowledge and skills emphasized in the instruction.

One of the first steps in constructing a specification chart is to itemize the content to be learned by students—the instructional content. For example, if you were planning a test for the unit of instruction "Finalizing My Ownership," the objectives should be identified from the content based on the learning outcomes that were sought. Table 14-1 shows a specification chart with the instructional content listed.

In addition to using a specification chart to select the content to be included in an examination, the teacher can assign weights to each objective based on its relative importance. The assignment of relative emphasis to objectives might appear as is shown in Table 14-1. The teacher who taught the material is best qualified to determine the relative emphasis.

The next decision faced by the teacher is to decide whether to develop a paper-and-pencil cognitive test, an affective rating scale, or a performance test. If a cognitive test is to be used, then the questions should reflect the level of desired learning. Were the students to learn facts, comprehend material, or apply information? The level of the evaluation should be consistent with the level of the objectives. Table 14-1 assumes fifty possible points to be awarded for an entire unit of instruction. Of these, nineteen would involve performance appraisals, with students actually doing the work and being evaluated on the result. Seven points are to be awarded on an affective rating scale. A 100-point written examination would be used to evaluate the students' knowledge, comprehension, and ability to apply their learning. The number of test items measuring each objective should be in proportion to the importance initially weighted to that objective and should be reflected in the design of the specification chart.

Many teachers fail to use enough items in developing a test. A longer test is more reliable. A longer test also can more comprehensively test the objectives. The length of the test also depends on how often a teacher gives exams. Similar results can be obtained from many short quizzes or a few long tests. In either case consider this. If a unit of instruction takes eighteen class meetings to complete, and each day the teacher provides a ten-minute (five to ten mixed items) quiz, the same amount of time would be used if students completed three one-hour exams, yet the learning was reinforced in smaller increments across the eighteen sessions. The teacher should feel confident that the scores have accurately measured what students have learned. This is more likely to occur if evaluations are of sufficient length and at appropriate intervals to provide consistent results.

Table 14–1 Specification Chart for "Finalizing My Ownership Supervised Agricultural Experience"

Instructional Content	Relative Emphasis	Level of Cognition			Area		Item Total	Number Points
		Knowing	Comprehending	Applying	Affective	Performance		
1. Name the advantages of being an owner-operator	2%	1					1	1
2. Explain how to establish ownership programs	6%	1	2				3	3
3. Develop growth objectives	10%		1			4	5	13
4. List the different types of business applications	5%	2					2	2
5. Develop written plans and goals	10%		1			4	5	13
6. Develop a financial statement	10%		1	2		2	5	13
7. Develop a budget	8%	2				2	4	6
8. Set up and enter inventory records	8%			4			4	6
9. Describe various efficiency measures	5%	1	2				3	3
10. Set up and keep expenses and receipt labor records	8%	1				3	4	6
11. Set up and keep nonfinancial records	6%	1				2	3	3
12. Prepare a financial summary	7%	1	1			2	4	5
13. Demonstrate acceptable personal appearance and hygiene	15%				7		7	26
TOTAL ITEMS		10	8	6	7	19	50	100

MULTIPLE-CHOICE TESTS

Multiple-choice items consist of a lead-in statement known as the stem, followed by a number of possible responses. Only one of the possible responses should be correct. The others, or the incorrect responses, are known as distractors. Most students and teachers have had considerable experience with this form of test question.

Advantages and Disadvantages

The popularity of the multiple-choice test is based on perceived advantages of this type of test over other types. Some benefits include:

- Adaptable for use in a wide range of testing situations
- Less ambiguous than completion or true–false questions
- Less susceptible to chance errors resulting from guessing than true-false items
- More objective than tests in which students supply the response
- Well adapted to machine scoring
- Easily understood by students
- Can measure recognition as well as recall

Multiple-choice tests do receive some criticism. Critics of this type of test item are often generally critical of all objective-type tests. Some of the criticism is legitimate because of the misuse or improper preparation of a test. Major disadvantages include:

- Failure to commit time to proper development of items
- Failure to develop items measuring higher-cognitive levels than recall or memorization
- Improper distractors, either making the correct choice obvious or causing some debate among experts as to whether only one choice is correct
- Space-consuming, causing the appearance of a long test
- Time-consuming to properly construct
- Conducive to guessing

Steps in Constructing Items

The previously discussed specification chart should be used as one begins to construct test items. Determine the major knowledge and skills to be tested relating to each objective, then prepare a question or an incomplete sentence that clearly implies a question. This is the stem. A good answer with a few well-chosen words should be written to the question. Several (usually three) incorrect answers should then be written. These distractors should be plausible to the student who lacks understanding of the question.

Table 14-2 Item Measuring a Specific Objective

What is a major advantage of a well-planned ownership program?
a. helps in improving the appearance and value of an agribusiness
b. helps in gaining experience in working for others
c. helps in accumulating earnings*
d. helps provide assurance of regular wage income
*Correct response

Source: From Amberson, Max L., and Anderson, B. Harold. *Learning through Experience in Agricultural Industry.* New York: McGraw-Hill, 1978.

Good multiple-choice items deal with important ideas, not with insignificant details. Too often teachers ask questions on illustrations or similar materials found in a text or discussed in class. These illustrations were designed to teach a concept, not in themselves to serve as the focus of the learning. Table 14-2 shows an item that could be used in partial measurement of the objective, "Name the advantages of being an owner-operator."

Teachers should avoid the use of opinion stems in multiple-choice items. Asking a student for an opinion as to a best answer will result in disagreement. Even though a student selects an answer the teacher believes is wrong, it is still the student's opinion and thus the answer as stated is correct.

It is sometimes helpful to illustrate a question. Student understanding can then be readily tested. For example, in Figure 14-1, one could easily test the application of what saw might be used in cutting across the grain of a board.

The type of saw used to make this cut is a:

a. crosscut saw*
b. rip saw
c. compass saw
d. coping saw

*Correct response.

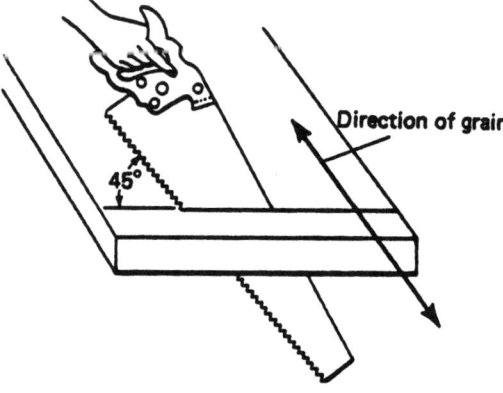

Figure 14-1 Illustrated test item.
Source: From Shinn, Glen C., and Weston, Curtis R. *Working in Agricultural Mechanics.* New York: McGraw-Hill, p. 80.

Teachers of agriculture should always attempt to word the stem positively, avoiding negative terms. When it is necessary to use negative terms, underline them for emphasis.

Avoid providing false clues or irrelevant information in the stem. The student should be able to quickly decipher the intended purpose of the question.

Careful wording of alternative responses is needed. All the responses should seem plausible, but only one should be clearly correct. Following are some suggestions for writing multiple-choice test items:

- All alternative answers should be appropriate to the question.
- All responses should be grammatically consistent with the stem.
- All responses should be similar in type of content, length, and complexity.
- There should be at least four choices.
- Each choice should be listed on a separate line.
- A series of figures should be listed in order.
- Correct responses should be scattered.
- Responses such as "all of the above" and "none of the above" should be avoided.
- Distractors must be plausible alternatives. Some ways of obtaining them include:
  - Using true statements that do not correctly answer the stem question.
  - Using common expressions or phrases that may seem attractive to students whose knowledge is only superficial.
- Avoid distractors that are unfamiliar to the students.
- Avoid distractors obviously incorrect, even to those not familiar with the material.

"All of the above" and "none of these" can be used when the responses deal with an absolutely correct or incorrect question such as an answer to an arithmetic calculation or spelling of words. They should not be used for answers that are "best" responses.

SHORT-ANSWER TESTS

A short-answer test is one in which a student is asked to supply a number, phrase, or word that answers a question or completes a thought. It is best suited to testing recall knowledge. The teacher forms the stem (as either a question or an incomplete statement), leaving space for the student to respond. A sample item is exhibited as Table 14-3.

Table 14-3 Sample Short-Answer Item

What is the term used to describe the end of a meeting?
Answer: Adjournment

Advantages and Disadvantages

The short-answer item is highly accurate in assessing knowledge. It has several additional advantages:

- Low chance of guessing correct answer
- Reasonably easy to write
- Requires less room on a page per item than multiple-choice items
- Easily understood by students

Some disadvantages of this type of item include

- Less objectivity in scoring than where students choose from among listed responses; therefore the test should be scored by a knowledgeable person
- Less adaptability to measuring higher cognitive items than those dealing with recall of knowledge
- Susceptible to improper use when hastily developed

Steps in Constructing Items

A problem many teachers have in writing short-answer items is that there may be several equally defendable answers to a question. For example, the question "Personal data sheets should be what?" [1] to which the intended answer was "neat, accurate, and typed," might also elicit such answers as "used to get a job," "reviewed by a friend to see if they are complete," or "printed on 8½" × 11" paper." Questions of this type should be worded to encourage a specific desired response. Following is a question designed to test recall of information for which there is only one correct response. "FFA makes a positive difference in the lives of students by developing their potential for premier _____ , personal _____ , and _____ success through agricultural education." Answer: leadership, growth, career. Be careful, however, depending on the nature of the question, to have fewer as opposed to more blanks in the item, and to carefully choose the placement of the blanks toward the end of the phrase so the students get the nature of the question prior to the blank(s). Additional considerations in developing short-answer items are as follows:

- Decide on the desired answer, then write a question to obtain the appropriate response.
- Avoid lifting items out of their context; the answer out of context may be difficult to defend as the only correct response.
- Make all blanks the same length.
- Arrange blanks at either the right or left margin of the page for ease of scoring.

---

[1] Written from material in Stewart, Bob R., *Leadership for Agricultural Industry*. New York: McGraw-Hill, 1978.

- Avoid unintended clues (example: "The leadership committee is responsible for developing student abilities in _____ ." Answer: leadership)
- Word items differently from in the reference material to elicit greater processing, thus, understanding

TRUE-FALSE TESTS

A popular form of test item among classroom teachers is the true-false item. This is probably because it is viewed as being easy to write and score. Well-constructed items do discriminate between students who have mastered the material and those who have not. However, there is general agreement that true-false tests, as written by many teachers, tend to confuse students. The most common uses of true-false tests are in identifying the correctness of definitions, statements of principles, and statements of facts. Complex and difficult problems can also be presented in this form.

Advantages and Disadvantages

True-false tests are efficient. Students can respond to many items in a short period of time. Therefore, a large body of knowledge may be tested in one examination. Other advantages are as follows:

- Relatively easy to write
- Scoring is objective and easy
- Well suited to force a choice between two options

Many teachers and students are critical of using true-false items because of their limitations. Some major disadvantages are as follows:

- Guessing is a factor; students have a reasonable chance of guessing the correct answer even though they have not learned the appropriate objectives.
- The item reliability is low, forcing a high number of items per test to accurately measure the content of a unit of instruction.
- Each item has less value in assessing how well students have achieved specific objectives; therefore, a high number of items per objective is needed.
- Students learn to make use of word clues, and some can score well with little knowledge of the material.
- It is difficult to construct a completely true or completely false statement without making the answer obvious.
- Be careful not to emphasize minor details for which items are easier to write.

Steps in Constructing Items

A true-false question should test an important concept. For example, if you were teaching the breeds of beef cattle, the country of origination of the Angus breed would be much less important than the fact that they are known for their excellent conformation (body type). However, a true-false item about origination is easier to write than an item about conformation.

The question should be designed to encourage students to think rather than only memorize. Following are two examples of items written to test at different cognitive levels. Item 3 was written to (1) reduce guessing on a true–false item and (2) raise the cognitive level at which a student responds to the item by forcing them to think through and write an explanation.

COMPARISON OF COGNITIVE LEVEL TEST ITEMS

a. Memorization
 - Softwood comes from *coniferous* trees. (T)
b. Thinking and Understanding
 - If one were to construct a nursery lath house, one would normally use poles made from *deciduous* trees. (F)

TRUE–FALSE ITEMS THAT ENCOURAGE THINKING

1. In the spring of the year pasture grasses are the main source of water for hindgut fermenters.

Explain _____

Following are some suggestions for developing true–false questions:

- Make approximately one-half the items true and one-half false; if a test is unbalanced, have more false items.
- Make the method of indicating responses simple; one might put a typed T and F to the left of each item for a student to circle.
- Do not make true statements consistently longer than false statements.
- Avoid negative statements.
- Use only a single idea in each statement.
- Word an item so superficial logic suggests a wrong answer.
- Do not lift sentences directly from a textbook.
- Do not organize true and false items in a predictable pattern.

MATCHING TESTS

A matching test contains two sets of items to be associated by the student on the basis of the set of directions supplied by the teacher. One set of items is a list of premises, the other a list of responses. In most cases, the lists could easily be interchanged, as, for example, one could either match dates to events or events to dates.

The matching item is really a form of multiple-choice testing. The student is expected to select a correct response from a list containing several distractors. A wide variety of types of lists may be matched. Examples include a chart of parts of an animal with names of the parts, plant common names with botanical names, dates with events, tools with names or use, and wildlife with area of habitation.

Advantages and Disadvantages

Matching items are efficient. Many items can be placed on a page. Students can answer items quickly. Thus, much material can be tested in a short period of time. Students are encouraged to integrate their knowledge as they attempt to match items from related sets of material.

Matching tests are generally limited to specific factual information. It is quite difficult to construct items to test for comprehension or application of material. Because matching requires choosing from a cluster of related items, some material in any given unit of instruction may be left out simply because it does not fit a cluster. Thus, matching items are seldom appropriate as the only method of testing in a unit. This method can, however, be a useful part of a test that also contains other types of items. Teachers must be careful to properly construct this type of test.

Steps in Constructing Items

Several basic practices are essential to developing good matching tests. Improperly constructed items allow the student to readily eliminate incorrect responses and correctly guess the correct response.

Related Sets. The premises and responses within a set must be related (homogeneous). Figure 14-2 illustrates how the items are related throughout the matching list.

Responses per Set. In most cases, there should be at least six but not more than ten responses per set. There may be exceptions for identification items. Shorter lists make it easier to keep items homogeneous. Normal practice might be to have three premises and five responses. Indicate to the students whether responses can be used more than one time. There should be at least three plausible choices for each premise.

Extra Choices. Use extra choices in the response list to reduce the chance of guessing the correct answer. This idea can be achieved by permitting the response alternatives to be used more than once or by adding extra responses. Again, when this is the case, indicate so to the students in the instructions. Note in Figure 14-2 that there are three incorrect responses in addition to the correct responses.

Logical Arrangement. Items should be arranged so students do not have to spend excessive time hunting for them. Put them in alphabetical or

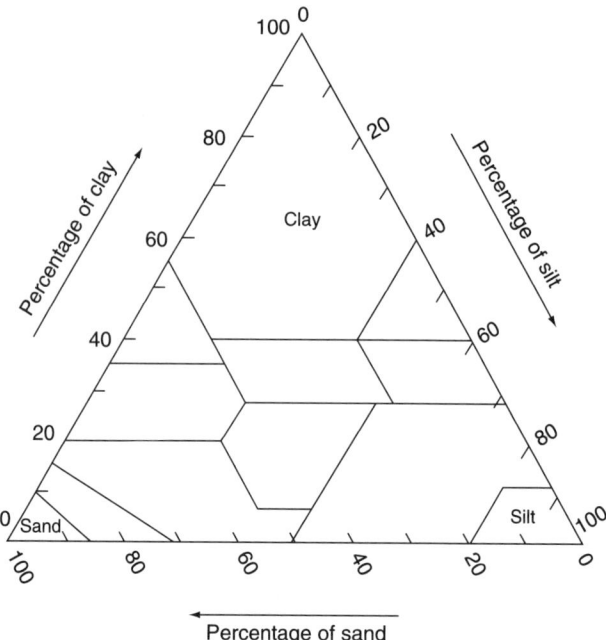

A soil triangle from which textures are determined

Directions: Place the correct letter in each of the nine unlabeled sections of the triangle.

    a. clay loam
    b. loam
    c. loamy sand
    d. sandy clay
    e. sandy clay loam
    f. sandy loam
    g. silty clay
    h. silty clay loam
    i. silty clay sand
    j. silty loam
    k. silty sand
    l. silty sand loam

Figure 14-2 Related set in a matching test item using a graphic.
Source: From Bishop, Douglas D., Chapman, Stephen R., and Carter, Lark P. *Working in Plant Science.* New York: McGraw-Hill.

chronological order unless some other system is preferred. Note the order of the responses in Figure 14-2.

One Page. Never carry a matching set over from one page to the next. A student becomes confused when he or she must flip pages while seeking a correct response.

ESSAY TESTS

The essay test is familiar to most teachers and students. It is a form of free response item that allows students to provide information in their own words.

Advantages and Disadvantages

The essay item is quite useful in assessing learning requiring the recall, organization, and presentation of ideas. The student must supply rather than choose a correct response. Proponents of essay exams suggest they:

- Develop abilities in written expression
- Require the student to use high mental processes
- Encourage deep study to prepare for the test
- Encourage creativity

They are much easier for the teacher to construct than most other test forms.

Although an essay exam does not provide opportunity for a student to guess the correct response, there is opportunity for bluffing by the student. The distribution of scores on an objective test is determined by the exam. On an essay test, the grader determines the distribution.

Essay tests are low in reliability. Different graders will score exams differently. Also, essay exams require a great deal of time for students to respond to each question; therefore, a smaller portion of the content can be sampled. When the questions are not clear to all the students, a wide range of responses will contribute to the low reliability.

Essay tests should not be a measure of the students' ability to write if the purpose of the test is to measure knowledge of some subject matter. Too often the score is a reflection of ability in written expression. This is not to say that grammar and spelling should be ignored. However, it must be remembered that the primary purpose for the exam is to assess the knowledge learned from the content that was taught.

The time saved in constructing essay exams is more than lost in grading them. They are difficult to grade uniformly. Students have a tendency to "pad" their answers. Misinterpretation of the question can cause a poor score even though the student may understand the material. Many of the deficiencies can, however, be kept in check with proper item preparation.

Writing Acceptable Questions

The teacher should primarily use essay items to test learning that short-answer items do not test as well. Always ask questions for which experts would agree there is a clearly correct answer. Do not simply ask students to discuss their opinions relating to a topic. Such questions are impossible to fairly grade.

Be as specific as possible in defining what is expected of the student. Give preference to more questions of a specific nature rather than fewer general questions. Do not give a student a choice among questions. Giving the student the option of answering any two of three essay questions will not differentiate between students who can answer all three and those that can answer only two of three. It is also impossible to equate the difficulty of optional questions.

After writing the question, the teacher should either write an ideal answer or, at a minimum, list the essential information expected of the student. Students should be informed before the test as to how their answers will be scored—whether the teacher will take off for grammatical errors or for incorrect information, or just determine if the relevant information is present.

Grading Student Answers

The value of essay exams largely depends on the quality of the grading process. The teacher must take extra care to be consistent and fair in rating the essay answers students have written.

Use a Consistent Scoring Method.
The teacher may choose to specify crucial elements of an ideal answer. The proportion of these crucial elements appearing in his or her answer determines a student's score. This type of grading provides clear justification to the student of the attained score. This system is best suited to questions likely to have uniformly structured answers.

Another method, perhaps better with questions having complex answers, is to simply read an answer and assign a numerical grade based on the rater's general impression of the response. A better, related system is for the teacher to sort the answers to a question into piles having corresponding levels of quality. Scores can then be assigned. Although this method may be as reliable, it is more difficult to justify scores to students.

Grade One Question at a Time.
The teacher should read the answers to one question on all the students' papers before going on to the next question. The teacher can concentrate on the answer to the one question and more fairly grade each of the students' responses to that question. It is suggested that teachers read through every response to one question without assigning a grade, and then reread the same items before awarding a grade.

Grade Items Anonymously. Answers can more nearly be graded on merit when the teacher examines an item without knowing who wrote it. Students can put their names in a place on the paper where the teacher need not see it until after the scores are assigned. This may be difficult in small classes where the teacher may recognize the handwriting. However, the teacher should make every effort to fairly and independently judge each response.

ORGANIZING AND ADMINISTERING A TEST

After the test items have been written, they should be arranged to assist the student in taking the test and for ease of scoring. If the teacher has followed the test plan, the items will represent a good example of the skills and knowledge to be tested. There should be a sufficient number of items to provide a reliable estimate of a student's knowledge and skills.

Organizing the Test

The first step in organizing a test is to group like items together. All the multiple-choice items should be in one section of the test, the true-false items in another section, and so on. In this way, the directions need be given for responding to the type of item only once.

Once the items are sorted by type, they should be arranged according to their level of difficulty. In other words, all the multiple-choice items should be arranged so the easier items appear first followed by increasingly difficult items. Classroom teachers need not be overly concerned as to whether items are in an exact order of difficulty. In general, however, the items should be arranged in this way.

Another important point in organizing the test is to arrange items for ease of scoring. Normally, the spaces for student responses should appear in a column running from the top to the bottom of the page. Even in fill-in-the-blank questions, items can have the answers appear in a vertical column (see Table 14–4). Some teachers prefer to use separate answer sheets for objective items that may be machine- or hand-scored more quickly.

Table 14–4 Fill-in-the-Blank Item with Choices Arranged Vertically

Answers in Right Column	
1. The rental charge for use of capital is ___(1)___	1. (interest)
2. Ownership capital is ___(2)___ capital and rented capital is ___(3)___ capital.	2. (investment)
	3. (loan or borrowed)
3. The portion of profits distributed to investor owners is known as ___(4)___.	4. (dividends)

Source: From Long, Don L., Oliver, J. Dale, and Coale, Charles W. *Introduction to Agribusiness Management.* New York: McGraw-Hill Book Co., 1979, pp. 47-50.

Directions should be prepared to advise the students concerning the test. The amount of detail in the directions should relate to whether the student is familiar with the type of exam being administered. General directions should cover

- Putting his or her name in the proper space
- Time allotted
- Whether guessing is encouraged or whether there are points off for guessing
- Whether students may ask questions for clarifying an item
- Necessity for following directions for each type of item
- Specific directions for each part of the test might include
  - Procedure for responding to each item
  - A sample item that has been completed

Examine the physical arrangement of items. Some hints are as follows:

- Leave sufficient space for all responses.
- Arrange items so responses will not form a particular pattern.
- Arrange items so students do not need to refer to more than one page to answer them.
- Number items consecutively from the beginning to the end of the test.
- On multiple-choice and similar items, arrange possible responses in a vertical rather than horizontal row.

A scoring key should be prepared prior to administering the test.

Administering a Test

This is probably the simplest phase of the testing process. Concerns in administering a test usually relate to time limits and cheating. However, it is also a responsibility of the teacher to prepare the students in test taking.

Students should be made aware of the importance of following the directions and the consequences of not doing so. They should understand how the test will be scored. Will points be subtracted for guessing? Will points be subtracted for errors in spelling? Will there be a premium for legibility and neatness? Students should be encouraged to pace themselves rather than spending excessive time on a difficult question. The teacher should provide guidelines for planning and organizing responses to essay questions.

If time limits are imposed, the teacher may make students aware of time constraints by writing the time remaining on the chalkboard periodically. However, know your students because indicating time could cause some students to panic.

The teacher has a definite responsibility to prevent cheating. The examination should be administered in such a way that cheating is not encouraged. Examination copies should be safeguarded prior to administration. Students

Chap. 14 Evaluation of Learning   345

and teachers must recognize that cheating requires consistent application of appropriate penalties such as failure, loss of credit, or suspension. The instructor must not allow cheating to be easy. Alternate forms of the exam may be prepared by arranging questions in a different order. An exam should be proctored at all times. Summary tips on administering a test follow:

- Announce the test well in advance of the test date so students have ample time to study, especially if a significant portion of the course grade will depend on the test score.
- Type the test and make copies of it; don't write questions on the writing surface or read them aloud.
- Produce clean, legible copy; have extra copies available.
- Fit the test to the available time; announce the time allotted; give at least ten- and five-minute warnings.
- Administer the test in a comfortable, familiar setting.
- Seat students so as to reduce any temptation to copy.
- Make sure all are ready to take the test: sharpened pencils, desks clear of books and notes, and so on.
- Distribute test papers face down and have all students start at the same time.
- Explain the instructions.
- Supervise the test; move quietly about the room; do not become involved in other duties.

AFFECTIVE ASSESSMENT

Affective assessment is concerned with measuring the interests, attitudes, values, and appreciations of students. In preparing students for work in agribusiness, these areas are quite important. In many cases, the affective area is as important in obtaining and keeping a job as the cognitive and psychomotor skills of a student.

The agriculture program has many aspects designed to develop the affective domain of students. The supervised agricultural experience program develops student initiative and responsibility. Work habits are developed in the agricultural mechanics laboratory, the greenhouse, and through other skill development avenues. The FFA organization provides opportunity for developing career success skills, citizenship, and leadership. Assessment must be designed and implemented to measure student progress in each of these aspects of the agricultural instruction program.

It is difficult to measure interests, attitudes, values, and appreciations. Also, some believe that education should be value-free. However, if the role of agricultural instruction is to prepare students for life beyond high school, then the affective development of students must be written as objectives and measured for accountability.

Some appropriate measures currently used by teachers to assess affect (interests, attitudes, etc.) include direct observation, observer checklists, progress charts, rating scales, and assessment rubrics.

Direct Observation

Many teachers of agriculture consider student "attitude" in assigning grades. Students who work hard and are cooperative are often given the benefit of the doubt when their scores are near a break between two grades. Some teachers average an "attitude" score with scores obtained on other measures of achievement in arriving at a final determination of grades. Such practices are often based on perceptions of the teacher who has directly observed student behavior.

The reliable observation of affect is difficult. Students who know they are being observed often respond differently from when observation is not apparent. Many students reflect socially acceptable behavior among peers but may exhibit different behavior in a work setting. Affective behavior is very difficult to quantify. Although most observations of affective behavior have been informal, teachers facing a need to justify evaluations of students have desired a more formal approach. Checklists and rating scales may be developed and used to measure affect and to communicate with students and parents about observed behavior.

Checklists

Checklists are a convenient way for teachers to rate student behavior. They can easily be developed simply by listing the desired traits to be observed. For example, a teacher might work with students to cooperatively develop a list prior to a field trip to encourage the affective expected behaviors. Some teachers have students rate themselves and then discuss differences between student and teacher ratings in individual conferences. Thus, the list is used for counseling as well as evaluation. An example checklist is shown in Figure 14-3. Teachers who use checklists of this nature must be sure that students have the opportunity to participate and practice the desired behavior.

Place a check mark (√) in the space next to each observed behavior.

_____ 1. Practices good manners
_____ 2. Greets people properly
_____ 3. Listens courteously
_____ 4. Is well groomed
_____ 5. Exhibits neat appearance
_____ 6. Is considerate of others
_____ 7. Practices good judgment

Figure 14-3 Checklist for social skills.
Source: From information in Stewart, Bob R. *Leadership for Agricultural Industry.* New York: McGraw-Hill, pp. 32–34.

Rating Scales

Rating scales are similar to checklists, except that students receive a quantitative score on each item rather than an all-or-none decision. Figure 14-4 is a rating scale containing items to assist students in developing cooperation, responsibility, leadership, work habits, and social habits. The specific items could be revised to meet the needs of any local program. Teachers could have students complete a weekly self-rating, using this or a modified form. Teachers then respond to the student with revisions as needed. Student improvement then becomes a cooperative venture rather than the teacher encouraging

Category	Indicators	Rating*
1. Cooperation	a. Works well with peer students	1 2 3 4 5 N
	b. Seeks help when needed	1 2 3 4 5 N
	c. Assists others	1 2 3 4 5 N
	d. Is receptive to suggestions	1 2 3 4 5 N
2. Responsibility	a. Completes assigned work	1 2 3 4 5 N
	b. Follows policy and rules	1 2 3 4 5 N
	c. Assumes responsibility for actions	1 2 3 4 5 N
	d. Attends regularly	1 2 3 4 5 N
	e. Is on time	1 2 3 4 5 N
3. Leadership	a. Influences others	1 2 3 4 5 N
	b. Makes decisions	1 2 3 4 5 N
	c. Provides constructive criticism	1 2 3 4 5 N
	d. Develops and works toward personal goals	1 2 3 4 5 N
	e. Demonstrates initiative	1 2 3 4 5 N
	f. Demonstrates poise and confidence	1 2 3 4 5 N
4. Work Habits	a. Works well under pressure	1 2 3 4 5 N
	b. Works rapidly and accurately	1 2 3 4 5 N
	c. Demonstrates patience	1 2 3 4 5 N
	d. Uses available time	1 2 3 4 5 N
5. Social Habits	a. Meets people properly	1 2 3 4 5 N
	b. Uses appropriate dress	1 2 3 4 5 N
	c. Respects the rights of others	1 2 3 4 5 N
	d. Is well groomed	1 2 3 4 5 N
	e. Is concerned about others	1 2 3 4 5 N

*A rating of 5 means excellent; 4, very satisfactory; 3, satisfactory; 2, needs improvement; 1, poor; and N, not applicable.

Figure 14-4 Example of a personal development rating scale.
Source: Printed with permission of The Ohio Curriculum Materials Services. From McCracken, J. David, and Bartsch, Bruce P. *Developing Local Courses of Study in Vocational Agriculture.* Columbus: The Ohio Curriculum Materials Service, The Ohio State University, p 44.

a reluctant student. The scores on scales of this nature can be used as a measure of student achievement in the affective domain.

Assessment Rubrics[2]

Often in an attempt to grade students' work it is found that the assessment criteria are vague and the performance behavior was subjective. Rubrics are authentic assessment tools that are useful in assessing criteria that are complex and subjective (see Table 14-5 for an example assessment rubric).

Authentic assessment is geared toward assessment methods that correspond as closely as possible to real-world experience. The teacher observes the student in the process of working on something real, provides feedback, monitors the student's use of the feedback, and adjusts instruction and evaluation accordingly.

The rubric is one authentic assessment tool that is designed to simulate real-life activity where students are engaged in solving problems. It is a formative type of assessment because it becomes an ongoing part of the whole teaching and learning process. Students themselves are involved in the assessment process through both peer and self-assessment. The advantages of using rubrics in assessment are as follows:

- They allow assessment to be objective and consistent.
- They focus the teacher to clarify his or her criteria in specific terms.
- They clearly show the student how his or her work will be evaluated and what is expected.
- They promote student awareness of criteria to use in assessing peer performance.
- They provide useful feedback regarding the effectiveness of the instruction.
- They provide benchmarks against which to measure and document progress.

Rubrics can be created in a variety of forms and levels of complexity; however, they all contain common features that

- focus on measuring a stated objective (performance, behavior, or quality)
- use a range to rate performance
- contain specific performance characteristics arranged in levels indicating the degree to which a standard has been met

Teachers of agriculture may use direct observation, checklists, rating scales, and assessment rubrics as measures of affective behavior. Students

---

[2] Wiggins, G. *The Case for Authentic Assessment*. ERIC Document Reproduction Service: ED 328 611. Also see http://webquest.sdsu.edu/rubrics/weblessons.htm.

Table 14–5 Example Assessment Rubric
Apply Problem-Solving Techniques

Competency Builder (Criteria)	Mastery or above Proficient	Proficient	Below Proficient	Weight
Diagnosed Problem	The student has acquired knowledge to correctly diagnose the problem. In addition, he or she is able to accurately distinguish between 3 or more problem-solving or decision-making models, which better assists the student in solving the problem.	The student is capable of identifying the problem. In addition, he or she is able to accurately distinguish between 1 and 2 problem-solving or decision-making models, allowing the student to correctly solve the problem.	The student has no logical understanding or knowledge of how to solve the given problem.	
Gathering Data	The student demonstrated his or her ability to gather relevant information. For those sources that were not used, the student cited and explained why they were not applicable.	By providing other case studies or similar data, the student demonstrated his or her ability to gather relevant information.	Although the student has acquired data, there is little evidence provided that justifies the data.	
Analyzing and Comprehension of Data	The student used comprehension strategies to identify more than 2 advantages and disadvantages of each solution. This information allowed the student to form logical and accurate conclusions from his or her data collection.	The student used comprehension strategies to identify 2 advantages and disadvantages of each solution. This information allowed the students to form knowledgeable conclusions from his or her data collection.	The student lacked the comprehension skills necessary to form more than 1 logical or accurate conclusion from his or her data collection.	
Effectiveness of the Solution	The student has thoroughly examined the problem's progress, and has provided information confirming that the solution has had substantially positive results.	The student has thoroughly examined the problem's progress, and has provided information confirming that the problem is being maintained.	The student has provided little to no evidence on the progress of the problem, or the evidence that was provided suggests that the rates are continually decreasing.	

(continued)

Table 14-5 Continued

Follow-Up	The student has continuously monitored the problem to prevent a reoccurrence in the future. In addition, the student inspects for any additional problems that may have arisen.	The student assesses the problem once a month to prevent a reoccurrence in the future.	After solving the problem, the student has not maintained any further assessment practices.
Written Communication Skills	The plan is clearly written and is free of grammar, style, or formatting errors.	The plan is written so that formatting and/or grammatical errors do not distract the reader; style and formatting are within standard guidelines.	The plan is poorly written. It may contain many grammar, formatting, and style errors.

Source: From The Ohio Curriculum Materials Service.

should be encouraged and rewarded as they develop the interests, attitudes, values, and appreciations needed for successful entry and advancement opportunities after high school.

PERFORMANCE ASSESSMENT

The primary use of performance evaluations is to assess student achievement. Three basic methods of measuring performance have been used:

1. Observing and rating the procedures or process used to accomplish a task.
2. Observing and rating the end product or project resulting from the performance of specified skills.
3. Rating a student in a general assessment of overall performance.

Learning by doing has been an important theme of agricultural instruction throughout its history. Students have been involved in applying learning in the school laboratory, in their supervised experience programs, and by participating in FFA activities. Skill development is an important part of all career and technical education curricula. To only measure achievement through cognitive testing would not provide an adequate measure of learning.

Assessments of Process

Performance tests can be designed to evaluate whether students follow a correct procedure in performing a task. A performance test could be designed for a student to demonstrate the proper procedures for setting up an oxyacetylene torch, lighting the torch, and turning it off. The list of procedures could be developed according to the process described by Shinn and Weston.[3] Either the teacher or a committee of students could rate each individual student using a checklist. Students could be required to correctly perform the procedures, thus achieving a perfect score on the test, prior to using the oxyacetylene torch in the school laboratory.

A skill sheet (see Chapter 7) or learning activity package might help the student follow correct procedures. A skill sheet for preventative maintenance might include step-by-step procedures as shown in Figure 14-5. Figure 14-5 illustrates a performance test designed to ensure correct procedures are used. In this example, the students must master critical tasks. They must achieve a prespecified level on the overall test.

Another example of a performance test is the individual safety test given by teachers of agriculture. Students must perform, to a prespecified standard, a series of tasks on each piece of lab equipment to successfully pass the test.

Performance measures of process are time-consuming to administer. They are needed, however, as measures of student mastery of the more important procedures.

Assessment of Product

Evaluation of a finished project completed by a student is a form of performance assessment. The teacher evaluates a finished product against a standard. A teacher might develop sample welding beads, with different beads used to illustrate varying quality of work. The quality of the work done by the student is compared with the standard to assign a score to the student project.

Some teachers use a skill chart to record the projects that have been satisfactorily completed by students. With a skill chart, students can check the competencies that have been satisfactorily completed. They can then compare their progress with others in the class. Quality scores, however, should be recorded in a grade book for privacy rather than on a skill chart that is displayed publicly.

Combination of Methods

Grades on laboratory work should be based on:

- The procedures used by the students
- The quality of the finished project
- The productivity or number of finished projects

---

[3] Shinn, Glen C., and Weston, Curtis R. *Working in Agricultural Mechanics*. New York: McGraw-Hill, 1978.

LESSON: PREVENTATIVE MAINTENANCE FOR THE ELECTROLYTIC BATTERY

COURSE: AGRICULTURE RESOURCE CONSERVATION

TERMINAL PERFORMANCE OBJECTIVE: The student will be able to accurately determine the state of charge of a battery by measuring the specific gravity of the electrolyte with a hydrometer.

GIVEN: An electrolytic battery of predetermined charge, hydrometer, comparative chart of specific gravity readings and charge.

PERFORMANCE: The student will be able to determine the state of charge of an electrolytic battery.

STANDARD: The student will be able to determine the specific gravity within 0.010 adjusted for temperature.

POINTS AWARDED	ITEMS TO BE OBSERVED	SAT.	UNSAT.	CRITERIA
	1. Removes battery caps	+2	0	1. Wears safety glasses. Places caps where they will not be lost. Allows sediment to settle before taking reading.
	2. Inserts hydrometer into cell and draws electrolyte into barrel	+5	0	2. Holds the hydrometer vertically so that the float does not touch the sides. Does not read the hydrometer until the float rides freely. Leaves hydrometer nozzle in battery to prevent loss of electrolyte.
	3. Indicates in notes the specific gravity (uncorrected for temperature)	+2	0	3. Writes specific gravity down in notes.
	4. Returns electrolyte to cell	+2	0	4. Does not spill electrolyte.
	5. Repeats steps 2, 3, and 4 for each battery cell	+4	0	5. Same criteria as above.

*Indicates critical items.

Figure 14–5A Performance test to determine specific gravity of a battery.

POINTS AWARDED	ITEMS TO BE OBSERVED	SAT.	UNSAT.	CRITERIA
	6. Replaces battery caps and flushes hydrometer with water	+2	0	6. Cleans hydrometer. Returns hydrometer to proper storage.
	7. Adjusts specific gravity reading for temperature in notes	+5	0	7. Uses formula. If temperature is greater than 80°F, add (+) four gravity points (0.004) for each 10°F above 80°F. If temperature is less than 80°F, subtract (−) four gravity points for each 10°F below 80°F. Writes corrected specific gravity in notes.
	8. Compares specific gravity with state of charge chart	+3	0	8. Indicates in equipment log the state of charge of the battery. Charges battery if charge is less than 80%. Indicates in log if there is more than 0.050 variation in specific gravity of electrolyte between cells.

TOTAL SCORE: _____ Pts.   TOTAL POSSIBLE SCORE: __25__ Pts.

MINIMUM MASTERY LEVEL SCORE: __20__ Pts.

Figure 14–5B Performance test to determine specific gravity of a battery. Continued.

Figure 14-6 illustrates a method of scoring that includes the three areas of assessment. A student's score is determined by adding the values assigned for procedure and product. A total score for the unit of instruction can be obtained by adding the points for all completed work. This type of record is useful for communicating the reason for the grades that have been assigned.

Supervised Agricultural Experience Evaluation

In agricultural instruction, an integral component of the curriculum, and thus an expectation of students, is that they conduct supervised agricultural experience projects that combine across time into a total supervised agricultural experience program (see Chapter 10). An important area of performance assessment relates to measurement of student achievement in this area. Many teachers visit students once each grading period to review the status of the supervised experience program and to assist the student in further program planning and development. A form that teachers may use to evaluate the project during or following a home visit is illustrated in Figure 14-7A. The form takes into account the size or scope of the project, the student effort required, and the condition or quality. The scope score should reflect how well the student has made use of his or her opportunity for project work. A form that teachers may use to evaluate the total SAE program during or following a home visit is illustrated by Figure 14-7B.

The effort score should reflect the time and managerial commitment required of the student. The condition score should be an indication of whether approved practices are in use.

Another way of evaluating the supervised experience program is to rate the quality and condition of the student record book. The record book grade should be based on records being complete, accurate, up-to-date, and neat.

Teachers of agriculture ask employers to assist in rating students placed in cooperative education. Forms such as that shown in Figure 14-8 are designed primarily to measure total performance, including affective behavior. Teachers should advise employers on the use of the form to encourage consistency among employers in the rating of placed students. The final decision about the grades given as a result of the rating must rest with the teacher.

Identification Tests

Identification tests involving real or simulated materials may be classified as either a form of performance test or a form of cognitive paper-and-pencil test. Such tests usually involve specimens that must be identified by the student. Examples include breeds of dogs, plant materials, hand tools, and soil types. Specimens with identifying numbers can be arranged two to three feet apart around the perimeter of tables. Students stand next to a "station," with each specimen serving as a station. The time at each station should be limited, with

Profit Assessment I

Student Name: _____

Required Projects	Procedures	Product	Points for Project	Score	Teacher's Initials / Date
Sharpening and Fitting					
1. Plane bit or wood chisel	1.0	.75	1.75	B	/
2. Wood auger bit	.75	.25	1.00	C	/
3. Screwdriver	1.00	1.00	2.00	A	/
4. Twist drill bit	1.00	.75	1.75	B	/
5. Cold chisel	1.00	1.00	2.00	A	/
6. Punch	.50	.50	1.00	C	/
Optional Projects			Double		
Sharpening and Fitting					
1. Crosscut saw	1.00	1.00	4.00	A+	/
2. Rip saw	N/A	N/A			
3. Circular saw blade	N/A	N/A			

Procedures	Product	Score
A = 1.00	A = 1.00	A = 2.0
B = 0.75	B = 0.75	B = 1.5
C = 0.50	C = 0.50	C = 1.0
D = 0.25	D = 0.25	D = .5

Figure 14–6 Tool sharpening and fitting score sheet.
Source: From Wilson, Richard H. *Methods of Student Evaluation.* Columbus: The Ohio State University, Department of Agricultural Education, unpublished mimeo, p. 20.

Project	Scope	Effort	Condition	Total
Example				
Wheat production	2	4	3	9
Pheasant release	4	2	2	8
Grand Total: 17/24			Grade:	C+

Figure 14-7A Supervised agricultural experience project evaluation form.

every student having equal time at each station. The students rotate to the next station until they have come full circle. The test is then complete. The identification test can be matching, fill-in-the-blank, or multiple-choice.

GRADING STUDENTS

Marks or grades are designed to serve as comprehensive measures of student achievement. Most schools require that marks be assigned and reported for each course. It is the responsibility of the teacher of agriculture to report a grade for each student for each grading period.

Many teachers do not like to assign grades. Teachers desire to serve students as guides and counselors. Some believe the role of the counselor is threatened when they judge achievement in assigning grades. Any grading system can be made strong or weak by the extent of dedication of the teachers who use it.

Purpose of Grades

Grades serve as a means of recording and communicating student progress in learning. Students often are motivated to work for grades. There is no harm in this if the grades are a true indication of learning progress. Grades must be accurate, precise measures of achievement in order to minimize misinterpretation of them by students, faculty, parents, and others who use them.

Kinds of Marking Systems

There are two major types of marking systems. Most agriculture teachers report student grades on a system based on the use of a few letter marks. Some still use the more traditional system of marking in percentages.

The percentage system is usually on a scale of 0 to 100 percent. Often a score somewhere between 60 and 75 percent is regarded as the minimum needed to pass.

The letter system usually consists of five letters: A, B, C, D, and F. The letter system is a less precise system but one with which most people are familiar.

Assessment Form

Student Name:

1. _____ (0–10) Planning documents (e.g., training plans, agreements, plans of practice, procedures, and budgets).

Initial (1–2)	Basic (3–4)	Commendable (5–6)	Advanced (7–8)	Superior (9–10)
Explores and understands an SAE with supervision.	Selects and develops planning documents for their SAE with supervision.	Shares responsibility for selecting an SAE and develops planning documents with supervision.	Selects SAE independently and develops planning documents with minimal supervision.	Understands components and importance of planning documents; can complete documents independently and seeks input.

2. _____ (0–10) Record keeping system.

Initial (1–2)	Basic (3–4)	Commendable (5–6)	Advanced (7–8)	Superior (9–10)
Selects a record system with instructor assistance.	Keeps appropriate records in a timely fashion with supervision; begins resume.	Completes appropriate records with some supervision; understands the importance of records; has a current resume.	Maintains accurate records with minimal supervision; summarizes records; updates resume.	Analyzes records, evaluates practices, and identifies alternatives based on their records with little supervision.

3. _____ (0–40) Performance of OCAP (or an equivalent in occupations without OCAPs) competencies pertinent to the occupational goal of the student. This is a percentage calculation based upon the competencies in the applicable OCAP. Formula: Number of competencies experienced in real world setting as a part of an SAE/Number of core competencies 40.

Initial (1–8)	Basic (9–16)	Commendable (17–24)	Advanced (25–32)	Superior (33–40)
0–20% performance	21–40% performance	41–60% performance	61–80% performance	81–100% performance

4. _____ (0–40) Extent to which the student directs the supervised agricultural experience program.

Initial (1–8)	Basic (9–16)	Commendable (17–24)	Advanced (25–32)	Superior (33–40)
Task identified by others, student works with supervision, student does not make decisions.	Task identified by others, student works with minimal supervision.	Task identified by others, student works independently, identifies some problems, and seeks help with solutions.	Shared decision making by student and other persons, student works independently.	Student makes decisions based upon current conditions and works independently, identifies problems and solves them.

Figure 14–7B Supervised agricultural experience program evaluation form.
Source: From The Ohio Curriculum Materials Service.

Supervised Agricultural Experience Program—Performance Objective Planning Form

Student Name: _____ **Date:** _____

Objectives

Mid-course targets:	Standard	Planned	Actual	Grade
1st Semester level	8			
2nd Semester level	18			
End of 1st summer	34			
1st Semester level	62			
2nd Semester level	100			

Signatures:

Student: _____

Parent: _____

Cooperator/Mentor: _____

Figure 14-7C Supervised agricultural experience program evaluation form.

Chap. 14 Evaluation of Learning

Student Employee _____ Date _____

Placement Site _____

Instructions:
Please rate the student employee on each of the following skills. Rate the student employee by placing a check mark in the appropriate column to the right of each skill. Use the following key for ratings:

X—No chance to observe   1—Below expectation   2—Meets expectation   3—Above expectation

Teachers will find that evaluation of student progress must be a joint effort between themselves and the employer for students involved in placement programs. This will necessitate the use of a form by the teacher and employer to evaluate the progress of the student. The rating form used by the employer might be completed while the teacher is conducting a supervisory visit in order that they may discuss the evaluation point by point.

Figure 14–8A Employer–Teacher–Student Observation form.

General Skills	X	1	2	3
Accepts and carries out responsibilities				
Attitude toward work, use of work time				
Adaptability; ability to work under pressure				
Speed and accuracy of work				
Attentiveness to work being done				
Promptness in reporting for work				
Care of work space				
Care of materials and equipment				
Observing, imagination				
Attitude toward customers				
Attitude toward workers, supervisors				
Personal appearance, grooming fitness				
Initiative				
Enthusiasm				
Cheerfulness, friendliness				
Courtesy, tact, diplomacy, manners				
Helpfulness				
Honesty, fairness, loyalty				
Maturity, poise, self-confidence				
Patience, self-control				
Sense of humor				
Selling ability, personality for selling				
TOTAL				

Figure 14–8B Employer–Teacher–Student Observation form. Continued.

Job Skills*	X	1	2	3
Knowledge of merchandise				
Mathematical ability				
Handwriting				
Speech, ability to convey ideas				
Stock keeping ability, orderliness				
Use of proper English				
Desire to serve people				
Like people, not afraid of people				
Filling orders				
Check incoming freight				
Mark merchandise for sale				
Use computerized cash register				
Writing sales slips				
Making sales				
TOTAL				

*Job skills may be developed as a part of the development plan of each student.

Figure 14-8C Employer–Teacher–Student Observation form. Continued.

Rating for liabilities:

X—No chance to observe 1—Below expectation 2—Meets expectation 3—Above expectation

Liabilities	X	1	2	3
Annoying mannerisms				
Making excuses				
Tendency to argue				
Tendency to bluff or "know it all"				
Tendency to complain				
TOTAL				

Please feel free to write comments below:

Evaluated by: _____

Position: _____

Figure 14–8D Employer–Teacher–Student Observation form. Continued.

Most often a combination of the percentage system and letter system is used. In this situation grades are assigned using a numerical percent, and then a letter grade is assigned accordingly where grades in the 90s are A ranges, 80s are B ranges, 70s are C ranges, and so on. Each grading system can be used effectively by a teacher to report the degree to which students have mastered needed knowledge and skills.

Bases for Marking Systems

Teachers of agriculture must consider what information will contribute to decisions about grades. In the traditional sense, grades should reflect the extent to which students have achieved the objectives. Some teachers, however, believe top grades should be given to all students who are working up to their capabilities. Others believe a more difficult course should award a higher percentage of top grades. Classes in which outstanding students are enrolled are often graded higher on the average than classes where students with greater limitations are enrolled. Should better grades be given to reward a student who needs encouragement, or should poorer marks be given to a student who needs to be punished? These issues are difficult to resolve. However, one must keep in mind that if grades are to be valid indicators of achievement, they must be assigned based on acquiring knowledge and skills. The primary basis for assignment of marks is the extent to which objectives have been achieved. Consideration of other factors only lowers the validity of grades as measures of student achievement. If grades are to be measures of accomplishment of objectives, teachers must be sure that objectives have been specified for cognitive learning, affective development, and performance of essential skills.

Factors in Grading

Teachers should develop a grading policy and share it with their students. There is then a clear understanding as to what is expected.

Objectives in agricultural instruction suggest that students should be achieving in their supervised agricultural experience program, in the classroom and laboratory, and in personal development. Figure 14-9 is an example of a grading scale that incorporates these factors in arriving at a mark for a grading period. In the grading scale, weights have been assigned by the teacher to each of three major areas. Teachers may determine the weight for each area based on the objectives of a particular course.

Grading students requires making decisions about marks for each test, for various affective behavior ratings, for performance evaluations, for completed projects, and other factors. Because most classes in agriculture are relatively small, teachers should refrain from grading on a curve and should instead rate students on the extent to which they achieved what was expected of them. Grades should be combined so that a composite score is obtained which has assigned an appropriate weight to each factor.

Student Assessment

Student Name _____

Grading Period _____

Criteria	Points Possible	My Score

Supervised Agricultural Experience
  Improvement Project—points
  Placement Experience—points
  Records—points		

Subtotal*	20	

Learning Activities
  Classroom Achievement
  Laboratory Performance
  Notebook		

Subtotal*	60	

Personal Development
  Cooperation and Social Habits
  Responsibility and Initiative
  Leadership and FFA Participation		

Subtotal*	20	
Grand Total		

A = 90–100 points
B = 80–89 points
C = 70–79 points
D = 60–69 points
F = Below 60 points

LETTER GRADE _____

---

*Students must achieve at least 50% of the possible points in each of the three major areas to satisfactorily complete the course.

Figure 14-9 Grading scale for agricultural education.
Source: Adapted from Wilson, Richard H. *Methods of Student Evaluation.*

One method of grading is to use percents to score each item. Another method is to use letter grades and then average the letter grades based on the following scale: A = 4; B = 3; C = 2; D = 1; F = 0. When plus and minus grades are used, the following scale might be appropriate: A = 11; A− = 10; B+ = 9; B = 8; B− = 7; C+ = 6; C = 5; C− = 4; D+ = 3; D = 2; D− = 1; F = 0. Converting letters to numbers helps in averaging to obtain a composite score.

It is important for the teacher to record several grades during each grading period. A personal development grade should be recorded at least weekly.

Record books and notebooks should be evaluated at least twice per grading period. There should be at least one test score recorded per week. Because grades are more reliable when they have been calculated based on many independent measures of achievement, teachers of agriculture are obligated to use a minimum of three to four grades per week from a variety of measures, such as speeches, quizzes, written papers, static displays, tests, skills projects, group projects, and products.

SUMMARY

Evaluation is needed for pretesting, in giving feedback on instructional effectiveness, for motivating students to learn, in student self-appraisal, for better student learning, and for assigning grades to students. Evaluation should be based on objectives. A test plan will help teachers ensure that the evaluation measures the objectives. Types of tests a teacher should be able to use include multiple-choice, short-answer, true–false, matching, and essay tests. Tests need to be properly organized and administered for fair evaluations. Affective and performance assessments should be conducted because these areas relate more closely to the ability of students to secure and maintain employment. A defendable grading system using many independent measures of achievement must be developed and used by teachers.

FOR FURTHER STUDY

1. Develop a grading system for use within a class of agriculture students. The system should provide a means of evaluating objectives relating to the supervised agricultural experience program, personal development activities, and classroom and laboratory activities.
2. Develop a specification chart that will aid in planning a test for a specific unit of instruction.
3. Write a ten-item multiple-choice test.
4. Write a five-item short-answer test.
5. Write a ten-item true–false test.
6. Develop six to ten premise/response matching test items.
7. Write an essay question with an ideal response. Underline key concepts in the response. Describe how the item would be scored.
8. Write test directions for students for one of the preceding tests.
9. Develop a performance test to measure student performance of a skill.

Appendix A

Sample Unit of Instruction

I. Performing Simple Wiring Jobs, prepared by Tim Miller
II. Situation
    A. Sophomore class of Ag-Ed II students (18 students).
    B. Two-thirds of the class has a farm background.
    C. All students are from homes using electricity, and they use it every day.
    D. The students have had no previous experience or training in electricity in class.
    E. Students can add, subtract, multiply, and divide. They can work simple equations with practice when shown how.
    F. It is not uncommon for simple wiring to be done on the farm or in the home.
    G. I need to find out if any students work for an electrician or have special abilities, needs, or interests in this area.
III. Teacher Objectives
    The student will be able to
    A. Draw wiring diagrams for simple wiring jobs.
    B. Match fuse size to load size and explain how delayed fuses work.
    C. List three safety rules to follow when using electricity.
    D. Wire two-way switches, three-way switches, outlets, and light fixtures.

IV. Teaching Procedure
   A. Interest Approach
      - Write on the board: "Performing Simple Wiring Jobs."
      - How many of you have a situation on your farm or in your home where you can turn a light on or off from two different places? (Have a demonstration board wired this way to illustrate.)
      - How many of you have a situation where it would be more convenient to be able to do this? (Examples: Hay loft? Stairway?)
      - How would you go about wiring two switches and a light so that they would work in this manner?
      - Call two students forward to demonstrate to the class how to wire a three-way switch.
      - Once the students realize they do not know how to do this, move to group objectives.
   B. Anticipated Group Objectives
      - What are some reasons why you think we should be able to perform simple wiring jobs?
      - Why is it important to know something about electricity?
         1. So we can wire switches.
         2. So we can use electricity safely.
         3. So we can save money.
         4. So we can do our own wiring.
   C. Anticipated Problems and Concerns
      - What do we need to know and be able to do in order to perform simple wiring jobs?
      - What are some questions that we need to answer before we can do this?
         1. How do I draw wiring diagrams?
         2. How do I select the proper fuse size for a given load, and how does a time-delay fuse work?
         3. How do I perform wiring tasks safely?
         4. How do I wire lights?
         5. How do I wire two-way switches?
         6. How do I wire three-way switches?
         7. How do I wire outlets?
   D. Plans for Solving Each Problem
      Problem 1. How do I draw wiring diagrams?
         1. Trial Discussion

Appendix A

- Why would we want to draw wiring diagrams before we do the actual wiring?
- Pull from the class the following:
  a. So we have a plan to follow.
  b. To save time and wasted energy when wiring.
  c. To avoid making mistakes in the actual wiring.
  d. To determine the supplies you will need ahead of time.

2. In our diagrams, we will be including such things as black, white, green, and red wires; lights; switches; and outlets (show an example of each item).
   - We must be able to represent these items on paper.
   - So we will need a code.
   - Here are the codes we will use in this class (see page A).
   - Put page A on overhead and uncover one at a time (show actual items also).
   - Make sure students get the codes and the following in their notes:
   - Write on board:
     a. Black wire = Hot wire (always hot).
     b. White wire = Neutral (only carries current when the load is on if wired properly).
     c. Green wire = Ground.
        - should not carry current
        - used as a safety
        - will be discussed later
     d. Red wire = Hot

After finishing page A, put the circuit on page B on the board and lead a discussion on it. (Have the students draw it in their notes also.)
- What color are the wires?
- What is this?
- Why is this switch here?
- Is the outlet always on?
- Are the lights always on?

Make sure they have the following in their notes:
a. Always run the neutral directly to the loads unswitched. This should be the first step.
b. Always switch the hot or black wire. We cover this in more detail in safety.

c. Notice that the hot wire is on the gold terminals.

d. You must have a complete circuit.

3. Summary

- Have each person draw a diagram of a circuit with four lights and one switch. Two lights are always on and two are switched.
- Also have them draw a diagram of a switched light with an unswitched grounded outlet on the other side of it.
- Hint: Need three wires for part of it (red wire).
- Help individual students as needed (see solutions in the accompanying diagram).

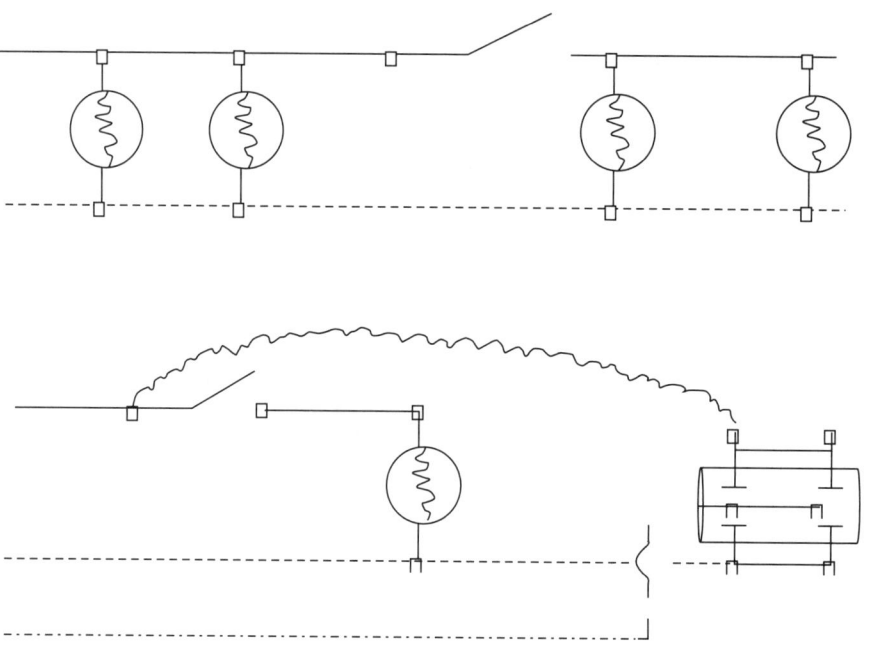

Problem 2. How do I select the proper fuse size for a given load, and how does a time-delay fuse work?

1. Trial Discussion
   - Why do we need to use the proper fuse size?
   - What happens if the fuse we use is too little? Too big?
   - What happens to a fuse when we have an overload? A dead short?

2. Demonstration/Experiment:
   a. Show the students a cord with a dead short wired across a micro wire.

Appendix A

    b. Ask them what they think will happen.

    c. Plug the cord in and blow up the wire.

    This was a dead short. The micro wire acted as a fuse or as the wire if it had not been fused. Place a micro wire in series with a circuit operating on an iron.

    a. Ask what will happen.

    b. Turn the iron on and wire gets red hot.

    c. Have a student touch the wire with paper.

    This is what can happen to a wire if it is overloaded and the fuse is too big (similar to 1,000 gpm through a quarter-inch hose). Time-delay fuses are made to protect a circuit from both an overload and a short circuit.

3. Here's how fuses work (put up a transparency of page C and explain).

    a. If overloaded for a period of about one minute, the solder melts and opens the circuit.

    b. The fuse handles short overloads.

    c. Short circuit blows the link.

    Now that we know how a fuse works, how do we know what size fuse to use?

    Anticipated answer:

    a. By the load.

    b. $V \times A = W$ or $A = W/v$.

      Where V = volts
           A = Amps
           W = Watts

    Example: A 2,300-W heater at 120 v? A 20-amp fuse.

    Note: For motors, figure the amps at full load and then multiply by 1.25 because 25 percent over current is allowed.

4. Summary

    Pull from the class:

    a. With a time-delay fuse, what happens if you have a slight overload for thirty seconds?

    b. What happens if there is a short circuit?

    c. How do we figure fuse size? For motors?

Problem 3. How do I perform wiring tasks safely?

1. Electricity can be dangerous, but so can a car, a match, or an unprotected power take-off shaft. If used properly, they will not hurt you. Electricity is one of the safest forms of energy if used properly. If not, it can kill you. What is it about

electricity that causes electrical shock or death? Volts? Amps? Watts?

- It is the current flow or amps.

    1/50 amp: You can't let go—get in notes and put on board.

    1/10 amp = Death—get in notes and put on board.

You must have about 100 v to push it through, but it is the amps that kill. You could stand 100,000 v if they were forcing only 1/1,000 amp (it has been done). So if we are to prevent electrical shock, what must we prevent?

- Current flow through the body.

    How do we do this?

2. I'm going to relate to you a true story and you see if you can figure out why it happened.

    - Tell story of man on a camper who got shocked when he touched the aluminum ladder while holding a drill.
    - Why? The case of the drill was hot because of a short, and he completed the circuit to ground. How could this have been corrected (in notebooks)?

    a. Double insulated tools (explain and show examples).

    b. Grounded.

    - Draw the circuit of (a) ungrounded and (b) grounded, and explain how each works.

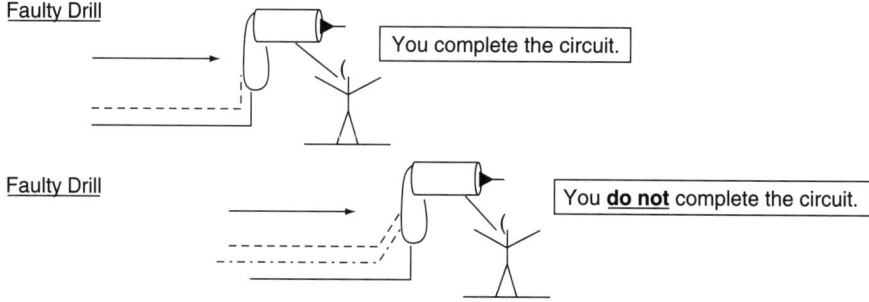

All wiring jobs should be polarized. Put on the board:

Polarity—using specific color, shape, or location of parts to identify the hot and neutral conductors of a circuit.

- We have already talked about this some (color codes for wire and screws), but let's take a closer look at why we need to do this.
- Use page D to help explain, along with an actual light and fixture.

Appendix A

   a. Show what is hot and what is neutral.
   b. Show potential shock hazard if not polarized in order for it to work.
   c. The entire system must be polarized in order for it to work.
   d. Point out why switches and fuses should be in the hot wire on any load.

   When you are performing wiring jobs or making electrical repairs, what is the surest way to keep from getting shocked?
   - Turn off all electricity by pulling the main fuse.
   - Always make sure the power is disconnected.
   - Some electricians who make repairs on live circuits say that they are safe, but it only takes one mistake.
     a. Wet ladder
     b. Dirty ladder
     c. Faulty insulation
     d. Wrong move or slip

3. Summary
   - If you had to take an electric shock, what would you want to make sure it was low in? How low?
   - amps: < 1/1,000 amp.
   - How does grounding help to prevent shock?
   - How does polarizing help to prevent shock?
   - When you are shocked, you are actually completing an electrical _____ ?
   - Circuit.
   - What should you always do to a circuit before working on it?
   - Disconnect it (pull main fuse).
   - Let's keep these things in mind when we are wiring.
   - Give a closed-book quiz over this problem.

Problems 4 and 5. How do I wire two-way switches and lights?
   - We have been studying electricity for _____ days now, but we have been doing it all on paper.
   - We have been studying circuits and safety, and drawing wiring diagrams on paper, but the real test is to see if you can actually wire the real thing.
   - Do you think you are ready?
   - Just to be sure, let's do an example together as a class and then you can try it on your own.

- Let's draw the circuit on the board for a switched light (have class tell how to draw it).

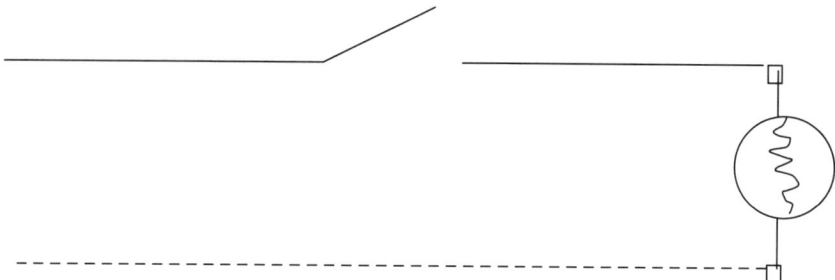

- Now, how do we wire it? If you all will come up here, we will wire this circuit and you can see.
- See page E for needed materials.
- See page F for steps to follow when conducting the demonstration.
- Divide students into groups of about three or four, depending on the number of wiring boards available, and have them wire two lights hooked in parallel controlled by a two-way switch.
- Before the students in a group may begin wiring, they must show me a wiring diagram of the circuit.
- Supervise the wiring and help where needed.
- For all wiring projects, the students will be working on tables away from outlets. They must come to me and have me check their work before going to the assigned area to test it.

Problem 6. How do I wire three-way switches?

1. Trial Discussion
   - If you remember back to when we first started this unit, several of you said that you had situations where you could operate a light from two different locations.
   - What kind of switch must we use to do this?
   - How does it work?
   - How do we wire it?
2. Discussion.
   - Have a three-way switch to show.
   - Notice there are three screws instead of just two.
   - Here is what it looks like on the inside (draw on board).

Appendix A

- Explain how it works.
- Draw two three-way switches and a light fixture on the board and see if students can figure out where to connect the wires (see page G). Explain why.
- Erase the first example from the board and have students diagram:

  Source → Switch → Switch → Light

  and then wire it.
- Check boards before they are plugged in.

Problem 7. How do I wire outlets?

1. Trial Discussion
   - How many of you are using multiple plugs such as this one (show one) at home so you can plug more things into an outlet?
   - What is wrong with these situations?

     a. Circuit overload.

     b. Tripping hazards.

     c. Fire hazard, and so on.
   - How could we correct these problems?

     a. Install more outlets.
   - Here is a duplex (will take two plugs) outlet. Do you notice anything different about the screws? Yes—it has a green one. What goes there? (The ground.)
   - Let's see if we can wire outlets.
   - First, I want you to show me a wiring diagram for the following circuit (one for each group) and then you can wire it (see page H). Source → Switch → Outlet → Outlet
   - You can use your notes if you need to. You may want to look back to see how to indicate an outlet on paper.

APPROVED PRACTICES

1. Draw a wiring diagram before wiring.
2. Use proper symbols when making wiring diagrams.

3. Protect all circuits with a fuse or circuit breaker with hot wire.
4. Always switch the hot wire.
5. Always run a neutral wire to the load unswitched.
6. Determine fuse size from this equation: volts x amps = watts.
7. Polarize all circuits.
8. Ground all metal-cased appliances and tools; use ones that are double encased.
9. Always pull the main fuse before making wiring repairs.
10. Put six inches of wire into each electrical box to work with.
11. Put wire around screw in the direction in which it will be tightened.
12. Use wire nuts to connect wires together, or solder them.
13. Always make connections inside a metal box.
14. If using a wire as a hot wire, paint two inches on each end black.

GENERAL PRINCIPLES

1. Electricity is created by the flowing of electrons.
2. Electrical pressure known as volts causes the electrons to flow.
3. A conductor allows electrons to flow through it.
4. Copper is a better conductor than aluminum.
5. An insulator does not allow electrons to easily flow through it.
6. An amp (ampere) is a measure of current flow.
7. An ohm is a measure of electrical resistance.
8. Electrical resistance causes heat.
9. Resistance is affected by
    a. Material size
    b. Size
    c. Length
10. Volts/ohms = amps *or* amps x volts = watts.
11. A watt is a measure of electrical power. Watts = amps x volts.
12. The larger the wire gauge number is, the smaller the wire is.
13. One thousand watts operated for one hour equals one kilowatt-hour (kwh).

Appendix A

ELECTRICITY TEST

Name _____

TRUE OR FALSE

Directions: Put a T in the blank before the item if it is true; put an F if the statement is false.

____ 1. Less than one-half amp can kill you if the voltage is high enough to push the current through your body.

____ 2. A large wire can carry more amps at a given voltage than a small one.

____ 3. A large wire has a larger wire gauge number than a small one.

____ 4. When a metal-cased electrical tool is not grounded and the case becomes "hot," the person holding it completes the circuit if he or she is grounded.

MULTIPLE-CHOICE

Directions: Select the best answer and place the letter in the blank to the left of the statement or question.

____ 5. When installing a switch in a simple lamp circuit, which wire should be attached to the switch?

    A. white   B. green

    C. black   D. yellow

____ 6. Metal materials in which electricity flows easily are called:

    A. transformers   B. insulators

    C. short circuits   D. conductors

____ 7. The flow of current is measured by:

    A. a voltmeter   B. an ammeter

    C. an ohmmeter   D. a micrometer

____ 8. Using color, shape, or position of electrical circuit parts to distinguish the hot from the neutral parts of the circuit is called:

    A. harmonizing   B. labeling

    C. grounding   D. polarizing

____ 9. Heat is generated from wires because of electrical:

    A. resistance   B. conductivity

    C. magnetic fields   D. insulators

____10. The longer an electrical wire is, the greater the _____ will be.

A. volts   B. ohms   C. amps   D. watts

____11. When making wiring repairs, the best way to keep from getting shocked is:

A. Never touch the hot wire while you are grounded.

B. Use rubber gloves.

C. Pull the main fuse.

D. Stand on wood (dry).

MATCHING

Directions: Put the proper letter from the selection of letters at the bottom in the blanks of the numbers below. There is only one answer for each blank, and you should not use the same answer twice.

____12. Neutral wire color

____13. Hot wire color

____14. Neutral screw color

____15. Ground wire color

____16. Hot screw color

a. green   e. silver
b. black   f. orange
c. blue   g. white
d. gold

FILL-IN

Directions: Fill in the blanks with the best word or words.

17. Electric companies charge for electrical energy in units of _____.

18. Electrical pressure is measured in _____.

19. Electrical resistance is measured in _____.

20. The amount of electrical flow of current in a circuit is measured in _____.

21. Electrical power is measured in _____.

PROBLEMS

Directions: Work the following questions. Show all work and circle your answer for full credit.

22. & 23. An electrical motor runs for twelve hours. The motor is 120 volts and pulls 115 amps.

Appendix A **379**

   a. How many watts is the motor using?
   b. How many kilowatts is the motor using?
   c. How many kilowatt-hours are used in the twelve hours?
   d. If you pay fourteen cents for each kilowatt-hour, how much does it cost to use it for the twelve hours?
24. Problem 2. What size fuse should be used to protect a 2,200-watt heater at 115 volts?

PAGE A

Black Wire ----------------------------
White Wire ----------------------------
Green Wire ----------------------------
Red Wire ~~~~~~~~~~~~~~~~~~~~~~~~~~~~

Light Fixture →

Gold Screw ⟶ ⊓ = Hot
Silver Screw ⟶ ⊓ = Neutral

Two-Way Switch

Three-Way Switch

OUTLET

PAGE B

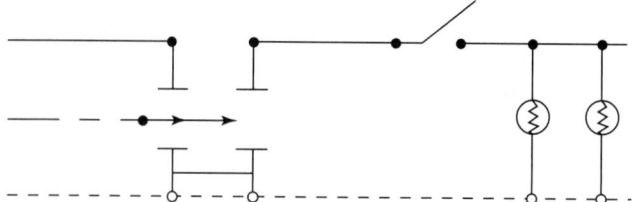

PAGE C

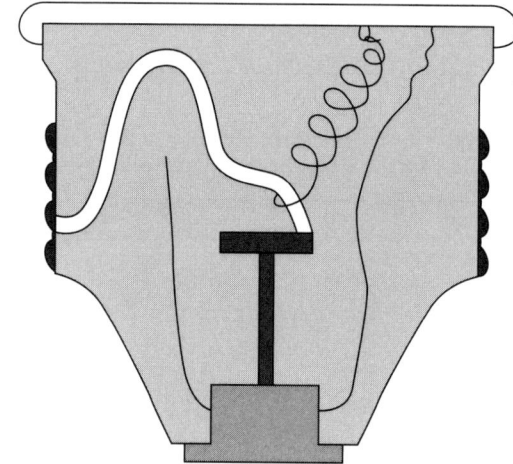

Time-delay fuse.

Appendix A

PAGE D

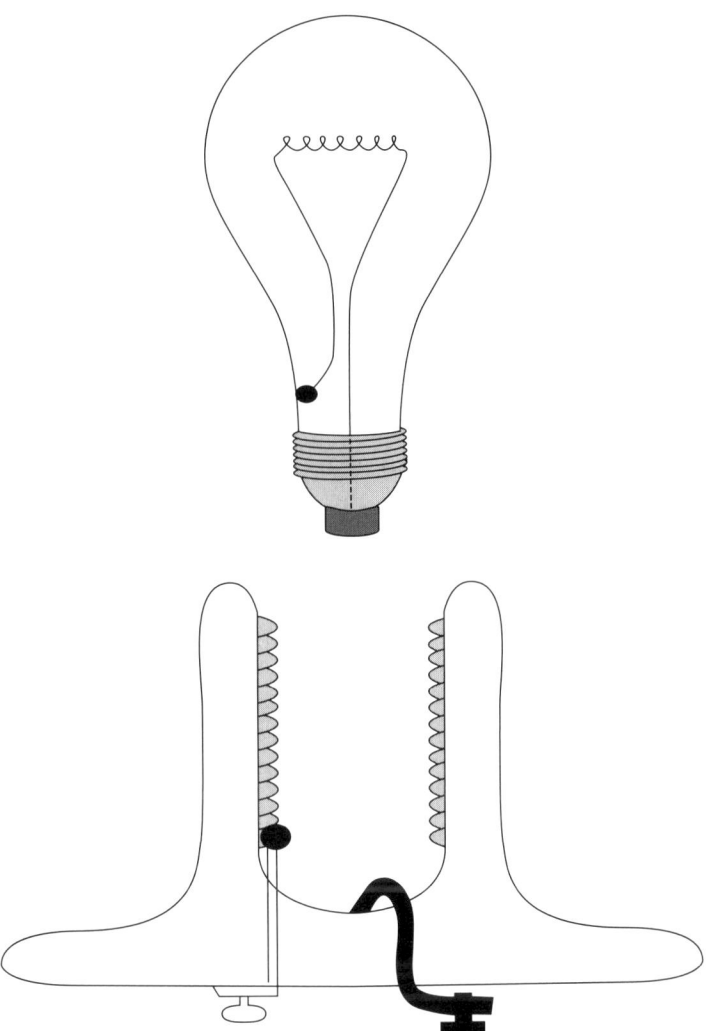

PAGE E

Materials Needed for Demonstration of Switch and Light

1. 5"+ of 14-2 wire
2. Pocket knife
3. Wire strippers
4. Wire pliers and cutters
5. Needle nose pliers
6. Wiring board
7. Wire nuts
8. Cable clamps
9. Screwdriver
10. Hammer
11. Switch
12. Light fixture
13. Light bulbs

PAGE F

Steps and Key Points

1. Disconnect power.
   - Pull main fuse usually.
2. Strip wire.
   - Use knife.
   - Use wire stripper.
3. Cut wire.
   - Long enough for six inches into each box.
4. Attach cable.
   - Use hammer and screwdriver to tighten.
5. Attach wires.
   - Follow color coding.
   - Put each wire around the screw in the direction in which the screw will be tightened.
   - Don't tighten too tightly or you will have trouble the next time.
6. Make connection of wires using wire nuts.
7. In class, let instructor check before proceeding.
8. Push switch and wire into outlet box.
9. Fasten switch to the box with screws.
10. Replace switch plate.
11. Fasten fixture to box.
12. Turn power on.
13. See if it works.

Appendix A **383**

PAGE G

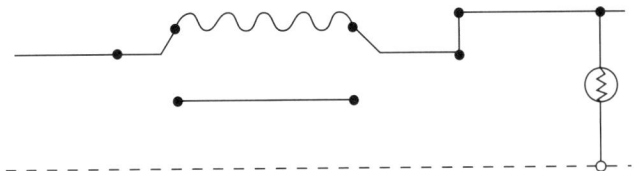

PAGE H

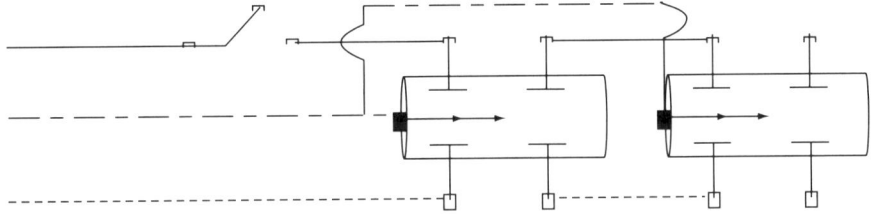

Appendix B

Methods in Teaching Agriculture

*The Ohio State University
Department of Human and Community Resource Development*

Vocabulary Useful in Developing Objectives and Test Items at Various Cognitive Levels.

Remembering	Processing		Creating	Evaluating
acquire	analyze	infer	combine	appraise
cite	apply	interpolate	compose	argue
define	associate	interpret	constitute	assess
identify	categorize	outline	construct	conclude
label	change	paraphrase	create	consider
list	choose	point out	derive	decide
name	classify	predict	design	evaluate
recall	compare	prepare	develop	judge
recite	compute	rearrange	devise	standardize
recognize	contrast	relate	document	validate
reproduce	convert	reorder	formulate	weigh
state	deduce	rephrase	integrate	
	demonstrate	represent	modify	

Appendix B

Remembering	Processing		Creating	Evaluating
	describe	restate	organize	
	detect	restructure	originate	
	determine	summarize	plan	
	differentiate	transfer	produce	
	discriminate	transform	propose	
	distinguish	translate	reorganize	
	draw	use	revise	
	estimate		rewrite	
	explain		specify	
	extend		synthesize	
	extrapolate		tell	
	generalize		transmit	
	illustrate		write	

Sources:

Chamberlain, V. M., & Kelly, J. M. (1981). *Creative home economics instruction.* New York: McGraw-Hill.

Clegg, A. A. (1968). Teacher strategies of questioning for eliciting selected cognitive student responses. (Research Report No. 1). Applied Research Training Program. University of Massachusetts.

Hall, K. S. (1983). Content structuring and question asking for computer-based instructional objectives classification scheme. *Educational Technology, 24* (7), 21–26.

Index

Adult education, 309-323
  adult farmer, 310-311
  agricultural business, 311
  avocational, 311
  evaluating instruction, 321-322
  planning for, 315-318
  reasons for, 311-313
  techniques for instruction, 318-321
  young farmer, 310
Adult farmer education
  enterprise classes, 310-311
  farm business management, 311
  seminars, 310
Adult instruction, 218-321
  discussion, 319
  individual instruction, 320-321
  panel discussion, 319
  resource persons, use of, 319-320
Adult learners, characteristics of, 313-315
Advisory committee, 15, 59

Affective assessment, 345-350
  checklists, 346
  direct observation, 346
  rating scales, 347-348
  rubrics, 348-350
Agricultural education, clientele, 11
  (*See also* Clientele)
Agricultural literacy, 6-8
Application of learning, 211
Assignment sheets, 177-181

Behavior, student, 186-209
Brainstorming, 120
Buzz groups, 120-121

Career development event, 274-275, 282
Clientele, 11, 17-21
Committee on Agricultural Education, 7
Community service, 283-284
Cooperative programs, 249

Course of study, 49–71
  allocation of time, 63–66
  factors influencing decisions, 52–57
  information influencing content, 57–62
  problem areas, 63–66
  reasons for, 50–51
  sequencing, 66–71
  teacher responsibility, 51–52
Cooperative learning, 145–146

Daily plan, 110–111
Deaf/hearing impaired students, 295–296
Decisions about teaching, 3–24
Deficit hyperactivity disorder, 296–297
Demonstrations, 127–135
Dewey, John, 73
"Dewey's Steps in Reflective Thinking," 74
Disability conditions, 291–292
Discussion, 120–127

Employer/employee visits, 258
Entrepreneurship projects, 248–249
Equipment, management of, 236–237
Evaluating adult instruction, 321–322
Evaluating laboratory performance, 232–236
Evaluation of learning
  achievement assessment, 333–345
  affective assessment, 345–350
  grading, 356–365
  performance assessment, 350–356
  reasons for, 328–330
Experiments, 164–171
Essay tests, 341–343

FFA as a laboratory, 268–285
  career development events, 274–275, 282
  chapter banquet, 282–283
  chapter meetings, 279–281
  committee work, 281–282
  community service, 283–284
  degree program, 275–277
  parliamentary procedure, 283
  planning for practice, 270–272
  proficiency awards, 275
  program of activities, 273–274
  public speaking, 283
  state and national activities, 284
  structured learning, 268–269
  teacher's responsibility, 272–273
Field trips, 135–139

Grading, students, 356–365
  factors in, 363–365
  marking systems, 356–363
  purpose of, 356
Group enterprise projects, 248
Group teaching techniques, 112–148
  cooperative learning, 145–146
  demonstrations, 127–135
  discussion, 120–127
  factors in selection of, 146–148
  field trips, 135–139
  lecture, 116–120
  resource people, use of, 142–145
  role playing, 139–142

Hearing impaired students, 295–296

Improvement projects, 250
Independent study, 171–174
Individualized educational plan (IEP), 304–307
Individualized teaching techniques, 150–184
  assignment sheets, 177–181
  experiments, 164–171
  factors in selection of, 181–184
  independent study, 171–174
  information sheets, 177–181
  skill sheets, 177–181
  student notebooks, use of, 174–177
  supervised study, 155–164
Information sheets, 177–181
Instructional objectives, 87, 91–94
Interest approach, 94–97

Laboratory
  clean-up, 238–239
  evaluating performance, 232–236
  FFA as, 268–285
  managing tools, 236–237
  planning instruction, 217–220
  progress chart, 237

Index

rationale for, 214-216
rotation system, 220-226
safety instruction, 226-227
sales policy, 239-241
supervising instruction, 227-232
Learners with special needs
deaf/hard of hearing, 295-296
deficit hyperactivity disorder, 296-297
definition of, 291
disability conditions, 291-292
individualized educational plan, 304-307
learning disabilities, 293
legal mandates, 290-291
mobility/medical impairments, 297-298
peer instruction, 307
psychiatric disabilities, 299
student personnel development, 303-304
student safety, 304
supervised agricultural experience, 304
teacher attitude, 300
teaching practices, 300-303
visually impaired, 294-295
Learning, principles of, 25-48 (Table 2-1)
Learning application, 107-109
Learning centers, rotation of students, 220-226
Learning process, 74-76
Lecture, 116-120
Lesson plan (sample), Appendix

Matching tests, 338-341
Meetings, FFA chapter, 279-281
Misbehavior, dealing with, 192-207
Mobility/medical impairments, 297-298
Motivation, 31-34
Multiple choice tests, 333-335

National strategic plan, 7
Notebooks, student, 174-177

Objectives of instruction, 6-10
advanced study, 9-10
agricultural literacy, 6-8
exploration and orientation, 8-9
occupational competence, 9

Parliamentary procedure, 283
Performance assessment, 350-356
identification tests, 354-356
product assessment, 351
process assessment, 350-351
supervised agricultural experience, 354
Placement programs, 249
Planning for instruction, 85-111
daily planning, 110-111
outline (format) for, 89-109
answering questions, 101-107
application of learning, 107-109
evaluation procedures, 109-110
questions to be answered, 99-101
instructional objectives, 91-94
interest approach, 94-97
problems to be solved, 99-101
reasons of studying unit, 98-99
references and teaching aids, 109
situation, 89-90
solving problems, 101-107
title, 89
rationale for written plans, 87
relationship to problem-solving, 88
relationship to teaching techniques, 88
revising and updating, 110
selecting the unit, 86-87
Principles of teaching and learning, 25-48 (Table 2-1)
motivation, 31-34
organization and structure, 27-30
readiness, 30
reward and reinforcement, 34-37
techniques of teaching, 37-41
transfer of learning, 42-43
Problem solving, 72-82
Problem-solving approach, 76-81
evaluation of solutions, 81
interest approach, 76-77
objectives, 78-79
problem solution, 79
testing solutions, 80-81
Production projects, 247-248
Proficiency awards, 275
Program of agricultural education, 12-16
Program of activities, 273-274
Progress chart, 237

Projects
　entrepreneurship, 248–249
　group enterprise, 248
　production, 247–248
Psychiatric disabilities, 299
Public speaking, 283

Questioning, 122–124

Records instruction, 258–265
Resource people, use of, 142–145
Reward and reinforcement, 34–37
Role playing, 139–142
Rubric, 348–350

Safety instruction, 226–227
Skill development projects, 250
Skill sheets, 177–181
Special needs learners, 289–308
Student behavior, managing, 186–209
Student misbehavior, 192–207
　dealing with, 200–207
　individual conferences, 201–205
　nonverbal communication, 205–206
　preventing, 192–200
Student notebooks, use of, 174–177
Supervised agricultural experience, 243–266
　developing individual plans, 250–256
　employer/employee visits, 258
　entrepreneurship projects, 248–249
　improvement projects, 250
　need for, 244–246
　placement programs, 249
　production projects, 247–248
　records instruction, 258–265
　relationship to teaching, 246–247
　skill development projects, 250
　supervisory visits, 256–258
　types of, 247–250
Supervised study, 155–164
Supervising laboratory instruction, 227–232
Supervisory visits, 256–258

Teaching students with special needs, 300–303
Teaching techniques, 37–41, 45–48 (Table 2-1)
Tests
　administration of, 344
　essay, 341–343
　identification, 354–356
　matching, 338–341
　multiple-choice, 333–335
　organization of, 343–344
　planning of, 330–332
　short-answer, 335–337
　true-false, 337–338
Transfer of learning, 42–43
True-false tests, 337–338

Visits
　employer/employee, 258
　instructional, 257
　orientation, 256
　program planning, 257
　supervisory, 256–258
Visually impaired students, 294–295

Young farmer education, 310